Biomedical Natural Language Processing Shared Task Workshop 2011

(BioNLP Shared Task 2011)

Held at ACL HTL 2011

Portland, Oregon, USA
24 June 2011

ISBN: 978-1-61839-234-3

Printed from e-media with permission by:

Curran Associates, Inc.
57 Morehouse Lane
Red Hook, NY 12571

Some format issues inherent in the e-media version may also appear in this print version.

Printed by Curran Associates, Inc. (2011)

Association for Computational Linguistics
209 N. Eighth Street
Stroudsburg, Pennsylvania 18360

Phone: 1-570-476-8006
Fax: 1-570-476-0860

acl@aclweb.org

Additional copies of this publication are available from:

Curran Associates, Inc.
57 Morehouse Lane
Red Hook, NY 12571 USA
Phone: 845-758-0400
Fax: 845-758-2634
Email: curran@proceedings.com
Web: www.proceedings.com

Introduction

The requirements of improved access to the massive amount of scientific literature in biomedical domain - through applications such as semantic search, assisted pathway annotation, and the automatic identification of specific biomolecular reactions for database curation support - place continuing demands on the development of methods and resources for advanced biomedical information extraction and text mining. The BioNLP Shared Task series seeks to advance this development through an increased focus on detailed structured representations of extracted information, novel corpus resources with fully text-bound annotation, and precise task definitions, support and evaluation.

The BioNLP Shared Task 2011 is the second in the series, following up on the first event organized in 2009. Seeking to build on the success of the previous event, the task was organized as a collaboration between several groups in Asia, Europe and the US who defined in total eight specific tasks involving diverse challenges, including in addition to structured event extraction also relation extraction and supporting tasks such as coreference resolution. The main theme of the 2011 event was generalization, and the main tasks further broadened on the 2009 setup in three aspects: text types, subject domains, and novel event extraction targets.

The task attracted broad interest from the community, and a total of 46 final submissions were received from 24 groups, maintaining the 2009 task participation numbers while nearly doubling its number of submissions. In addition to the continued interest from the biomedical text mining community, we were glad to welcome the participation of many new groups from academia and industry. The submissions demonstrated substantial progress at the established event extraction task and showed that event extraction methods generalize well, among other aspects, to full papers, new subject domains such as infectious diseases and bacterial interactions, and new sets of events such as protein post-translational modifications.

Thanks to the many excellent manuscripts received from participants and the efforts of the programme committee, it is our pleasure to present these proceedings describing the task and the participating systems.

BioNLP Shared Task 2011 chairs

Jun'ichi Tsujii
Jin-Dong Kim
Sampo Pyysalo

Scientific Advisory Committee:

Jun'ichi Tsujii (Chair), Microsoft Research Asia
Sophia Ananiadou, National Centre for Text Mining (NaCTeM)
Kevin Cohen, University of Colorado, and MITRE
Claire Nédellec, French National Institute for Agricultural Research (INRA)
Andrey Rzhetsky, University of Chicago
Bruno Sobral, Virginia Bioinformatics Institute
Tapio Salakoski, University of Turku
Toshihisa Takagi, Database Center for Life Science (DBCLS)

Organizing Committee:

Jin-Dong Kim (Chair), Database Center for Life Science (DBCLS)
Sampo Pyysalo (Chair), University of Tokyo
Tomoko Ohta, University of Tokyo
Robert Bossy, French National Institute for Agricultural Research (INRA)
Chunhong Mao, Virginia Bioinformatics Institute
Dan Sullivan, Virginia Bioinformatics Institute
Rafal Rak, National Centre for Text Mining (NaCTeM)
Ngan Nguyen, University of Tokyo

Program Committee:

Timothy Baldwin, University of Melbourne
Sabine Bergler, Concordia University
Olivier Bodenreider, National Library of Medicine (NLM)
Wendy Chapman, University of California, San Diego (UCSD)
Kevin Cohen, University of Colorado, and MITRE
Nigel Collier, National Institute of Informatics (NII)
Filip Ginter, University of Turku
Jörg Hakenberg, Arizona State University
Minlie Huang, Tsinghua University
Su Jian, Institute for Infocomm Research
Min-Yen Kan, National University of Singapore
Jung-Jae Kim, Nanyang Technological University
Martin Krallinger, National Biotechnology Center (CNB)
Zhiyong Lu, National Library of Medicine (NLM)
David McClosky, Stanford University
Roser Morante, University of Antwerp
Claire Nédellec, French National Institute for Agricultural Research (INRA)
Serguei Pakhomov, University of Minnesota
Thierry Poibeau, French Institute for Fundamental Research (CNRS)
Hoifung Poon, University of Washington

Sebastian Riedel, University of Massachusetts
Fabio Rinaldi, University of Zurich
Thomas Rindflesch, National Library of Medicine (NLM)
Yvan Saeys, Ghent University
Tapio Salakoski, University of Turku
Hagit Shatkay, University of Delaware
Rune Sætre, Norwegian University of Science and Technology (NTNU)
Yuka Tateisi, Kogakuin University
Yoshimasa Tsuruoka, Japan Advanced Institute of Science and Technology (JAIST)
Karin Verspoor, University of Colorado
Xinglong Wang, National Centre for Text Mining
Hong Yu, University of Wisconsin-Milwaukee
Pierre Zweigenbaum, French National Center for Scientific Research (CNRS)

Table of Contents

Conference Program

Friday, June 24, 2011

Session 1: Oral presentations and discussion

09:00–09:25 Overview of BioNLP Shared Task 2011 (I) - *Overall Organization*, and *GE*, *IDE* and *ID* Tasks
Shared Task Organizers

09:25–09:40 *Biomedical Event Extraction from Abstracts and Full Papers using Search-based Structured Prediction*
Andreas Vlachos and Mark Craven

09:40–09:55 *Event Extraction as Dependency Parsing for BioNLP 2011*
David McClosky, Mihai Surdeanu and Christopher Manning

09:55–10:10 *Robust Biomedical Event Extraction with Dual Decomposition and Minimal Domain Adaptation*
Sebastian Riedel and Andrew McCallum

10:10–10:25 *Model Combination for Event Extraction in BioNLP 2011*
Sebastian Riedel, David McClosky, Mihai Surdeanu, Andrew McCallum and Christopher D. Manning

10:25–10:30 Discussion

10:30–11:00 Morning break

Friday, June 24, 2011 (continued)

Session 2: Oral presentations and discussion

11:00–11:20 Overview of BioNLP Shared Task 2011 (II) - Bacteria Track (*BB*, *BI*) and Supporting Tasks (*CO*, *REL* and REN)
Shared Task Organizers

11:20–11:35 *The Taming of Reconcile as a Biomedical Coreference Resolver*
Youngjun Kim, Ellen Riloff and Nathan Gilbert

11:35–11:55 *Extracting Bacteria Biotopes with Semi-supervised Named Entity Recognition and Coreference Resolution*
Nhung T. H. Nguyen and Yoshimasa Tsuruoka

11:55–12:15 *BioNLP 2011 Task Bacteria Biotope – The Alvis system*
Zorana Ratkovic, Wiktoria Golik, Pierre Warnier, Philippe Veber and Claire Nédellec

12:15–12:30 Discussion

12:30–14:00 Lunch break

Friday, June 24, 2011 (continued)

Session 4: Oral presentations and discussion

Overview of BioNLP Shared Task 2011

Jin-Dong Kim
Database Center for Life Science
2-11-16 Yayoi, Bunkyo-ku, Tokyo
jdkim@dbcls.rois.ac.jp

Sampo Pyysalo
University of Tokyo
7-3-1 Hongo, Bunkyo-ku, Tokyo
smp@is.s.u-tokyo.ac.jp

Tomoko Ohta
University of Tokyo
7-3-1 Hongo, Bunkyo-ku, Tokyo
okap@is.s.u-tokyo.ac.jp

Robert Bossy
National Institute for Agricultural Research
78352 Jouy en Josas, Cedex
Robert.Bossy@jouy.inra.fr

Ngan Nguyen
University of Tokyo
7-3-1 Hongo, Bunkyo-ku, Tokyo
nltngan@is.s.u-tokyo.ac.jp

Jun'ichi Tsujii
Microsoft Research Asia
5 Dan Ling Street, Haiian District, Beijing
jtsujii@microsoft.com

Abstract

The BioNLP Shared Task 2011, an information extraction task held over 6 months up to March 2011, met with community-wide participation, receiving 46 final submissions from 24 teams. Five main tasks and three supporting tasks were arranged, and their results show advances in the state of the art in fine-grained biomedical domain information extraction and demonstrate that extraction methods successfully generalize in various aspects.

1 Introduction

The BioNLP Shared Task (BioNLP-ST, hereafter) series represents a community-wide move toward fine-grained information extraction (IE), in particular biomolecular event extraction (Kim et al., 2009; Ananiadou et al., 2010). The series is complementary to BioCreative (Hirschman et al., 2007); while BioCreative emphasizes the short-term *applicability* of introduced IE methods for tasks such as database curation, BioNLP-ST places more emphasis on the *measurability* of the state-of-the-art and *traceability* of challenges in extraction through an approach more closely tied to text.

These goals were pursued in the first event, BioNLP-ST 2009 (Kim et al., 2009), through *high quality benchmark data* provided for system development and *detailed evaluation* performed to identify remaining problems hindering extraction performance. Also, as the complexity of the task was high and system development time limited, we encouraged *focus on fine-grained IE* by providing gold annotation for named entities as well as various supporting resources. BioNLP-ST 2009 attracted wide attention, with 24 teams submitting final results. The task setup and data since have served as the basis for numerous studies (Miwa et al., 2010b; Poon and Vanderwende, 2010; Vlachos, 2010; Miwa et al., 2010a; Björne et al., 2010).

As the second event of the series, BioNLP-ST 2011 preserves the general design and goals of the previous event, but adds a new focus on *variability* to address a limitation of BioNLP-ST 2009: the benchmark data sets were based on the Genia corpus (Kim et al., 2008), restricting the community-wide effort to resources developed by a single group for a small subdomain of molecular biology. BioNLP-ST 2011 is organized as a joint effort of several groups preparing various tasks and resources, in which variability is pursued in three primary directions: *text types*, *event types*, and *subject domains*. Consequently, *generalization* of fine grained bio-IE in these directions is emphasized as the main theme of the second event.

This paper summarizes the entire BioNLP-ST 2011, covering the relationships between tasks and similar broad issues. Each task is presented in detail in separate overview papers and extraction systems in papers by participants.

Proceedings of BioNLP Shared Task 2011 Workshop, pages 1–6,
Portland, Oregon, USA, 24 June, 2011. ©2011 Association for Computational Linguistics

2 Main tasks

BioNLP-ST 2011 includes four main tracks (with five tasks) representing fine-grained bio-IE.

2.1 Genia task (GE)

The GE task (Kim et al., 2011) preserves the task definition of BioNLP-ST 2009, arranged based on the Genia corpus (Kim et al., 2008). The data represents a focused domain of molecular biology: *transcription factors in human blood cells*. The purpose of the GE task is two-fold: to measure the progress of the community since the last event, and to evaluate generalization of the technology to full papers. For the second purpose, the provided data is composed of two collections: the *abstract collection*, identical to the BioNLP-ST 2009 data, and the new *full paper collection*. Progress on the task is measured through the unchanged task definition and the abstract collection, while generalization to full papers is measured on the full paper collection. In this way, the GE task is intended to connect the entire event to the previous one.

2.2 Epigenetics and post-translational modification task (EPI)

The EPI task (Ohta et al., 2011) focuses on IE for protein and DNA modifications, with particular emphasis on events of epigenetics interest. While the basic task setup and entity definitions follow those of the GE task, EPI extends on the extraction targets by defining 14 new event types relevant to task topics, including major protein modification types and their reverse reactions. For capturing the ways in which different entities participate in these events, the task extends the GE argument roles with two new roles specific to the domain, *Sidechain* and *Contextgene*. The task design and setup are oriented toward the needs of pathway extraction and curation for domain databases (Wu et al., 2003; Ongenaert et al., 2008) and are informed by previous studies on extraction of the target events (Ohta et al., 2010b; Ohta et al., 2010c).

2.3 Infectious diseases task (ID)

The ID task (Pyysalo et al., 2011a) concerns the extraction of events relevant to biomolecular mechanisms of infectious diseases from full-text publications. The task follows the basic design of BioNLP-ST 2009, and the ID entities and extraction targets are a superset of the GE ones. The task extends considerably on core entities, adding to PROTEIN four new entity types, including CHEMICAL and ORGANISM. The events extend on the GE definitions in allowing arguments of the new entity types as well as in introducing a new event category for high-level biological processes. The task was implemented in collaboration with domain experts and informed by prior studies on domain information extraction requirements (Pyysalo et al., 2010; Ananiadou et al., 2011), including the support of systems such as PATRIC (`http://patricbrc.org`).

2.4 Bacteria track

The bacteria track consists of two tasks, BB and BI.

2.4.1 Bacteria biotope task (BB)

The aim of the BB task (Bossy et al., 2011) is to extract the habitats of bacteria mentioned in textbook-level texts written for non-experts. The texts are Web pages about the state of the art knowledge about bacterial species. BB targets general relations, *Localization* and *PartOf*, and is challenging in that texts contain more coreferences than usual, habitat references are not necessarily named entities, and, unlike in other BioNLP-ST 2011 tasks, all entities need to be recognized by participants. BB is the first task to target phenotypic information and, as habitats are yet to be normalized by the field community, presents an opportunity for the BioNLP community to contribute to the standardization effort.

2.4.2 Bacteria interaction task (BI)

The BI task (Jourde et al., 2011) is devoted to the extraction of bacterial molecular interactions and regulations from publication abstracts. Mainly focused on gene transcriptional regulation in *Bacillus subtilis*, the BI corpus is provided to participants with rich semantic annotation derived from a recently proposed ontology (Manine et al., 2009) defining ten entity types such as gene, protein and derivatives as well as DNA sites/motifs. Their interactions are described through ten relation types. The BI corpus consists of the sentences of the LLL corpus (Nédellec, 2005), provided with manually checked linguistic annotations.

Task	Text	Focus	#
GE	abstracts, full papers	domain (HT)	9
EPI	abstracts	event types	15
ID	full papers	domain (TCS)	10
BB	web pages	domain (BB)	2
BI	abstracts	domain (BS)	10

Table 1: Characteristics of BioNLP-ST 2011 main tasks. '#': number of event/relation types targeted. Domains: HT = human transcription factors in blood cells, TCS = two-component systems, BB = bacteria biology, BS = Bacillus subtilis

2.5 Characteristics of main tasks

The main tasks are characterized in Table 1. From the text type perspective, BioNLP-ST 2011 generalizes from abstracts in 2009 to full papers (GE and ID) and web pages (BB). It also includes data collections for a variety of specific subject domains (GE, ID, BB an BI) and a task (EPI) whose scope is not defined through a domain but rather event types. In terms of the target event types, ID targets a superset of GE events and EPI extends on the representation for PHOSPHORYLATION events of GE. The two bacteria track tasks represent an independent perspective relatively far from other tasks in terms of their target information.

3 Supporting tasks

BioNLP-ST 2011 includes three supporting tasks designed to assist in primary the extraction tasks. Other supporting resources made available to participants are presented in (Stenetorp et al., 2011).

3.1 Protein coreference task (CO)

The CO task (Nguyen et al., 2011) concerns the recognition of coreferences to protein references. It is motivated from a finding from BioNLP-ST 2009 result analysis: coreference structures in biomedical text hinder the extraction results of fine-grained IE systems. While finding connections between event triggers and protein references is a major part of event extraction, it becomes much harder if one is replaced with a coreferencing expression. The CO task seeks to address this problem. The data sets for the task were produced based on MedCO annotation (Su et al., 2008) and other Genia resources (Tateisi et al., 2005; Kim et al., 2008).

Event	Date	Note
Sample Data	31 Aug. 2010	
Support. Tasks		
Train. Data	27 Sep. 2010	7 weeks for development
Test Data	15 Nov. 2010	4 days for submission
Submission	19 Nov. 2010	
Evaluation	22 Nov. 2010	
Main Tasks		
Train. Data	1 Dec. 2010	3 months for development
Test Data	1 Mar. 2011	9 days for submission
Submission	10 Mar. 2011	extended from 8 Mar.
Evaluation	11 Mar. 2011	extended from 10 Mar.

Table 2: Schedule of BioNLP-ST 2011

3.2 Entity relations task (REL)

The REL task (Pyysalo et al., 2011b) involves the recognition of two binary part-of relations between entities: PROTEIN-COMPONENT and SUBUNIT-COMPLEX. The task is motivated by specific challenges: the identification of the components of proteins in text is relevant e.g. to the recognition of *Site* arguments (cf. GE, EPI and ID tasks), and relations between proteins and their complexes relevant to any task involving them. REL setup is informed by recent semantic relation tasks (Hendrickx et al., 2010). The task data, consisting of new annotations for GE data, extends a previously introduced resource (Pyysalo et al., 2009; Ohta et al., 2010a).

3.3 Gene renaming task (REN)

The REN task (Jourde et al., 2011) objective is to extract renaming pairs of *Bacillus subtilis* gene/protein names from PubMed abstracts, motivated by discrepancies between nomenclature databases that interfere with search and complicate normalization. REN relations partially overlap several concepts: explicit renaming mentions, synonymy, and renaming deduced from biological proof. While the task is related to synonymy relation extraction (Yu and Agichtein, 2003), it has a novel definition of renaming, one name permanently replacing the other.

4 Schedule

Table 2 shows the task schedule, split into two phases to allow the use of supporting task results in addressing the main tasks. In recognition of their higher complexity, a longer development period was arranged for the main tasks (3 months vs 7 weeks).

Team	GE	EPI	ID	BB	BI	CO	REL	REN
UTurku	1	1	1	1	1	1	1	1
ConcordU	1	1	1			1	1	1
UMass	1	1	1					
Stanford	1	1	1					
FAUST	1	1	1					
MSR-NLP	1	1						
CCP-BTMG	1	1						
Others	8	0	2	2	0	4	2	1
SUM	15	7	7	3	1	6	4	3

Table 3: Final submissions to BioNLP-ST 2011 tasks.

5 Participation

BioNLP-ST 2011 received 46 submissions from 24 teams (Table 3). While seven teams participated in multiple tasks, only one team, UTurku, submitted final results to all the tasks. The remaining 17 teams participated in only single tasks. Disappointingly, only two teams (UTurku, and ConcordU) performed both supporting and main tasks, and neither used supporting task analyses for the main tasks.

6 Results

Detailed evaluation results and analyses are presented in individual task papers, but interesting observations can be obtained also by comparisons over the tasks. Table 4 summarizes best results for various criteria (Note that the results shown for e.g. GEa, GEf and GEp may be from different teams).

The community has made a significant improvement in the repeated GE task, with an over 10% reduction in error from '09 to GEa. Three teams achieved better results than M10, the best previously reported individual result on the '09 data. This indicates a beneficial role from focused efforts like BioNLP-ST. The GEf and ID results show that generalization to full papers is feasible, with very modest loss in performance compared to abstracts (GEa). The results for PHOSPHORYLATION events in GE and EPI are comparable (GEp vs EPIp), with the small drop for the EPI result, suggesting that the removal of the GE domain specificity does not compromise extraction performance. EPIc results indicate some challenges in generalization to similar event types, and EPIf suggest substantial further challenges in additional argument extraction. The complexity of ID is comparable to GE, also reflected to their final results, which further indicate success-

Task	Evaluation Results
BioNLP-ST 2009 ('09)	*46.73 / 58.48 / 51.95*
Miwa et al. (2010b) (M10)	*48.62 / 58.96 / 53.29*
LLL 2005 (LLL)	*53.00 / 55.60 / 54.30*
GE abstracts (GEa)	50.00 / 67.53 / 57.46
GE full texts (GEf)	47.84 / 59.76 / 53.14
GE PHOSPHORYLATION (GEp)	79.26 / 86.99 / 82.95
GE LOCALIZATION (GEl)	37.88 / 77.42 / 50.87
EPI full task (EPIf)	52.69 / 53.98 / 53.33
EPI core task (EPIc)	68.51 / 69.20 / 68.86
EPI PHOSPHORYLATION (EPIp)	86.15 / 74.67 / 80.00
ID full task (IDf)	48.03 / 65.97 / 55.59
ID core task (IDc)	50.62 / 66.06 / 57.32
BB	45.00 / 45.00 / 45.00
BB PartOf (BBp)	32.00 / 83.00 / 46.00
BI	71.00 / 85.00 / 77.00
CO	22.18 / 73.26 / 34.05
REL	50.10 / 68.00 / 57.70
REN	79.60 / 95.90 / 87.00

Table 4: Best results for various (sub)tasks (recall / precision / f-score (%)). GEl: task 2 without trigger detection.

ful generalization to a new subject domain as well as to new argument (entity) types. The BB task is in part comparable to GEl and involves a representation similar to REL, with lower results likely in part because BB requires entity recognition. The BI task is comparable to LLL Challenge, though BI involves more entity and event types. The BI result is 20 points above the LLL best result, indicating a substantial progress of the community in five years.

7 Discussion and Conclusions

Meeting with wide participation from the community, BioNLP-ST 2011 produced a wealth of valuable resources for the advancement of fine-grained IE in biology and biomedicine, and demonstrated that event extraction methods can successfully generalize to new text types, event types, and domains. However, the goal to observe the capacity of supporting tasks to assist the main tasks was not met. The entire shared task period was very long, more than 6 months, and the complexity of the task was high, which could be an excessive burden for participants, limiting the application of novel resources. There have been ongoing efforts since BioNLP-ST 2009 to develop IE systems based on the task resources, and we hope to see continued efforts also following BioNLP-ST 2011, especially exploring the use of supporting task resources for main tasks.

References

Sophia Ananiadou, Sampo Pyysalo, Junichi Tsujii, and Douglas B. Kell. 2010. Event extraction for systems biology by text mining the literature. *Trends in Biotechnology*.

Sophia Ananiadou, Dan Sullivan, William Black, Gina-Anne Levow, Joseph J. Gillespie, Chunhong Mao, Sampo Pyysalo, BalaKrishna Kolluru, Junichi Tsujii, and Bruno Sobral. 2011. Named entity recognition for bacterial type IV secretion systems. *PLoS ONE*, 6(3):e14780.

Jari Björne, Filip Ginter, Sampo Pyysalo, Jun'ichi Tsujii, and Tapio Salakoski. 2010. Complex event extraction at PubMed scale. *Bioinformatics*, 26(12):i382–390.

Robert Bossy, Julien Jourde, Philippe Bessières, Marteen van de Guchte, and Claire Nédellec. 2011. BioNLP Shared Task 2011 - Bacteria Biotope. In *Proceedings of the BioNLP 2011 Workshop Companion Volume for Shared Task*, Portland, Oregon, June. Association for Computational Linguistics.

Iris Hendrickx, Su Nam Kim, Zornitsa Kozareva, Preslav Nakov, Diarmuid Ó. Séaghdha, Sebastian Padó, Marco Pennacchiotti, Lorenza Romano, and Stan Szpakowicz. 2010. Semeval-2010 task 8: Multi-way classification of semantic relations between pairs of nominals. In *Proceedings of the 5th International Workshop on Semantic Evaluation*, SemEval '10, pages 33–38, Stroudsburg, PA, USA. Association for Computational Linguistics.

Lynette Hirschman, Martin Krallinger, and Alfonso Valencia, editors. 2007. *Proceedings of the Second BioCreative Challenge Evaluation Workshop*. CNIO Centro Nacional de Investigaciones Oncológicas.

Julien Jourde, Alain-Pierre Manine, Philippe Veber, Karën Fort, Robert Bossy, Erick Alphonse, and Philippe Bessières. 2011. BioNLP Shared Task 2011 - Bacteria Gene Interactions and Renaming. In *Proceedings of the BioNLP 2011 Workshop Companion Volume for Shared Task*, Portland, Oregon, June. Association for Computational Linguistics.

Jin-Dong Kim, Tomoko Ohta, and Jun'ichi Tsujii. 2008. Corpus annotation for mining biomedical events from lterature. *BMC Bioinformatics*, 9(1):10.

Jin-Dong Kim, Tomoko Ohta, Sampo Pyysalo, Yoshinobu Kano, and Jun'ichi Tsujii. 2009. Overview of BioNLP'09 Shared Task on Event Extraction. In *Proceedings of Natural Language Processing in Biomedicine (BioNLP) NAACL 2009 Workshop*, pages 1–9.

Jin-Dong Kim, Yue Wang, Toshihisa Takagi, and Akinori Yonezawa. 2011. Overview of the Genia Event task in BioNLP Shared Task 2011. In *Proceedings of the BioNLP 2011 Workshop Companion Volume for Shared Task*, Portland, Oregon, June. Association for Computational Linguistics.

A.P. Manine, E. Alphonse, and Bessières P. 2009. Learning ontological rules to extract multiple relations of genic interactions from text. *International Journal of Medical Informatics*, 78(12):e31–38.

Makoto Miwa, Sampo Pyysalo, Tadayoshi Hara, and Jun'ichi Tsujii. 2010a. A comparative study of syntactic parsers for event extraction. In *Proceedings of BioNLP'10*, pages 37–45.

Makoto Miwa, Rune Sætre, Jin-Dong Kim, and Jun'ichi Tsujii. 2010b. Event extraction with complex event classification using rich features. *Journal of Bioinformatics and Computational Biology (JBCB)*, 8(1):131–146, February.

Nédellec. 2005. Learning Language in Logic – Genic Interaction Extraction Challenge. In *Proceedings of 4th Learning Language in Logic Workshop (LLL'05)*, pages 31–37.

Ngan Nguyen, Jin-Dong Kim, and Jun'ichi Tsujii. 2011. Overview of the Protein Coreference task in BioNLP Shared Task 2011. In *Proceedings of the BioNLP 2011 Workshop Companion Volume for Shared Task*, Portland, Oregon, June. Association for Computational Linguistics.

Tomoko Ohta, Sampo Pyysalo, Jin-Dong Kim, and Jun'ichi Tsujii. 2010a. A re-evaluation of biomedical named entity-term relations. *Journal of Bioinformatics and Computational Biology (JBCB)*, 8(5):917–928.

Tomoko Ohta, Sampo Pyysalo, Makoto Miwa, Jin-Dong Kim, and Jun'ichi Tsujii. 2010b. Event extraction for post-translational modifications. In *Proceedings of BioNLP'10*, pages 19–27.

Tomoko Ohta, Sampo Pyysalo, Makoto Miwa, and Jun'ichi Tsujii. 2010c. Event extraction for dna methylation. In *Proceedings of SMBM'10*.

Tomoko Ohta, Sampo Pyysalo, and Jun'ichi Tsujii. 2011. Overview of the Epigenetics and Post-translational Modifications (EPI) task of BioNLP Shared Task 2011. In *Proceedings of the BioNLP 2011 Workshop Companion Volume for Shared Task*, Portland, Oregon, June. Association for Computational Linguistics.

Maté Ongenaert, Leander Van Neste, Tim De Meyer, Gerben Menschaert, Sofie Bekaert, and Wim Van Criekinge. 2008. PubMeth: a cancer methylation database combining text-mining and expert annotation. *Nucleic Acids Research*, 36(suppl_1):D842–846.

Hoifung Poon and Lucy Vanderwende. 2010. Joint inference for knowledge extraction from biomedical literature. In *Proceedings of NAACL-HLT'10*, pages 813–821.

Sampo Pyysalo, Tomoko Ohta, Jin-Dong Kim, and Jun'ichi Tsujii. 2009. Static Relations: a Piece in the Biomedical Information Extraction Puzzle. In *Proceedings of Natural Language Processing in Biomedicine (BioNLP) NAACL 2009 Workshop*, pages 1–9, Boulder, Colorado. Association for Computational Linguistics.

Sampo Pyysalo, Tomoko Ohta, Han-Cheol Cho, Dan Sullivan, Chunhong Mao, Bruno Sobral, Jun'ichi Tsujii, and Sophia Ananiadou. 2010. Towards event extraction from full texts on infectious diseases. In *Proceedings of BioNLP'10*, pages 132–140.

Sampo Pyysalo, Tomoko Ohta, Rafal Rak, Dan Sullivan, Chunhong Mao, Chunxia Wang, Bruno Sobral, Jun'ichi Tsujii, and Sophia Ananiadou. 2011a. Overview of the Infectious Diseases (ID) task of BioNLP Shared Task 2011. In *Proceedings of the BioNLP 2011 Workshop Companion Volume for Shared Task*, Portland, Oregon, June. Association for Computational Linguistics.

Sampo Pyysalo, Tomoko Ohta, and Jun'ichi Tsujii. 2011b. Overview of the Entity Relations (REL) supporting task of BioNLP Shared Task 2011. In *Proceedings of the BioNLP 2011 Workshop Companion Volume for Shared Task*, Portland, Oregon, June. Association for Computational Linguistics.

Pontus Stenetorp, Goran Topić, Sampo Pyysalo, Tomoko Ohta, Jin-Dong Kim, and Jun'ichi Tsujii. 2011. BioNLP Shared Task 2011: Supporting Resources. In *Proceedings of the BioNLP 2011 Workshop Companion Volume for Shared Task*, Portland, Oregon, June. Association for Computational Linguistics.

Jian Su, Xiaofeng Yang, Huaqing Hong, Yuka Tateisi, and Jun'ichi Tsujii. 2008. Coreference Resolution in Biomedical Texts: a Machine Learning Approach. In *Ontologies and Text Mining for Life Sciences'08*.

Yuka Tateisi, Akane Yakushiji, Tomoko Ohta, and Jun'ichi Tsujii. 2005. Syntax Annotation for the GENIA corpus. In *Proceedings of the IJCNLP 2005, Companion volume*, pages 222–227.

Andreas Vlachos. 2010. Two strong baselines for the bionlp 2009 event extraction task. In *Proceedings of BioNLP'10*, pages 1–9.

Cathy H. Wu, Lai-Su L. Yeh, Hongzhan Huang, Leslie Arminski, Jorge Castro-Alvear, Yongxing Chen, Zhangzhi Hu, Panagiotis Kourtesis, Robert S. Ledley, Baris E. Suzek, C.R. Vinayaka, Jian Zhang, and Winona C. Barker. 2003. The Protein Information Resource. *Nucleic Acids Research*, 31(1):345–347.

H. Yu and E. Agichtein. 2003. Extracting synonymous gene and protein terms from biological literature. *Bioinformatics*, 19(suppl 1):i340.

Overview of Genia Event Task in BioNLP Shared Task 2011

Jin-Dong Kim
Database Center for Life Science
2-11-16 Yayoi, Bunkyo-ku, Tokyo
jdkim@dbcls.rois.ac.jp

Yue Wang
Database Center for Life Science
2-11-16 Yayoi, Bunkyo-ku, Tokyo
wang@dbcls.rois.ac.jp

Toshihisa Takagi
University of Tokyo
5-1-5 Kashiwa-no-ha, Kashiwa, Chiba
tt@k.u-tokyo.ac.jp

Akinori Yonezawa
Database Center for Life Science
2-11-16 Yayoi, Bunkyo-ku, Tokyo
yonezawa@dbcls.rois.ac.jp

Abstract

The Genia event task, a bio-molecular event extraction task, is arranged as one of the main tasks of BioNLP Shared Task 2011. As its second time to be arranged for community-wide focused efforts, it aimed to measure the advance of the community since 2009, and to evaluate generalization of the technology to full text papers. After a 3-month system development period, 15 teams submitted their performance results on test cases. The results show the community has made a significant advancement in terms of both performance improvement and generalization.

1 Introduction

The BioNLP Shared Task (BioNLP-ST, hereafter) is a series of efforts to promote a community-wide collaboration towards fine-grained information extraction (IE) in biomedical domain. The first event, BioNLP-ST 2009, introducing a bio-molecular event (bio-event) extraction task to the community, attracted a wide attention, with 42 teams being registered for participation and 24 teams submitting final results (Kim et al., 2009).

To establish a community effort, the organizers provided the task definition, benchmark data, and evaluations, and the participants competed in developing systems to perform the task. Meanwhile, participants and organizers communicated to develop a better setup of evaluation, and some provided their tools and resources for other participants, making it a collaborative competition.

The final results enabled to observe the state-of-the-art performance of the community on the bio-event extraction task, which showed that the automatic extraction of simple events - those with unary arguments, e.g. gene expression, localization, phosphorylation - could be achieved at the performance level of 70% in F-score, but the extraction of complex events, e.g. binding and regulation, was a lot more challenging, having achieved 40% of performance level.

After BioNLP-ST 2009, all the resources from the event were released to the public, to encourage continuous efforts for further advancement. Since then, several improvements have been reported (Miwa et al., 2010b; Poon and Vanderwende, 2010; Vlachos, 2010; Miwa et al., 2010a; Björne et al., 2010). For example, Miwa et al. (Miwa et al., 2010b) reported a significant improvement with binding events, achieving 50% of performance level.

The task introduced in BioNLP-ST 2009 was renamed to *Genia event (GE) task*, and was hosted again in BioNLP-ST 2011, which also hosted four other IE tasks and three supporting tasks (Kim et al., 2011). As the sole task that was repeated in the two events, the GE task was referenced during the development of other tasks, and took the role of connecting the results of the 2009 event to the main tasks of 2011. The GE task in 2011 received final submissions from 15 teams. The results show the community made a significant progress with the task, and also show the technology can be generalized to full papers at moderate cost of performance.

This paper presents the task setup, preparation, and discusses the results.

Proceedings of BioNLP Shared Task 2011 Workshop, pages 7–15,
Portland, Oregon, USA, 24 June, 2011. ©2011 Association for Computational Linguistics

Event Type	Primary Argument	Secondary Argument
Gene_expression	Theme(Protein)	
Transcription	Theme(Protein)	
Protein_catabolism	Theme(Protein)	
Phosphorylation	Theme(Protein)	Site(Entity)
Localization	Theme(Protein)	AtLoc(Entity), ToLoc(Entity)
Binding	Theme(Protein)+	Site(Entity)+
Regulation	Theme(Protein/Event), Cause(Protein/Event)	Site(Entity), CSite(Entity)
Positive_regulation	Theme(Protein/Event), Cause(Protein/Event)	Site(Entity), CSite(Entity)
Negative_regulation	Theme(Protein/Event), Cause(Protein/Event)	Site(Entity), CSite(Entity)

Table 1: Event types and their arguments for Genia event task. The type of each filler entity is specified in parenthesis. Arguments that may be filled more than once per event are marked with "+".

2 Task Definition

The GE task follows the task definition of BioNLP-ST 2009, which is briefly described in this section. For more detail, please refer to (Kim et al., 2009).

Table 1 shows the event types to be addressed in the task. For each event type, the primary and secondary arguments to be extracted with an event are defined. For example, a *Phosphorylation* event is primarily extracted with the protein to be phosphorylated. As secondary information, the specific site to be phosphorylated may be extracted.

From a computational point of view, the event types represent different levels of complexity. When only primary arguments are considered, the first five event types in Table 1 are classified as *simple event types*, requiring only unary arguments. The *Binding* and *Regulation* types are more complex: *Binding* requires detection of an arbitrary number of arguments, and *Regulation* requires detection of recursive event structure.

Based on the definition of event types, the entire task is divided to three sub-tasks addressing event extraction at different levels of specificity:

Task 1. Core event extraction addresses the extraction of typed events together with their primary arguments.

Task 2. Event enrichment addresses the extraction of secondary arguments that further specify the events extracted in Task 1.

Task 3. Negation/Speculation detection addresses the detection of negations and speculations over the extracted events.

Task 1 serves as the backbone of the GE task and is mandatory for all participants, while the other two are optional.

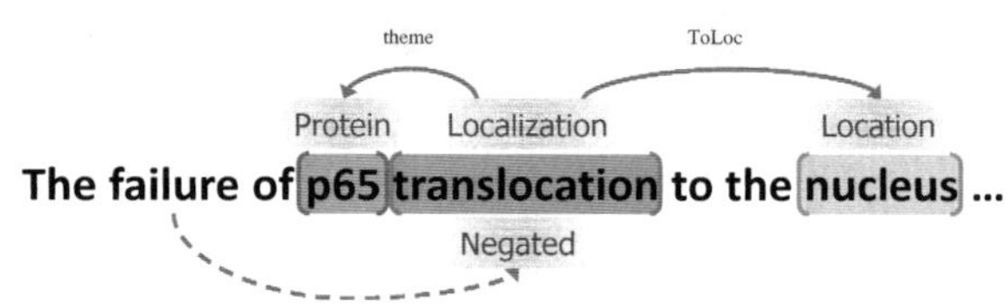

Figure 1: Event annotation example

Figure 1 shows an example of event annotation. The event encoded in the text is represented in a standoff-style annotation as follows:

```
T1   Protein        15 18
T2   Localization   19 32
T3   Entity         40 46
E1   Localization:T2 Theme:T1 ToLoc:T1
M1   Negation  E1
```

The annotation `T1` identifies the entity referred to by the string (*p65*) between the character offsets, *15* and *18* to be a *Protein*. `T2` identifies the string, *translocation*, to refer to a *Localization* event. Entities other than proteins or event type references are classified into a default class *Entity*, as in `T3`. `E1` then represents the event defined by the three entities, as defined in Table 1. Note that for Task 1, the entity, `T3`, does not need to be identified, and the event, `E1`, may be identified without specification of the secondary argument, `ToLoc:T1`:

```
E1'  Localization:T2 Theme:T1
```

Finding the full representation of `E1` is the goal of Task 2. In the example, the localization event, `E1`, is negated as expressed in *the failure of*. Finding the negation, `M1` is the goal of Task 3.

Item	Training Abs.	Training Full	Devel Abs.	Devel Full	Test Abs.	Test Full
Articles	*800*	*5*	*150*	*5*	*260*	*4*
Words	*176146*	*29583*	*33827*	*30305*	*57256*	*21791*
Proteins	9300	2325	2080	2610	3589	1712
Events	8615	1695	1795	1455	3193	1294
Gene_expression	1738	527	356	393	722	280
Transcription	576	91	82	76	137	37
Protein_catabolism	110	0	21	2	14	1
Phosphorylation	169	23	47	64	139	50
Localization	265	16	53	14	174	17
Binding	887	101	249	126	349	153
Regulation	961	152	173	123	292	96
Positive_regulation	2847	538	618	382	987	466
Negative_regulation	1062	247	196	275	379	194

Table 2: Statistics of annotations in training, development, and test sets

3 Data preparation

The data sets are prepared in two collections: the abstract and the full text collections. The *abstract collection* includes the same data used for BioNLP-ST 2009, and is meant to be used to measure the progress of the community. The *full text collection* includes full papers which are newly annotated, and is meant to be used to measure the generalization of the technology to full papers. Table 2 shows the statistics of the annotations in the GE task data sets. Since the training data from the full text collection is relatively small despite of the expected rich variety of expressions in full text, it is expected that 'generalization' of a model from the abstract collection to full papers would be a key technique to get a reasonable performance.

A full paper consists of several sections including the title, abstract, introduction, results, conclusion, methods, and so on. Different sections would be written with different purposes, which may affect the type of information that are found in the sections. Table 3 shows the distribution of annotations in different sections. It indicates that event mentions, according to the event definition in Table 1, in *Methods* and *Caption*s are much less frequent than in the other *TIAB*, *Intro.* and *R/D/C* sections. Figure 2 illustrates the different distribution of annotated event types in the five sections. It is notable that the *Methods* section (depicted in blue) shows very different distribution compared to others: while *Regulation* and *Positive_regulation* events are not as frequent as in other sections, *Negative_regulation* is relatively much more frequent. It may agree with an intuition that experimental devices, which will be explained in *Methods* sections, often consists of artificial processes that are designed to cause a negative regulatory effect, e.g. mutation, addition of inhibitor proteins, etc. This observation suggests a different event annotation scheme, or a different event extraction strategy would be required for *Methods* sections.

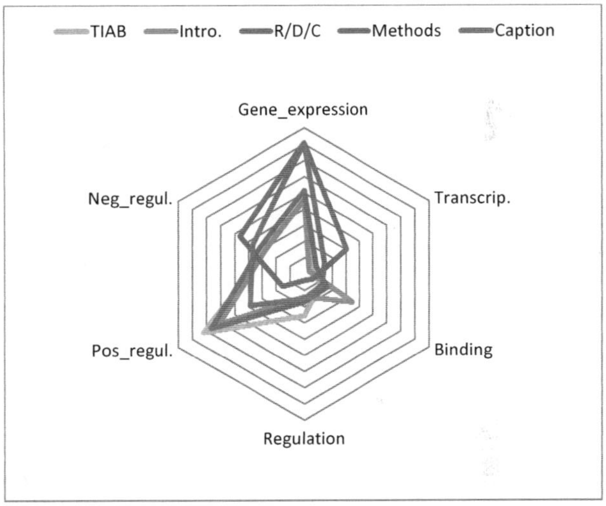

Figure 2: Event distribution in different sections

Item	Abstract	Full Paper					
		Whole	TIAB	Intro.	R/D/C	Methods	Caption
Words	*267229*	80962	3538	7878	43420	19406	6720
Proteins	*14969*	6580	336	597	3980	916	751
(Density: P / W)	*(5.60%)*	(8.13%)	(9.50%)	(7.58%)	(9.17%)	(4.72%)	(11.18%)
Events	*13603*	4436	272	427	3234	198	278
(Density: E / W)	*(5.09%)*	(5.48%)	(7.69%)	(5.42%)	(7.51%)	(1.02%)	(4.14%)
(Density: E / P)	*(90.87%)*	(67.42%)	(80.95%)	(71.52%)	(81.93%)	(21.62%)	(37.02%)
Gene_expression	*2816*	1193	62	98	841	80	112
Transcription	*795*	204	7	7	140	30	20
Protein_catabolism	*145*	3	0	0	3	0	0
Phosphorylation	*355*	137	12	12	101	10	2
Localization	*492*	47	3	15	22	7	0
Binding	*1485*	380	16	74	266	6	18
Regulation	*1426*	371	35	30	281	4	21
Positive_regulation	*4452*	1385	98	131	1087	15	54
Negative_regulation	*1637*	716	39	60	520	46	51

Table 3: Statistics of annotations in different sections of text: the *Abstract* column is of the abstraction collection (1210 titles and abstracts), and the following columns are of full paper collection (14 full papers). *TIAB* = title and abstract, *Intro.* = introduction and background, *R/D/C* = results, discussions, and conclusions, *Methods* = methods, materials, and experimental procedures. Some minor sections, supporting information, supplementary material, and synopsis, are ignored. *Density* = relative density of annotation (P/W = Protein/Word, E/W = Event/Word, and E/P = Event/Protein).

4 Participation

In total, 15 teams submitted final results. All 15 teams participated in the mandatory Task 1, four teams in Task 2, and two teams in Task 3. Only one team, `UTurku`, completed all the three tasks.

Table 4 shows the profile of the teams, excepting three who chose to remain anonymous. A brief examination on the team organization (the **People** column) suggests the importance of a computer science background, `C` and `BI`, to perform the GE task, which agrees with the same observation made in 2009. It is interpreted as follows: the role of computer scientists may be emphasized in part due to the fact that the task requires complex computational modeling, demanding particular efforts in framework design and implementation and computational resources. The **'09** column suggests that previous experience in the task may have affected to the performance of the teams, especially in a complex task like the GE task.

Table 5 shows the profile of the systems. A notable observation is that four teams developed their systems based on the model of `UTurku09` (Björne et al., 2009) which was the winning system of BioNLP-ST 2009. It may show an influence of the BioNLP-ST series in the task. For syntactic analyses, the prevailing use of Charniak Johnson re-ranking parser (Charniak and Johnson, 2005) using the self-trained biomedical model from McClosky (2008) (*McCCJ*) which is converted to Stanford Dependency (de Marneffe et al., 2006) is notable, which may also be an influence from the results of BioNLP-ST 2009. The last two teams, `XABioNLP` and `HCMUS`, who did not use syntactic analyses could not get a performance comparable to the others, which may suggest the importance of using syntactic analyses for a complex IE task like GE task.

5 Results

5.1 Task 1

Table 6 shows the final evaluation results of Task 1. For reference, the reported performance of the two systems, `UTurku09` and `Miwa10` is listed in the top. `UTurku09` was the winning system of Task 1 in 2009 (Björne et al., 2009), and `Miwa10` was the best system reported after BioNLP-ST 2009 (Miwa et al., 2010b). Particularly, the latter made

Team	'09	Task	People	reference
FAUST	√	12-	3C	(Riedel et al., 2011)
UMASS	√	12-	1C	(Riedel and McCallum, 2011)
UTurku	√	123	1BI	(Bjrne and Salakoski, 2011)
MSR-NLP		1--	4C	(Quirk et al., 2011)
ConcordU	√	1-3	2C	(Kilicoglu and Bergler, 2011)
UWMadison	√	1--	2C	(Vlachos and Craven, 2011)
Stanford		1--	3C+1.5L	(McClosky et al., 2011)
BMI@ASU	√	12-	3C	(Emadzadeh et al., 2011)
CCP-BTMG	√	1--	3BI	(Liu et al., 2011)
TM-SCS		1--	1C	(Bui and Sloot, 2011)
XABioNLP		1--	4C	(Casillas et al., 2011)
HCMUS		1--	6L	(Minh et al., 2011)

Table 4: Team profiles: The **'09** column indicates whether at least one team member participated in BioNLP-ST 2009. In **People** column, C=Computer Scientist, BI=Bioinformatician, B=Biologist, L=Linguist

	NLP		Task			Other resources	
Team	**Lexical Proc.**	**Syntactic Proc.**	**Trig.**	**Arg.**	**group**	**Dictionary**	**Other**
FAUST	SnowBall, CNLP	McCCJ+SD	Stacking (UMASS + Stanford)				
UMASS	SnowBall, CNLP	McCCJ+SD	Joint infer., Dual Decomposition				
UTurku	Porter	McCCJ+SD	SVM	SVM	SVM	S. cues	
MSR-NLP	Porter	McCCJ+SD, Enju	SVM	MaxEnt	rules		Coref(Hobbs)
ConcordU	-	McCCJ+SD	dic	rules	rules	S./N. cues	
UWMadison	Morpha, Porter	MCCCJ+SD	Joint infer., SEARN				
Stanford	Morpha, CNLP	McCCJ+SD	MaxEnt	MSTParser			word clusters
BMI@ASU	Porter, WordNet	Stanford+SD	SVM	SVM	-		MeSH
CCP-BTMG	Porter, WordNet	Stanford+SD	Subgraph Isomorphism				
TM-SCS	Stanford	Stanford	dic	rules	rules		
XABioNLP	KAF	-	rules				
HCMUS	OpenNLP	-	dic, rules	rules			UIMA

Table 5: System profiles: SnowBall=SnowBall Stemmer, CNLP=Stanford CoreNLP (tokenization), KAF=Kyoto Annotation Format McCCJ=McClosky-Charniak-Johnson Parser, Stanford=Stanford Parser, SD=Stanford Dependency Conversion, S.=Speculation, N.=Negation

an impressive improvement with Binding events (44.41%→52.62%).

The best performance in Task 1 this time is achieved by the `FAUST` system, which adopts a combination model of `UMass` and `Stanford`. Its performance on the *abstract collection*, 56.04%, demonstrates a significant improvement of the community in the repeated GE task, when compared to both `UTurku09`, 51.95% and `Miwa10`, 53.29%. The biggest improvement is made to the Regulation events (40.11%→46.97%) which requires a complex modeling for recursive event structure - an event may become an argument of another event. The second ranked system, `UMass`, shows the best performance on the *full paper collection*. It suggests that what `FAUST` obtained from the model combination might be a better optimization to abstracts.

The `ConcordU` system is notable as it is the sole rule-based system that is ranked above the average. It shows a performance optimized for precision with relatively low recall. The same tendency is roughly replicated by other rule-based systems, `CCP-BTMG`, `TM-SCS`, `XABioNLP`, and `HCMUS`. It suggests that a rule-based system might not be a good choice if a high coverage is desired. However, the performance of `ConcordU` for simple events suggests that a high precision can be achieved by a rule based system with a modest loss of recall. It might be more true when the task is less complex.

This time, three teams achieved better results than `Miwa10`, which indicates some role of focused efforts like BioNLP-ST. The comparison between the performance on abstract and full paper collections shows that generalization to full papers is feasible with very modest loss in performance.

5.2 Task 2

Tables 7 shows final evaluation results of Task 2. For reference, the reported performance of the task-winning system in 2009, `UT+DBCLS09` (Riedel et al., 2009), is shown in the top. The first and second ranked system, `FAUST` and `UMass`, which share a same author with `Riedel09`, made a significant improvement over `Riedel09` in the *abstract collection*. `UTurku` achieved the best performance in finding sites arguments but did not produce location arguments. In table 7, the performance of all the systems in *full text collection* suggests that finding secondary arguments in full text is much more challenging.

In detail, a significant improvement was made for *Location* arguments (36.59%→50.00%). A further breakdown of the results of *site* extraction, shown in table 8, shows that finding *site* arguments for *Phosphorylation*, *Binding* and *Regulation* events are all significantly improved, but in different ways. The extraction of protein sites to be phosphorylated is approaching a practical level of performance (84.21%), while protein sites to be bound or to be regulated remains challenging to be extracted.

5.3 Task 3

Table 9 shows final evaluation results of Task 3. For reference, the reported performance of the task-winning system in 2009, `Kilicoglu09`(Kilicoglu and Bergler, 2009), is shown in the top. Among the two teams participated in the task, `UTurku` showed a better performance in extracting negated events, while `ConcordU` showed a better performance in extracting speculated events.

6 Conclusions

The Genia event task which was repeated for BioNLP-ST 2009 and 2011 took a role of measuring the progress of the community and generalization IE technology to full papers. The results from 15 teams who made their final submissions to the task show that a clear advance of the community in terms of the performance on a focused domain and also generalization to full papers. To our disappointment, however, an effective use of supporting task results was not observed, which thus remains as future work for further improvement.

Acknowledgments

This work is supported by the "Integrated Database Project" funded by the Ministry of Education, Culture, Sports, Science and Technology of Japan.

References

Jari Björne, Juho Heimonen, Filip Ginter, Antti Airola, Tapio Pahikkala, and Tapio Salakoski. 2009. Extracting complex biological events with rich graph-based feature sets. In *Proceedings of the BioNLP 2009 Workshop Companion Volume for Shared Task*, pages 10–18, Boulder, Colorado, June. Association for Computational Linguistics.

Jari Björne, Filip Ginter, Sampo Pyysalo, Jun'ichi Tsujii, and Tapio Salakoski. 2010. Complex event extraction at PubMed scale. *Bioinformatics*, 26(12):i382–390.

Jari Bjrne and Tapio Salakoski. 2011. Generalizing Biomedical Event Extraction. In *Proceedings of the BioNLP 2011 Workshop Companion Volume for Shared Task*, Portland, Oregon, June. Association for Computational Linguistics.

Quoc-Chinh Bui and Peter. M.A. Sloot. 2011. Extracting biological events from text using simple syntactic patterns. In *Proceedings of the BioNLP 2011 Workshop Companion Volume for Shared Task*, Portland, Oregon, June. Association for Computational Linguistics.

Arantza Casillas, Arantza Daz de Ilarraza, Koldo Gojenola, Maite Oronoz, and German Rigau. 2011. Using Kybots for Extracting Events in Biomedical Texts. In *Proceedings of the BioNLP 2011 Workshop Companion Volume for Shared Task*, Portland, Oregon, June. Association for Computational Linguistics.

Eugene Charniak and Mark Johnson. 2005. Coarse-to-Fine n-Best Parsing and MaxEnt Discriminative Reranking. In *Proceedings of the 43rd Annual Meeting of the Association for Computational Linguistics (ACL'05)*, pages 173–180.

Marie-Catherine de Marneffe, Bill MacCartney, and Christopher D. Manning. 2006. Generating Typed Dependency Parses from Phrase Structure Parses. In *Proceedings of the Fifth International Conference on Language Resources and Evaluation (LREC'06)*, pages 449–454.

Ehsan Emadzadeh, Azadeh Nikfarjam, and Graciela Gonzalez. 2011. Double Layered Learning for Biological Event Extraction from Text. In *Proceedings of the BioNLP 2011 Workshop Companion Volume for Shared Task*, Portland, Oregon, June. Association for Computational Linguistics.

Halil Kilicoglu and Sabine Bergler. 2009. Syntactic dependency based heuristics for biological event extraction. In *Proceedings of the BioNLP 2009 Workshop Companion Volume for Shared Task*, pages 119–127, Boulder, Colorado, June. Association for Computational Linguistics.

Halil Kilicoglu and Sabine Bergler. 2011. Adapting a General Semantic Interpretation Approach to Biological Event Extraction. In *Proceedings of the BioNLP 2011 Workshop Companion Volume for Shared Task*, Portland, Oregon, June. Association for Computational Linguistics.

Jin-Dong Kim, Tomoko Ohta, Sampo Pyysalo, Yoshinobu Kano, and Jun'ichi Tsujii. 2009. Overview of BioNLP'09 Shared Task on Event Extraction. In *Proceedings of Natural Language Processing in Biomedicine (BioNLP) NAACL 2009 Workshop*, pages 1–9.

Jin-Dong Kim, Sampo Pyysalo, Tomoko Ohta, Robert Bossy, and Jun'ichi Tsujii. 2011. Overview of BioNLP Shared Task 2011. In *Proceedings of the BioNLP 2011 Workshop Companion Volume for Shared Task*, Portland, Oregon, June. Association for Computational Linguistics.

Haibin Liu, Ravikumar Komandur, and Karin Verspoor. 2011. From graphs to events: A subgraph matching approach for information extraction from biomedical text. In *Proceedings of the BioNLP 2011 Workshop Companion Volume for Shared Task*, Portland, Oregon, June. Association for Computational Linguistics.

David McClosky and Eugene Charniak. 2008. Self-Training for Biomedical Parsing. In *Proceedings of the 46th Annual Meeting of the Association for Computational Linguistics - Human Language Technologies (ACL-HLT'08)*, pages 101–104.

David McClosky, Mihai Surdeanu, and Christopher Manning. 2011. Event Extraction as Dependency Parsing for BioNLP 2011. In *Proceedings of the BioNLP 2011 Workshop Companion Volume for Shared Task*, Portland, Oregon, June. Association for Computational Linguistics.

Quang Le Minh, Son Nguyen Truong, and Quoc Ho Bao. 2011. A pattern approach for Biomedical Event Annotation . In *Proceedings of the BioNLP 2011 Workshop Companion Volume for Shared Task*, Portland, Oregon, June. Association for Computational Linguistics.

Makoto Miwa, Sampo Pyysalo, Tadayoshi Hara, and Jun'ichi Tsujii. 2010a. A comparative study of syntactic parsers for event extraction. In *Proceedings of BioNLP'10*, pages 37–45.

Makoto Miwa, Rune Sætre, Jin-Dong Kim, and Jun'ichi Tsujii. 2010b. Event extraction with complex event classification using rich features. *Journal of Bioinformatics and Computational Biology (JBCB)*, 8(1):131–146, February.

Hoifung Poon and Lucy Vanderwende. 2010. Joint inference for knowledge extraction from biomedical literature. In *Proceedings of NAACL-HLT'10*, pages 813–821.

Chris Quirk, Pallavi Choudhury, Michael Gamon, and Lucy Vanderwend. 2011. MSR-NLP Entry in BioNLP Shared Task 2011. In *Proceedings of the BioNLP 2011 Workshop Companion Volume for Shared Task*, Portland, Oregon, June. Association for Computational Linguistics.

Sebastian Riedel and Andrew McCallum. 2011. Robust Biomedical Event Extraction with Dual Decomposition and Minimal Domain Adaptation. In *Proceedings of the BioNLP 2011 Workshop Companion Volume for Shared Task*, Portland, Oregon, June. Association for Computational Linguistics.

Sebastian Riedel, Hong-Woo Chun, Toshihisa Takagi, and Jun'ichi Tsujii. 2009. A markov logic approach to bio-molecular event extraction. In *Proceedings of the BioNLP 2009 Workshop Companion Volume for Shared Task*, pages 41–49, Boulder, Colorado, June. Association for Computational Linguistics.

Sebastian Riedel, David McClosky, Mihai Surdeanu, Andrew McCallum, and Christopher Manning. 2011. Model Combination for Event Extraction in BioNLP 2011. In *Proceedings of the BioNLP 2011 Workshop Companion Volume for Shared Task*, Portland, Oregon, June. Association for Computational Linguistics.

Andreas Vlachos and Mark Craven. 2011. Biomedical Event Extraction from Abstracts and Full Papers using Search-based Structured Prediction. In *Proceedings of the BioNLP 2011 Workshop Companion Volume for Shared Task*, Portland, Oregon, June. Association for Computational Linguistics.

Andreas Vlachos. 2010. Two strong baselines for the bionlp 2009 event extraction task. In *Proceedings of BioNLP'10*, pages 1–9.

Team		Simple Event	Binding	Regulation	All
UTurku09	A	*64.21 / 77.45 / 70.21*	*40.06 / 49.82 / 44.41*	*35.63 / 45.87 / 40.11*	*46.73 / 58.48 / 51.95*
Miwa10	A	*70.44*	*52.62*	*40.60*	*48.62 / 58.96 / 53.29*
	W	**68.47 / 80.25 / 73.90**	44.20 / 53.71 / 48.49	**38.02 / 54.94 / 44.94**	**49.41 / 64.75 / 56.04**
FAUST	A	**66.16 / 81.04 / 72.85**	**45.53 / 58.09 / 51.05**	**39.38 / 58.18 / 46.97**	**50.00 / 67.53 / 57.46**
	F	75.58 / 78.23 / 76.88	40.97 / 44.70 / 42.75	**34.99 / 48.24 / 40.56**	47.92 / 58.47 / 52.67
	W	67.01 / 81.40 / 73.50	**42.97 / 56.42 / 48.79**	37.52 / 52.67 / 43.82	48.49 / 64.08 / 55.20
UMass	A	64.21 / 80.74 / 71.54	43.52 / 60.89 / 50.76	38.78 / 55.07 / 45.51	48.74 / 65.94 / 56.05
	F	**75.58 / 83.14 / 79.18**	**41.67 / 47.62 / 44.44**	34.72 / 47.51 / 40.12	**47.84 / 59.76 / 53.14**
	W	68.22 / 76.47 / 72.11	42.97 / 43.60 / 43.28	38.72 / 47.64 / 42.72	49.56 / 57.65 / 53.30
UTurku	A	64.97 / 76.72 / 70.36	45.24 / 50.00 / 47.50	40.41 / 49.01 / 44.30	50.06 / 59.48 / 54.37
	F	78.18 / 75.82 / 76.98	37.50 / 31.76 / 34.39	34.99 / 44.46 / 39.16	48.31 / 53.38 / 50.72
	W	68.99 / 74.30 / 71.54	42.36 / 40.47 / 41.39	36.64 / 44.08 / 40.02	48.64 / 54.71 / 51.50
MSR-NLP	A	65.99 / 74.71 / 70.08	43.23 / 44.51 / 43.86	37.14 / 45.38 / 40.85	48.52 / 56.47 / 52.20
	F	78.18 / 73.24 / 75.63	40.28 / 32.77 / 36.14	35.52 / 41.34 / 38.21	48.94 / 50.77 / 49.84
	W	59.99 / **85.53** / 70.52	29.33 / 49.66 / 36.88	35.72 / 45.85 / 40.16	43.55 / 59.58 / 50.32
ConcordU	A	56.51 / **84.56** / 67.75	29.97 / 49.76 / 37.41	36.24 / 47.09 / 40.96	43.09 / 60.37 / 50.28
	F	70.65 / **88.03** / 78.39	27.78 / 49.38 / 35.56	34.58 / 43.22 / 38.42	44.71 / 57.75 / 50.40
	W	59.67 / 80.95 / 68.70	29.33 / 49.66 / 36.88	34.10 / 49.46 / 40.37	42.56 / 61.21 / 50.21
UWMadison	A	54.99 / 79.85 / 65.13	34.87 / 56.81 / 43.21	34.54 / 50.67 / 41.08	42.17 / 62.30 / 50.30
	F	74.03 / 83.58 / 78.51	15.97 / 29.87 / 20.81	33.11 / 46.87 / 38.81	43.53 / 58.73 / 50.00
	W	65.79 / 76.83 / 70.88	**39.92 / 49.87 / 44.34**	27.55 / 48.75 / 35.21	42.36 / 61.08 / 50.03
Stanford	A	62.61 / 77.57 / 69.29	**42.36 / 54.24 / 47.57**	28.25 / 49.95 / 36.09	42.55 / 62.69 / 50.69
	F	75.58 / 75.00 / 75.29	**34.03 / 40.16 / 36.84**	26.01 / 46.08 / 33.25	41.88 / 57.36 / 48.41
	W	62.09 / 76.55 / 68.57	27.90 / 44.92 / 34.42	22.30 / 40.26 / 28.70	36.91 / 56.63 / 44.69
BMI@ASU	A	58.71 / 78.51 / 67.18	26.22 / 47.40 / 33.77	22.99 / 40.47 / 29.32	36.61 / 57.82 / 44.83
	F	72.47 / 72.09 / 72.28	31.94 / 40.71 / 35.80	20.78 / 39.74 / 27.29	37.65 / 53.93 / 44.34
	W	53.61 / 75.13 / 62.57	22.61 / 49.12 / 30.96	19.01 / 43.80 / 26.51	31.57 / 58.99 / 41.13
CCP-BTMG	A	50.93 / 74.50 / 60.50	25.65 / 53.29 / 34.63	19.54 / 43.47 / 26.96	31.87 / 59.02 / 41.39
	F	61.82 / 76.77 / 68.49	15.28 / 37.29 / 21.67	17.83 / 44.63 / 25.48	30.82 / 58.92 / 40.47
	W	57.33 / 71.34 / 63.57	34.01 / 44.77 / 38.66	16.39 / 25.37 / 19.91	32.73 / 45.84 / 38.19
TM-SCS	A	53.65 / 71.66 / 61.36	36.02 / 49.41 / 41.67	18.29 / 27.07 / 21.83	33.36 / 47.09 / 39.06
	F	68.57 / 70.59 / 69.57	29.17 / 35.00 / 31.82	12.20 / 21.02 / 15.44	31.14 / 42.83 / 36.06
	W	43.71 / 47.18 / 45.38	05.30 / 50.00 / 09.58	05.79 / 26.94 / 09.54	19.07 / 42.08 / 26.25
XABioNLP	A	39.76 / 45.90 / 42.61	06.34 / 56.41 / 11.40	04.72 / 23.21 / 07.84	17.91 / 40.74 / 24.89
	F	55.84 / 50.23 / 52.89	02.78 / 30.77 / 05.10	08.18 / 33.89 / 13.17	21.96 / 45.09 / 29.54
	W	24.82 / 35.14 / 29.09	04.68 / 12.92 / 06.88	01.63 / 10.40 / 02.81	10.12 / 27.17 / 14.75
HCMUS	A	22.42 / 37.38 / 28.03	04.61 / 10.46 / 06.40	01.69 / 10.37 / 02.91	09.71 / 27.30 / 14.33
	F	32.21 / 31.16 / 31.67	04.86 / 28.00 / 08.28	01.47 / 10.48 / 02.59	11.14 / 26.89 / 15.75

Table 6: Evaluation results (recall / precision / f-score) of Task 1 in (W)hole data set, (A)bstracts only, and (F)ull papers only. Some notable figures are emphasized in bold.

Team		Sites (222)	Locations (66)	All (288)
UT+DBCLS09	A		*23.08 / 88.24 / 36.59*	*32.14 / 72.41 / 44.52*
	W	32.88 / 70.87 / 44.92	36.36 / 75.00 / 48.98	**33.68 / 71.85 / 45.86**
FAUST	A	43.51 / 71.25 / 54.03	**36.92 / 77.42 / 50.00**	**41.33 / 72.97 / 52.77**
	F	17.58 / 69.57 / 28.07	-	17.39 / 66.67 / 27.59
	W	31.98 / 71.00 / 44.10	**36.36 / 77.42 / 49.48**	32.99 / 72.52 / 45.35
UMass	A	42.75 / 70.00 / 53.08	36.92 / 77.42 / 50.00	40.82 / 72.07 / 52.12
	F	16.48 / 75.00 / 27.03	-	16.30 / 75.00 / 26.79
	W	32.88 / 62.93 / 43.20	22.73 / 83.33 / 35.71	30.56 / 65.67 / 41.71
BMI@ASU	A	37.40 / 67.12 / 48.04	23.08 / 83.33 / 36.14	32.65 / 70.33 / 44.60
	F	26.37 / 55.81 / 35.82	-	26.09 / 55.81 / 35.56
	W	**40.09 / 65.44 / 49.72**	00.00 / 00.00 / 00.00	30.90 / 65.44 / 41.98
UTurku	A	**48.09 / 69.23 / 56.76**	00.00 / 00.00 / 00.00	32.14 / 69.23 / 43.90
	F	**28.57 / 57.78 / 38.24**	-	**28.26 / 57.78 / 37.96**

Table 7: Evaluation results of Task 2 in (W)hole data set, (A)bstracts only, and (F)ull papers only

Team		Phospho. (67)	Binding (84)	Reg. (71)
Riedel'09	A	*71.43 / 71.43 / 71.43*	*04.76 / 50.00 / 08.70*	*12.96 / 58.33 / 21.21*
	W	71.64 / 84.21 / 77.42	05.95 / 38.46 / 10.31	**28.17 / 60.61 / 38.46**
FAUST	A	71.43 / 81.63 / 76.19	04.76 / 14.29 / 07.14	**29.63 / 66.67 / 41.03**
	F	**72.73 / 100.0 / 84.21**	06.35 / 66.67 / 11.59	23.53 / 44.44 / 30.77
	W	76.12 / 79.69 / 77.86	04.76 / 36.36 / 08.42	22.54 / 64.00 / 33.33
UMass	A	76.79 / 76.79 / 76.79	04.76 / 14.29 / 07.14	22.22 / 70.59 / 33.80
	F	**72.73 / 100.0 / 84.21**	04.76 / 75.00 / 08.96	**23.53 / 50.00 / 32.00**
	W	52.24 / 97.22 / 67.96	20.24 / 53.12 / 29.31	29.58 / 43.75 / 35.29
BMI@ASU	A	53.57 / 96.77 / 68.97	**09.52 / 22.22 / 13.33**	31.48 / 51.52 / 39.08
	F	45.45 / 100.0 / 62.50	23.81 / 65.22 / 34.88	23.53 / 26.67 / 25.00
	W	**76.12 / 91.07 / 82.93**	**21.43 / 51.43 / 30.25**	28.17 / 44.44 / 34.48
UTurku	A	**78.57 / 89.80 / 83.81**	09.52 / 18.18 / 12.50	31.48 / 54.84 / 40.00
	F	63.64 / 100.0 / 77.78	**25.40 / 66.67 / 36.78**	17.65 / 21.43 / 19.35

Table 8: Evaluation results of Site information for different event types in (A)bstracts

Team		Negation	Speculation	All
Kilicoglu09	A	*14.98 / 50.75 / 23.13*	*16.83 / 50.72 / 25.27*	*15.86 / 50.74 / 24.17*
	W	**22.87 / 48.85 / 31.15**	17.86 / 32.54 / 23.06	**20.30 / 39.67 / 26.86**
UTurku	A	**22.03 / 49.02 / 30.40**	19.23 / 38.46 / 25.64	**20.69 / 43.69 / 28.08**
	F	**25.76 / 48.28 / 33.59**	15.00 / 23.08 / 18.18	19.28 / 30.85 / 23.73
	W	18.77 / 44.26 / 26.36	**21.10 / 38.46 / 27.25**	19.97 / 40.89 / 26.83
ConcordU	A	18.06 / 46.59 / 26.03	**23.08 / 40.00 / 29.27**	20.46 / 42.79 / 27.68
	F	21.21 / 38.24 / 27.29	**17.00 / 34.69 / 22.82**	**18.67 / 36.14 / 24.63**

Table 9: Evaluation results of Task 3 in (W)hole data set, (A)bstracts only, and (F)ull papers only

Overview of the Epigenetics and Post-translational Modifications (EPI) task of BioNLP Shared Task 2011

Tomoko Ohta* **Sampo Pyysalo*** **Jun'ichi Tsujii**†
*Department of Computer Science, University of Tokyo, Tokyo, Japan
†Microsoft Research Asia, Beijing, China
{okap,smp}@is.s.u-tokyo.ac.jp, jtsujii@microsoft.com

Abstract

This paper presents the preparation, resources, results and analysis of the Epigenetics and Post-translational Modifications (EPI) task, a main task of the BioNLP Shared Task 2011. The task concerns the extraction of detailed representations of 14 protein and DNA modification events, the catalysis of these reactions, and the identification of instances of negated or speculatively stated event instances. Seven teams submitted final results to the EPI task in the shared task, with the highest-performing system achieving 53% F-score in the full task and 69% F-score in the extraction of a simplified set of core event arguments.

1 Introduction

The Epigenetics and Post-translational Modifications (EPI) task is a shared task on event extraction from biomedical domain scientific publications, first introduced as a main task in the BioNLP Shared Task 2011 (Kim et al., 2011a).

The EPI task focuses on events relating to epigenetic change, including DNA methylation and histone methylation and acetylation (see e.g. (Holliday, 1987; Jaenisch and Bird, 2003)), as well as other common protein post-translational modifications (PTMs) (Witze et al., 2007). PTMs are chemical modifications of the amino acid residues of proteins, and DNA methylation a parallel modification of the nucleotides on DNA. While these modifications are chemically simple reactions and can thus be straightforwardly represented in full detail, they have a crucial role in the regulation of gene expression and protein function: the modifications can alter the conformation of DNA or proteins and thus control their ability to associate with other molecules, making PTMs key steps in protein biosynthesis for introducing the full range of protein functions. For instance, protein phosphorylation – the attachment of phosphate – is a common mechanism for activating or inactivating enzymes by altering the conformation of protein active sites (Stock et al., 1989; Barford et al., 1998), and protein ubiquitination – the post-translational attachment of the small protein ubiquitin – is the first step of a major mechanism for the destruction (breakdown) of many proteins (Glickman and Ciechanover, 2002).

Many of the PTMs targeted in the EPI task involve modification of histone, a core protein that forms an octameric complex that has a crucial role in packaging chromosomal DNA. The level of methylation and acetylation of histones controls the tightness of the chromatin structure, and only "unwound" chromatin exposes the gene packed around the histone core to the transcriptional machinery. Since histone modification is of substantial current interest in epigenetics, we designed aspects of the EPI task to capture the full detail in which histone modification events are stated in text. Finally, the DNA methylation of gene regulatory elements controls the expression of the gene by altering the affinity with which DNA-binding proteins (including transcription factors) bind, and highly methylated genes are not transcribed at all (Riggs, 1975; Holliday and Pugh, 1975). DNA methylation can thus "switch off" genes, "removing" them from the genome in a way that is reversible through DNA demethylation.

Proceedings of BioNLP Shared Task 2011 Workshop, pages 16–25,
Portland, Oregon, USA, 24 June, 2011.

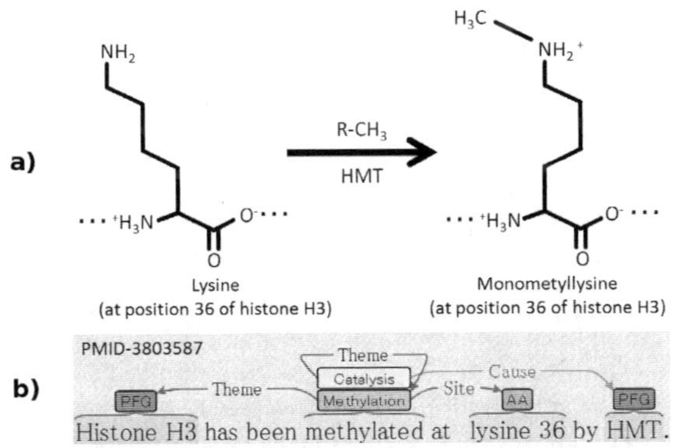

c)

Protein ID	Feature Type	Feature Description	Feature Position	Pubmed ID
HSKW3	Modified site	N6-methyllysine(Lys)	36	3803587

Figure 1: Three views of protein methylation. a) chemical formula b) event representation c) modification database entry.

The BioNLP'09 Shared Task on Event Extraction (Kim et al., 2009), the first task in the present shared task series, involved the extraction of nine event types including one PTM type, PHOSPHORYLATION. The results of the shared task showed this PTM event to be the single most reliably extracted event type in the task, with the best-performing system for the type achieving 91% precision and 76% recall (83% F-score) in its extraction (Buyko et al., 2009). The results suggest both that the event representation is well applicable to PTM extraction and that current extraction methods are capable of reliable PTM extraction. The EPI task follows up on these opportunities, introducing specific, strongly biologically motivated extraction targets that are expected to be both feasible for high-accuracy event extraction, relevant to the needs of present-day molecular biology, and closely applicable to biomolecular database curation needs (see Figure 1) (Ohta et al., 2010a).

2 Task Setting

The EPI task is an *event extraction task* in the sense popularized by a number of recent domain resources and challenges (e.g. (Pyysalo et al., 2007; Kim et al., 2008; Thompson et al., 2009; Kim et al., 2009; Ananiadou et al., 2010)). In broad outline, the task focuses on the extraction of information on statements regarding change in the state or properties of (physical) entities, modeled using an *event representation*.

Figure 2: Illustration of the event representation. An event of type METHYLATION (expressed through the text "methylation") with two participants of the types PROTEIN ("histone H3") and ENTITY ("Lys9"), participating in the event in *Theme* and *Site* roles, respectively.

In this representation, events are typed n-ary associations of participants (entities or other events) in specific roles. Events are bound to specific expressions in text (the *event trigger* or *text binding*) and are primary objects of annotation, allowing them to be marked in turn e.g. as negated or as participants in other events. Figure 2 illustrates these concepts.

In its specific formulation, EPI broadly follows the definition of the BioNLP'09 shared task on event extraction. Basic modification events are defined similarly to the PHOSPHORYLATION event type targeted in the '09 and the 2011 GE and ID tasks (Kim et al., 2011b; Pyysalo et al., 2011b), with the full task extending previously defined arguments with two additional ones, *Sidechain* and *Contextgene*.

2.1 Entities

The EPI task follows the general policy of the BioNLP Shared Task in isolating the basic task of named entity recognition from the event extraction task by providing task participants with manually annotated gene and gene product entities as a starting point for extraction. The entity types follow the BioNLP'09 Shared Task scheme, where genes and their products are simply marked as PROTEIN.[1]

In addition to the given PROTEIN entities, some events involve other entities, such as the modification *Site*. These entities are not given and must thus be identified by systems targeting the full task (see Section 4). In part to reduce the demands of this entity recognition component of the task, these additional entities are not given specific types but are generically marked as ENTITY.

[1]While most of the modifications targeted in the task involve proteins, this naming is somewhat inaccurate for the *Themes* of DNA METHYLATION and DNA DEMETHYLATION events and for *Contextgene* arguments, which refer to genes. Despite this inaccuracy, we chose to follow this naming scheme for consistency with other tasks.

Type	Core arguments	Additional arguments
HYDROXYLATION	*Theme*(PROTEIN)	*Site*(ENTITY)
PHOSPHORYLATION	*Theme*(PROTEIN)	*Site*(ENTITY)
UBIQUITINATION	*Theme*(PROTEIN)	*Site*(ENTITY)
DNA METHYLATION	*Theme*(PROTEIN)	*Site*(ENTITY)
GLYCOSYLATION	*Theme*(PROTEIN)	*Site*(ENTITY), *Sidechain*(ENTITY)
ACETYLATION	*Theme*(PROTEIN)	*Site*(ENTITY), *Contextgene*(PROTEIN)
METHYLATION	*Theme*(PROTEIN)	*Site*(ENTITY), *Contextgene*(PROTEIN)
CATALYSIS	*Theme*(Event), *Cause*(PROTEIN)	

Table 1: Event types and their arguments. The type of entity allowed as argument is specified in parenthesis. For each event type except CATALYSIS, the reverse reaction (e.g. DEACETYLATION for ACETYLATION) is also defined, with identical arguments. The total number of event types in the task is thus 15.

2.2 Relations

The EPI task does not define any explicit relation extraction targets. However, the task annotation involves one relation type, EQUIV. This is a binary, symmetric, transitive relation between entities that defines two entities to be equivalent (Hoehndorf et al., 2010). The relation is used in the gold annotation to mark local aliases such as the full and abbreviated forms of a protein name as referring to the same real-world entity. While the '09 task only recognized equivalent PROTEIN entities, EPI extends on the scope of EQUIV annotations in allowing entities of any type to be marked equivalent. In evaluation, references to any of a set of equivalent entities are treated identically.

2.3 Events

While the EPI task entity definition closely follows that of the previous shared task, the task introduces considerable novelty in the targeted events, adding a total of 14 novel event types and two new participant roles. Table 1 summarizes the targeted event types and their arguments.

As in the BioNLP'09 shared task, *Theme* arguments identify the entity that the event is *about*, such as the protein that is acetylated in an acetylation event. A *Theme* is always mandatory for all EPI task events. *Site* arguments identify the modification site on the *Theme* entity, such as a specific residue on a modified protein or a specific region on a methylated gene. The *Sidechain* argument, specific to GLYCOSYLATION and DEGLYCOSYLATION among the targeted events, identifies the moiety attached or removed in the event (in glycosylation, the sugar).[2] Finally, the *Contextgene* argument, specific to ACETYLATION and METHYLATION events and their reverse reactions, identifies the gene whose expression is controlled by these modifications. This argument applies specifically for histone protein modification: the modification of the histones that form the nucleosomes that structure DNA are key to the epigenetic control of gene expression. The *Site*, *Sidechain* and *Contextgene* arguments are not mandatory, and should only be extracted when explicitly stated.

For CATALYSIS events, representing the catalysis of protein or DNA modification by another protein, both *Theme* and *Cause* are mandatory. While CATALYSIS is a new event type, it is related to the '09 POSITIVE_REGULATION type by a class-subclass relation: any CATALYSIS event is a POSITIVE_REGULATION event in the '09 task terms (but not vice versa).

2.4 Event modifications

In addition to events, the EPI task defines two *event modification* extraction targets: NEGATION and SPECULATION. Both are represented as simple binary "flags" that apply to events, marking them as being explicitly negated (e.g. *H2A is not methylated*) or stated in a speculative context (e.g. *H2A may be methylated*). Events may be both negated and speculated.

[2]Note that while arguments similar to *Sidechain* could be defined for other event types also, their extraction would provide no additional information: the attached molecule is always acetyl in acetylation, methyl in methylation, etc.

3 Data

The primary EPI task data were annotated specifically for the BioNLP Shared Task 2011 and are not based on any previously released resource. Before starting this annotation effort, we performed two preparatory studies using in part previously released related datasets: in (Ohta et al., 2010a) we considered the extraction of four protein post-translational modifications event types with reference to annotations originally created for the Protein Information Resource[3] (PIR) (Wu et al., 2003), and in (Ohta et al., 2010b) we studied the annotation and extraction of DNA methylation events with reference to annotations created for the PubMeth[4] (Ongenaert et al., 2008) database. The corpus text selection and annotation scheme were then defined following the understanding formed in these studies.

3.1 Document selection

The texts for the EPI task corpus were drawn from PubMed abstracts. In selecting the primary corpus texts, we aimed to gather a representative sample of all PubMed documents relevant to selected modification events, avoiding bias toward, for example, specific genes/proteins, species, forms of event expression, or subdomains. We primarily targeted DNA methylation and the "prominent PTM types" identified in (Ohta et al., 2010a). We defined the following document selection protocol: for each of the targeted event types, 1) Select a random sample of PubMed abstracts annotated with the MeSH term corresponding to the target event (e.g. `Acetylation`) 2) Automatically tag protein/gene entities in the selected abstracts, removing ones where fewer than a specific cutoff are found 3) Perform manual filtering removing documents not relevant to the targeted topic (optional).

MeSH is a controlled vocabulary of over 25,000 terms that is used to manually annotate each document in PubMed. By performing initial document retrieval using MeSH terms it is possible to select relevant documents without bias toward specific expressions in text. While search for documents tagged with e.g. the `Acetylation` MeSH term is sufficient to select documents relevant to the modification, not all such documents necessarily concern specifically protein modification, necessitating a filtering step. Following preliminary experiments, we chose to apply the BANNER named entity tagger (Leaman and Gonzalez, 2008) trained on the GENETAG corpus (Tanabe et al., 2005) and to filter documents where fewer than five entities were identified. Finally, for some modification types this protocol selected also a substantial number of non-relevant documents. In these cases a manual filtering step was performed prior to full annotation to avoid marking large numbers of non-relevant abstracts.

This primary corpus text selection protocol does not explicitly target reverse reactions such as deacetylation, and the total number of these events in the resulting corpus was low for many types. To be able to measure the extraction performance for these types, we defined a secondary selection protocol that augmented the primary protocol with a regular expression-based filter removing documents that did not (likely) contain mentions of reverse reactions. This protocol was used to select a secondary set of test abstracts enriched in mentions of reverse reactions. Performance on this secondary test set was also evaluated, but is not part of the primary task evaluation. Due to space considerations, we only present the primary test set results in this paper, referring to the shared task website for the secondary results.

3.2 Annotation

Annotation was performed manually. The gene/protein entities automatically detected in the document selection step were provided to annotators for reference for creating PROTEIN annotations, but all entity annotations were checked and revised to conform to the specific guidelines for the task.[5] For the annotation of PROTEIN entities, we adopted the GENIA gene/gene product (GGP) annotation guidelines (Ohta et al., 2009), adding one specific exception: while the primary guidelines require that only specific individual gene or gene product names are annotated, we allowed also the annotation of mentions of groups of histones or

[3] `http://pir.georgetown.edu`

[4] `http://www.pubmeth.org/`

[5] This revision was substantial: only approximately 65% of final PROTEIN annotations exactly match an automatically predicted one due to differences in annotation criteria (Wang et al., 2009).

the entire histone protein family to capture histone modification events also in cases where only the group is mentioned.

All event annotations were created from scratch without automatic support to avoid bias toward specific automatic extraction methods or approaches. The event annotation follows the GENIA event corpus annotation guidelines (Kim et al., 2008) as they apply to protein modifications, with CATALYSIS being annotated following the criteria for the POSITIVE_REGULATION event type with the additional constraints that the *Cause* of the event is a gene or gene product entity and the form of regulation is catalysis of a modification reaction.

The manual annotation was performed by three experienced annotators with a molecular biology background, with one chief annotator with extensive experience in domain event annotation organizing and supervising the annotator training and the overall process. After completion of primary annotation, we performed a final check targeting simple human errors using an automatic extraction system.[6] This correction process resulted in the revision of approximately 2% of the event annotations. To evaluate the consistency of the annotation, we performed independent event annotation (taking PROTEIN annotations as given) for a random sample of 10% of the corpus documents. Comparison of the two manually created sets of event annotations under the primary task evaluation criteria gave an F-score of 82% for the full task and 89% for the core task.[7] We found that CATALYSIS events were particularly challenging, showing just 65% agreement for the core task.

Table 2 shows the statistics of the primary task data. We note that while the corpus is broadly comparable in size to the BioNLP'09 shared task dataset (Kim et al., 2009) in terms of the number of abstracts and annotated entities, the number of annotated events in the EPI corpus is approximately 20% of that in the '09 dataset, reflecting the more focused event types.

[6]High-confidence system predictions differing from gold annotations were provided to a human annotator, not used directly to change corpus data. To further reduce the risk of bias, we only informed the annotator of the entities involved, not of the predicted event structure.

[7]Due to symmetry of precision/recall and the applied criteria, this score was not affected by the choice of which set of annotations to consider as "gold" for the comparison.

Item	Training	Devel	Test
Abstract	600	200	400
Word	127,312	43,497	82,819
Protein	7,595	2,499	5,096
Event	1,852	601	1,261
Modification	173	79	117

Table 2: Statistics of the EPI corpus. Test set statistics shown only for the primary test data.

4 Evaluation

Evaluation is instance- and event-oriented and based on the standard precision/recall/F-score[8] metrics. The primary evaluation criteria are the same as in the BioNLP'09 shared task, incorporating the "approximate span matching" and "approximate recursive matching" variants to strict matching. In brief, under these criteria text-bound annotations (event triggers and entities) in a submission are considered to match a corresponding gold annotation if their span is contained within the (mildly extended) span of the gold annotation, and events that refer to other events as arguments are considered to match if the *Theme* arguments of the recursively referred events match, that is, non-*Theme* arguments are ignored in recursively referred events. For a detailed description of these evaluation criteria, we refer to (Kim et al., 2009).

In addition to the primary evaluation criteria, we introduced a new relaxed evaluation criterion we term *single partial penalty*. Under the primary criteria, when a predicted event matches a gold event in some of its arguments but lacks one or more arguments of the gold event, the submission is arguably given a double penalty: the predicted event is counted as a false positive (FP), and the gold event is counted as a false negative (FN). Under the single partial penalty evaluation criterion, predicted events that match a gold event in all their arguments are not counted as FP, although the corresponding gold event still counts as FN (the "single penalty"). Analogously, gold events that partially match a predicted event are not counted as FN, although the corresponding predicted event with "extra" arguments counts as FP. This criterion can give a more nuanced view of performance for partially correctly predicted events.

[8]Specifically F_1. F is used for short throughout.

			NLP		Events				Other resources	
Rank	Team	Org	word	parse	trigger	arg	group	modif.	corpora	other
1	UTurku	1BI	Porter	McCCJ + SD	SVM	SVM	SVM	SVM	-	hedge words
2	FAUST	3NLP	CoreNLP, SnowBall	McCCJ + SD	(UMass+Stanford as features)			-	-	word clusters
3	MSR-NLP	1SDE, 3NLP	Porter, custom	McCCJ + SD, Enju	SVM	SVM	SVM	-	-	triggers, word clusters
4	UMass	1NLP	CoreNLP, SnowBall	McCCJ + SD	Joint, dual decomposition			-	-	-
5	Stanford	3NLP	custom	McCCJ + SD	MaxEnt	Joint, MSTParser		-	-	word clusters
6	CCP-BTMG	3BI	Porter, WN-lemma	Stanford + SD	Graph extraction & matching			-	-	-
7	ConcordU	2NLP	-	McCCJ + SD	Dict	Rules	Rules	Rules	-	triggers and hedge words

Table 3: Participants and summary of system descriptions. Abbreviations: BI=Bioinformatician, NLP=Natural Language Processing researcher, SDE=Software Development Engineer, CoreNLP=Stanford CoreNLP, Porter=Porter stemmer, Snowball=Snowball stemmer, WN-lemma=WordNet lemmatization, McCCJ=McClosky-Charniak-Johnson parser, Charniak=Charniak parser, SD=Stanford Dependency conversion, Dict=Dictionary

The full EPI task involves many partially independent challenges, incorporating what were treated in the BioNLP'09 shared task as separate subtasks: the identification of additional non-*Theme* event participants (Task 2 in '09) and the detection of negated and speculated events (Task 3 in '09). The EPI task does not include explicit subtasks. However, we specifies minimal *core* extraction targets in addition to the *full* task targets. Results are reported separately for core targets and full task, allowing participants to choose to only extract core targets. The full task results are considered the primary evaluation for the task e.g. for the purposes of determining the ranking of participating systems.

5 Results

5.1 Participation

Table 3 summarizes the participating groups and the features of their extraction systems. We note that, similarly to the '09 task, machine learning-based systems remain dominant overall, although there is considerable divergence in the specific methods applied. In addition to domain mainstays such as support vector machines and maximum entropy models, we find increased application of joint models (Riedel et al., 2011; McClosky et al., 2011; Riedel and McCallum, 2011) as opposed to pure pipeline systems (Björne and Salakoski, 2011; Quirk et al., 2011) . Remarkably, the application of full parsing together with dependency-based representations of syntactic analyses is adopted by all participants, with the parser of Charniak and Johnson (2005) with the biomedical domain model of McClosky (2009) is applied in all but one system (Liu et al., 2011) and the Stanford Dependency representation (de Marneffe et al., 2006) in all. These choices may be motivated in part by the success of systems using the tools in the previous shared task and the availability of the analyses as supporting resources (Stenetorp et al., 2011).

Despite the availability of PTM and DNA methylation resources other than those specifically introduced for the task and the PHOSPHORYLATION annotations in the GE task (Kim et al., 2011b), no participant chose to apply other corpora for training. With the exception of externally acquired unlabeled data such as PubMed-derived word clusters applied by three groups, the task results thus reflect a closed task setting in which only the given data is used for training.

5.2 Evaluation results

Table 4 presents a the primary results by event type, and Table 5 summarizes these results. We note that only two teams, UTurku (Björne and Salakoski, 2011) and ConcordU (Kilicoglu and Bergler, 2011), predicted event modifications, and only UTurku predicted additional (non-core) event arguments (data not shown). The other five systems thus addressed

	UTurku	FAUST	MSR-NLP	UMass	Stanford	CCP-BTMG	Con-cordU	Size
HYDROXYLATION	**42.25**	10.26	10.20	12.80	9.45	12.84	6.32	139
DEHYDROXYLATION	-	-	-	-	-	-	-	1
PHOSPHORYLATION	**67.12**	51.61	50.00	49.18	40.98	47.06	44.44	130
DEPHOSPHORYLATION	0.00	0.00	0.00	0.00	0.00	**50.00**	0.00	3
UBIQUITINATION	**75.34**	72.95	67.88	72.94	67.44	70.87	69.97	340
DEUBIQUITINATION	**54.55**	40.00	0.00	31.58	0.00	42.11	14.29	17
DNA METHYLATION	**60.21**	31.21	34.54	23.82	31.02	15.65	8.22	416
DNA DEMETHYLATION	**26.67**	0.00	0.00	0.00	0.00	0.00	0.00	21
Simple event total	***63.05***	*45.17*	*44.97*	*43.01*	*40.96*	*40.62*	*37.84*	*1067*
GLYCOSYLATION	**49.43**	41.10	38.87	40.00	37.22	25.62	25.94	347
DEGLYCOSYLATION	**40.00**	35.29	0.00	38.10	30.00	35.29	26.67	27
ACETYLATION	**57.22**	40.00	41.42	40.25	35.12	37.50	38.19	337
DEACETYLATION	**54.90**	28.00	31.82	29.17	21.74	24.56	27.27	50
METHYLATION	**57.67**	24.82	19.57	23.67	18.54	16.99	15.50	374
DEMETHYLATION	**35.71**	0.00	0.00	0.00	0.00	0.00	0.00	13
Non-simple event total	***54.36***	*33.86*	*31.85*	*33.07*	*29.28*	*25.06*	*25.10*	*1148*
CATALYSIS	7.06	6.58	**7.75**	5.00	2.84	7.58	1.74	238
Subtotal	***55.02***	*36.93*	*36.17*	*35.30*	*32.85*	*30.58*	*28.92*	*2453*
NEGATION	18.60	0.00	0.00	0.00	0.00	0.00	**26.51**	149
SPECULATION	**37.65**	0.00	0.00	0.00	0.00	0.00	6.82	103
Modification total	***28.07***	*0.00*	*0.00*	*0.00*	*0.00*	*0.00*	*16.37*	*252*
Total	***53.33***	*35.03*	*34.27*	*33.52*	*31.22*	*28.97*	*27.88*	*2705*
Addition total	***59.33***	*40.27*	*39.05*	*38.65*	*36.03*	*32.75*	*31.50*	*2038*
Removal total	***44.29***	*22.41*	*15.73*	*22.76*	*14.41*	*23.53*	*17.48*	*132*

Table 4: Primary evaluation F-scores by event type. The "size" column gives the number of annotations of each type in the given data (training+development). Best result for each type shown in bold. For DEHYDROXYLATION, no examples were present in the test data and none were predicted by any participant.

Team	recall	prec.	F-score
UTurku	52.69	53.98	53.33
FAUST	28.88	44.51	35.03
MSR-NLP	27.79	44.69	34.27
UMass	28.08	41.55	33.52
Stanford	26.56	37.85	31.22
CCP-BTMG	23.44	37.93	28.97
ConcordU	20.83	42.14	27.88

Table 5: Primary evaluation results

only the core task. For the full task, this difference in approach is reflected in the substantial performance advantage for the UTurku system, which exhibits highest performance overall as well as for most individual event types.

Extraction performance for simple events taking only *Theme* and *Site* arguments is consistently higher than for other event types, with absolute F-score differences of over 10% points for many systems. Similar notable performance differences are seen between the *addition* events, for which ample training data was available, and the *removal* types for which data was limited. This effect is particularly noticeable for DEPHOSPHORYLATION, DNA DEMETHYLATION and DEMETHYLATION, for which the clear majority of systems failed to predict any correct events. Extraction performance for CATALYSIS events is very low despite a relatively large set of training examples, indicating that the extraction of nested event structures remains very challenging. This low performance may also be related to the fact that CATALYSIS events are often triggered by the same word as the catalysed modification (e.g. Figure 1b), requiring the assignment of multiple event labels to a single word in typical system architectures.

Table 6 summarizes the full task results with the addition of the single partial penalty criterion. The F-scores for the seven participants under this crite-

Team	recall	prec.	F-score	Δ
UTurku	54.79	58.42	56.55	3.22
FAUST	28.88	72.05	41.24	6.21
MSR-NLP	27.79	66.72	39.24	4.97
UMass	28.08	63.28	38.90	5.38
Stanford	26.56	56.83	36.20	4.98
CCP-BTMG	23.44	50.79	32.08	3.11
ConcordU	20.83	60.55	30.99	3.11

Table 6: Full task evaluation results for primary criteria and with single partial penalty. The Δ column gives F-score difference to the primary results.

Team	recall	prec.	F-score	Δ_1	Δ_2
UTurku	68.51	69.20	68.86	15.53	12.31
FAUST	59.88	80.25	68.59	33.56	27.35
MSR-NLP	55.70	77.60	64.85	30.58	25.61
UMass	57.04	73.30	64.15	30.63	25.25
Stanford	56.87	70.22	62.84	31.62	26.64
ConcordU	40.28	76.71	52.83	24.95	21.84
CCP-BTMG	45.06	63.37	52.67	23.70	20.59

Table 7: Core task evaluation results. The Δ_1 column gives F-score difference to primary full task results, Δ_2 to full task results with single partial penalty.

rion are on average over 4% points higher than under the primary criteria, with the most substantial increases seen for high-ranking participants only addressing the core task: for example, the precision of the FAUST system (Riedel et al., 2011) is nearly 30% higher under the relaxed criterion. These results provide new perspective deserving further detailed study into the question of what are the most meaningful criteria for event extraction system evaluation.

Table 7 summarizes the core task results. While all systems show notably higher performance than for the full task, high-ranking participants focusing on the core task gain most dramatically, with the FAUST system core task F-score essentially matching that of the top system (UTurku). For the core task, all participants achieve F-scores over 50% – a result achieved by only a single system in the '09 task – and the top four participants average over 65% F-score. These results confirm that current event extraction technology is well applicable to the core PTM extraction task even when the number of targeted event types is relatively high and may be ready to address the challenges of exhaustive PTM extraction (Pyysalo et al., 2011a). The best core tasks results, approaching 70% F-score, are particularly encouraging as the level of performance is comparable to or better than state-of-the-art results for many reference resources for protein-protein interaction extraction (see e.g. Tikk et al. (2010))) using the simple untyped entity pair representation, a standard task that has been extensively studied in the domain.

6 Discussion and Conclusions

This paper has presented the preparation, resources, results and analysis of the BioNLP Shared Task 2011 Epigenetics and Post-translational modifications (EPI) main task. The results demonstrate that the core extraction target of identifying statements of 14 different modification types with the modified gene or gene product can be reliably addressed by current event extraction methods, with two systems approaching 70% F-score at this task. Nevertheless, challenges remain in detecting statements regarding the catalysis of these events as well as in resolving the full detail of such modification events, a task attempted by only one participant and at which performance remains at somewhat above 50% in F-score.

Detailed evaluation showed that the highly competitive participating systems differ substantially in their relative strengths, indicating potential for further development at protein and DNA modification event detection. The task results are available in full detail from the shared task webpage, `http://sites.google.com/site/bionlpst/`.

In the future, we will follow the example of the BioNLP'09 shared task in making the data and resources of the EPI task open to all interested parties to encourage further study of event extraction for epigenetics and post-translational modification events, to facilitate system comparison on a well-defined standard task, and to support the development of further applications of event extraction technology in this important area of biomolecular science.

Acknowledgments

We would like to thank Yoshiro Okuda and Yo Shidahara of NalaPro Technologies for their efforts in producing the EPI task annotation. This work was supported by Grant-in-Aid for Specially Promoted Research (MEXT, Japan).

References

Sophia Ananiadou, Sampo Pyysalo, Jun'ichi Tsujii, and Douglas B. Kell. 2010. Event extraction for systems biology by text mining the literature. *Trends in Biotechnology*, 28(7):381–390.

D. Barford, A.K. Das, and M.P. Egloff. 1998. The structure and mechanism of protein phosphatases: insights into catalysis and regulation. *Annual review of biophysics and biomolecular structure*, 27(1):133–164.

Jari Björne and Tapio Salakoski. 2011. Generalizing biomedical event extraction. In *Proceedings of the BioNLP 2011 Workshop Companion Volume for Shared Task*, Portland, Oregon, June. Association for Computational Linguistics.

Ekaterina Buyko, Erik Faessler, Joachim Wermter, and Udo Hahn. 2009. Event extraction from trimmed dependency graphs. In *Proceedings of BioNLP Shared Task 2009*, pages 19–27.

Eugene Charniak and Mark Johnson. 2005. Coarse-to-Fine n-Best Parsing and MaxEnt Discriminative Reranking. In *Proceedings of ACL'05*, pages 173–180.

Marie-Catherine de Marneffe, Bill MacCartney, and Christopher D. Manning. 2006. Generating Typed Dependency Parses from Phrase Structure Parses. In *Proceedings of the Fifth International Conference on Language Resources and Evaluation (LREC'06)*, pages 449–454.

M.H. Glickman and A. Ciechanover. 2002. The ubiquitin-proteasome proteolytic pathway: destruction for the sake of construction. *Physiological reviews*, 82(2):373.

R. Hoehndorf, A.C.N. Ngomo, S. Pyysalo, T. Ohta, A. Oellrich, and D. Rebholz-Schuhmann. 2010. Applying ontology design patterns to the implementation of relations in GENIA. In *Proceedings of the Fourth Symposium on Semantic Mining in Biomedicine SMBM 2010*.

Robin Holliday and JE Pugh. 1975. Dna modification mechanisms and gene activity during development. *Science*, 187:226–232.

Robin Holliday. 1987. The inheritance of epigenetic defects. *Science*, 238:163–170.

Rudolf Jaenisch and Adrian Bird. 2003. Epigenetic regulation of gene expression: how the genome integrates intrinsic and environmental signals. *Nature Genetics*, 33:245–254.

Halil Kilicoglu and Sabine Bergler. 2011. Adapting a general semantic interpretation approach to biological event extraction. In *Proceedings of the BioNLP 2011 Workshop Companion Volume for Shared Task*, Portland, Oregon, June. Association for Computational Linguistics.

Jin-Dong Kim, Tomoko Ohta, and Jun'ichi Tsujii. 2008. Corpus annotation for mining biomedical events from literature. *BMC Bioinformatics*, 9(10).

Jin-Dong Kim, Tomoko Ohta, Sampo Pyysalo, Yoshinobu Kano, and Jun'ichi Tsujii. 2009. Overview of BioNLP'09 Shared Task on Event Extraction. In *Proceedings of Natural Language Processing in Biomedicine (BioNLP) NAACL 2009 Workshop*, pages 1–9.

Jin-Dong Kim, Sampo Pyysalo, Tomoko Ohta, Robert Bossy, and Jun'ichi Tsujii. 2011a. Overview of BioNLP Shared Task 2011. In *Proceedings of the BioNLP 2011 Workshop Companion Volume for Shared Task*, Portland, Oregon, June. Association for Computational Linguistics.

Jin-Dong Kim, Yue Wang, Toshihisa Takagi, and Akinori Yonezawa. 2011b. Overview of the Genia Event task in BioNLP Shared Task 2011. In *Proceedings of the BioNLP 2011 Workshop Companion Volume for Shared Task*, Portland, Oregon, June. Association for Computational Linguistics.

R. Leaman and G. Gonzalez. 2008. Banner: an executable survey of advances in biomedical named entity recognition. *Proceedings of the Pacific Symposium on Biocomputing (PSB'08)*, pages 652–663.

Haibin Liu, Ravikumar Komandur, and Karin Verspoor. 2011. From graphs to events: A subgraph matching approach for information extraction from biomedical text. In *Proceedings of the BioNLP 2011 Workshop Companion Volume for Shared Task*, Portland, Oregon, June. Association for Computational Linguistics.

David McClosky, Mihai Surdeanu, and Christopher Manning. 2011. Event extraction as dependency parsing for bionlp 2011. In *Proceedings of the BioNLP 2011 Workshop Companion Volume for Shared Task*, Portland, Oregon, June. Association for Computational Linguistics.

David McClosky. 2009. *Any Domain Parsing: Automatic Domain Adaptation for Natural Language Parsing*. Ph.D. thesis, Department of Computer Science, Brown University.

Tomoko Ohta, Jin-Dong Kim, Sampo Pyysalo, Yue Wang, and Jun'ichi Tsujii. 2009. Incorporating GENETAG-style annotation to GENIA corpus. In *Proceedings of BioNLP'09*, pages 106–107.

Tomoko Ohta, Sampo Pyysalo, Makoto Miwa, Jin-Dong Kim, and Jun'ichi Tsujii. 2010a. Event extraction for post-translational modifications. In *Proceedings of BioNLP'10*, pages 19–27.

Tomoko Ohta, Sampo Pyysalo, Makoto Miwa, and Jun'ichi Tsujii. 2010b. Event extraction for dna methylation. In *Proceedings of SMBM'10*.

Maté Ongenaert, Leander Van Neste, Tim De Meyer, Gerben Menschaert, Sofie Bekaert, and Wim

Van Criekinge. 2008. PubMeth: a cancer methylation database combining text-mining and expert annotation. *Nucl. Acids Res.*, 36(suppl_1):D842–846.

Sampo Pyysalo, Filip Ginter, Juho Heimonen, Jari Björne, Jorma Boberg, Jouni Järvinen, and Tapio Salakoski. 2007. BioInfer: A corpus for information extraction in the biomedical domain. *BMC Bioinformatics*, 8(50).

Sampo Pyysalo, Tomoko Ohta, Makoto Miwa, and Jun'ichi Tsujii. 2011a. Towards exhaustive protein modification event extraction. In *Proceedings of the BioNLP 2011 Workshop*, Portland, Oregon, June. Association for Computational Linguistics.

Sampo Pyysalo, Tomoko Ohta, Rafal Rak, Dan Sullivan, Chunhong Mao, Chunxia Wang, Bruno Sobral, Jun'ichi Tsujii, and Sophia Ananiadou. 2011b. Overview of the Infectious Diseases (ID) task of BioNLP Shared Task 2011. In *Proceedings of the BioNLP 2011 Workshop Companion Volume for Shared Task*, Portland, Oregon, June. Association for Computational Linguistics.

Chris Quirk, Pallavi Choudhury, Michael Gamon, and Lucy Vanderwende. 2011. Msr-nlp entry in bionlp shared task 2011. In *Proceedings of the BioNLP 2011 Workshop Companion Volume for Shared Task*, Portland, Oregon, June. Association for Computational Linguistics.

Sebastian Riedel and Andrew McCallum. 2011. Robust biomedical event extraction with dual decomposition and minimal domain adaptation. In *Proceedings of the BioNLP 2011 Workshop Companion Volume for Shared Task*, Portland, Oregon, June. Association for Computational Linguistics.

Sebastian Riedel, David McClosky, Mihai Surdeanu, Andrew McCallum, and Chris Manning. 2011. Model combination for event extraction in bionlp 2011. In *Proceedings of the BioNLP 2011 Workshop Companion Volume for Shared Task*, Portland, Oregon, June. Association for Computational Linguistics.

A.D. Riggs. 1975. X inactivation, differentiation, and dna methylation. *Cytogenetic and Genome Research*, 14:9–25.

Pontus Stenetorp, Goran Topić, Sampo Pyysalo, Tomoko Ohta, Jin-Dong Kim, and Jun'ichi Tsujii. 2011. BioNLP Shared Task 2011: Supporting Resources. In *Proceedings of the BioNLP 2011 Workshop Companion Volume for Shared Task*, Portland, Oregon, June. Association for Computational Linguistics.

JB Stock, AJ Ninfa, and AM Stock. 1989. Protein phosphorylation and regulation of adaptive responses in bacteria. *Microbiology and Molecular Biology Reviews*, 53(4):450.

Lorraine Tanabe, Natalie Xie, Lynne Thom, Wayne Matten, and John Wilbur. 2005. Genetag: a tagged corpus for gene/protein named entity recognition. *BMC Bioinformatics*, 6(Suppl 1):S3.

Paul Thompson, Syed Iqbal, John McNaught, and Sophia Ananiadou. 2009. Construction of an annotated corpus to support biomedical information extraction. *BMC Bioinformatics*, 10(1):349.

Domonkos Tikk, Philippe Thomas, Peter Palaga, Jörg Hakenberg, and Ulf Leser. 2010. A comprehensive benchmark of kernel methods to extract protein-protein interactions from literature. *PLoS Comput Biol*, 6(7):e1000837, 07.

Yue Wang, Jin-Dong Kim, Rune Sætre, Sampo Pyysalo, and Jun'ichi Tsujii. 2009. Investigating heterogeneous protein annotations toward cross-corpora utilization. *BMC Bioinformatics*, 10(403), Dec. ISSN: 1471-2105.

Eric S Witze, William M Old, Katheryn A Resing, and Natalie G Ahn. 2007. Mapping protein post-translational modifications with mass spectrometry. *Nature Methods*, 4:798–806.

Cathy H. Wu, Lai-Su L. Yeh, Hongzhan Huang, Leslie Arminski, Jorge Castro-Alvear, Yongxing Chen, Zhangzhi Hu, Panagiotis Kourtesis, Robert S. Ledley, Baris E. Suzek, C.R. Vinayaka, Jian Zhang, and Winona C. Barker. 2003. The Protein Information Resource. *Nucl. Acids Res.*, 31(1):345–347.

Overview of the Infectious Diseases (ID) task of BioNLP Shared Task 2011

Sampo Pyysalo*[*] Tomoko Ohta[*] Rafal Rak[‡§] Dan Sullivan[†] Chunhong Mao[†] Chunxia Wang[†] Bruno Sobral[†] Jun'ichi Tsujii[¶] Sophia Ananiadou[‡§]

[*]Department of Computer Science, University of Tokyo, Tokyo, Japan
[†]Virginia Bioinformatics Institute, Virginia Tech, Blacksburg, Virginia, USA
[‡]School of Computer Science, University of Manchester, Manchester, UK
[§]National Centre for Text Mining, University of Manchester, Manchester, UK
[¶]Microsoft Research Asia, Beijing, China

{smp,okap}@is.s.u-tokyo.ac.jp jtsujii@microsoft.com
{dsulliva,cmao,cwang,sobral}@vbi.vt.edu
{rafal.rak,sophia.ananiadou}@manchester.ac.uk

Abstract

This paper presents the preparation, resources, results and analysis of the Infectious Diseases (ID) information extraction task, a main task of the BioNLP Shared Task 2011. The ID task represents an application and extension of the BioNLP'09 shared task event extraction approach to full papers on infectious diseases. Seven teams submitted final results to the task, with the highest-performing system achieving 56% F-score in the full task, comparable to state-of-the-art performance in the established BioNLP'09 task. The results indicate that event extraction methods generalize well to new domains and full-text publications and are applicable to the extraction of events relevant to the molecular mechanisms of infectious diseases.

1 Introduction

The Infectious Diseases (ID) task of the BioNLP Shared Task 2011 (Kim et al., 2011a) is an information extraction task focusing on the biomolecular mechanisms of infectious diseases. The primary target of the task is event extraction (Ananiadou et al., 2010), broadly following the task setup of the BioNLP'09 Shared Task (BioNLP ST'09) (Kim et al., 2009).

The task concentrates on the specific domain of two-component systems (TCSs, or two-component regulatory systems), a mechanism widely used by bacteria to sense and respond to the environment (Thomason and Kay, 2000). Typical TCSs consist of two proteins, a membrane-associated sensor kinase and a cytoplasmic response regulator. The sensor kinase monitors changes in the environment while the response regulator mediates an adaptive response, usually through differential expression of target genes (Mascher et al., 2006). TCSs have many functions, but those of particular interest for infectious disease researchers include virulence, response to antibiotics, quorum sensing, and bacterial cell attachment (Krell et al., 2010). Not all TCS functions are well known: in some cases, TCSs are involved in metabolic processes that are difficult to precisely characterize (Wang et al., 2010). TCSs are of interest also as drugs designed to disrupt TCSs may reduce the virulence of bacteria without killing it, thus avoiding the potential selective pressure of antibiotics lethal to some pathogenic bacteria (Gotoh et al., 2010). Information extraction techniques may support better understanding of these fundamental systems by identifying and structuring the molecular processes underlying two component signaling.

The ID task seeks to address these opportunities by adapting the BioNLP ST'09 event extraction model to domain scientific publications. This model was originally introduced to represent biomolecular events relating to transcription factors in human blood cells, and its adaptation to a domain that centrally concerns both bacteria and their hosts involves a variety of novel aspects, such as events concerning whole organisms, the chemical environment of bacteria, prokaryote-specific concepts (e.g. regulons as elements of gene expression), as well as the effects of biomolecules on larger-scale processes involving hosts such as virulence.

Proceedings of BioNLP Shared Task 2011 Workshop, pages 26–35,
Portland, Oregon, USA, 24 June, 2011.

2 Task Setting

The ID task broadly follows the task definition and event types of the BioNLP ST'09, extending it with new entity categories, correspondingly broadening the scope of events, and introducing a new class of events, high-level biological processes.

2.1 Entities

The ID task defines five core types of entities: genes/gene products, two-component systems, regulons/operons, chemicals, and organisms. Following the general policy of the BioNLP Shared Task, the recognition of the core entities is not part of the ID task. As named entity recognition (NER) is considered in other prominent domain evaluations (Krallinger et al., 2008), we have chosen to isolate aspects of extraction performance relating to NER from the main task of interest, event extraction, by providing participants with human-created gold annotations for core entities. These annotations are briefly presented in the following.

Mentions of names of genes and their products (RNA and proteins) are annotated with a single type, without differentiating between subtypes, following the guidelines of the GENIA GGP corpus (Ohta et al., 2009). This type is named PROTEIN to maintain consistency with related tasks (e.g. BioNLP ST'09), despite slight inaccuracy for cases specifically referencing RNA or DNA forms. Two-component systems, consisting of two proteins, frequently have names derived from the names of the proteins involved (e.g. *PhoP-PhoR* or *SsrA/SsrB*). Mentions of TCSs are annotated as TWO-COMPONENT-SYSTEM, nesting PROTEIN annotations if present. Regulons and operons are collections of genes whose expression is jointly regulated. Like the names of TCSs, their names may derive from the names of the involved genes and proteins, and are annotated as embedding PROTEIN annotations when they do. The annotation does not differentiate between the two, marking both with a single type REGULON-OPERON.

In addition to these three classes relating to genes and proteins, the core entity annotation recognizes the classes CHEMICAL and ORGANISM. All mentions of formal and informal names of atoms, inorganic compounds, carbohydrates and lipids as well as organic compounds other than amino acid and nucleic acid compounds (i.e. gene/protein-related compounds) are annotated as CHEMICAL. Mentions of names of families, genera, species and strains as well as non-name references with comparable specificity are annotated as ORGANISM.

Finally, the non-specific type ENTITY[1] is defined for marking entities that specify additional details of events such as the binding site in a BINDING event or the location an entity moves to in a LOCALIZATION event. Unlike the core entities, annotations of the generic ENTITY type are not provided for test data and must be detected by participants addressing the full task.

2.2 Relations

The ID task involves one relation, EQUIV, defining entities (of any of the core types) to be equivalent. This relation is used to annotate abbreviations and local aliases and it is not a target of extraction, but provided for reference and applied in evaluation, where references to any of a set of equivalent entities are treated identically.

2.3 Events

The primary extraction targets of the ID task are the event types summarized in Table 1. These are a superset of those targeted in the BioNLP ST'09 and its repeat, the 2011 GE task (Kim et al., 2011b). This design makes it possible to study aspects of domain adaptation by having the same extraction targets in two subdomains of biomedicine, that of transcription factors in human blood cells (GE) and infectious diseases. The events in the ID task extend on those of GE in the inclusion of additional entity types as participants in previously considered event types and the introduction of a new type, PROCESS. We next briefly discuss the semantics of these events, defined (as in GE) with reference to the community-standard Gene Ontology (Ashburner et al., 2000). We refer to (Kim et al., 2008; Kim et al., 2009) for the ST'09/GE definitions.

[1]In terms of the GENIA ontology, ENTITY is used to mark e.g. PROTEIN DOMAIN OR REGION references. Specific types were applied in manual annotation, but these were replaced with the generic ENTITY in part to maintain consistency with BioNLP ST'09 data and to reduce the NER-related demands on participating systems by not requiring the assignment of detailed types.

Type	Core arguments	Additional arguments
GENE EXPRESSION	*Theme*(PROTEIN or REGULON-OPERON)	
TRANSCRIPTION	*Theme*(PROTEIN or REGULON-OPERON)	
PROTEIN CATABOLISM	*Theme*(PROTEIN)	
PHOSPHORYLATION	*Theme*(PROTEIN)	*Site*(ENTITY)
LOCALIZATION	*Theme*(Core entity)	*AtLoc*(ENTITY), *ToLoc*(ENTITY)
BINDING	*Theme*(Core entity)+	*Site*(ENTITY)+
PROCESS	*Participant*(Core entity)?	
REGULATION	*Theme*(Core entity / Event), *Cause*(Core entity / Event)?	*Site*(ENTITY), *CSite*(ENTITY)
POSITIVE REGULATION	*Theme*(Core entity / Event), *Cause*(Core entity / Event)?	*Site*(ENTITY), *CSite*(ENTITY)
NEGATIVE REGULATION	*Theme*(Core entity / Event), *Cause*(Core entity / Event)?	*Site*(ENTITY), *CSite*(ENTITY)

Table 1: Event types and their arguments. The type of entity allowed as argument is specified in parenthesis. "Core entity" is any of PROTEIN, TWO-COMPONENT-SYSTEM, REGULON-OPERON, CHEMICAL, or ORGANISM. Arguments that can be filled multiple times marked with "+", non-mandatory core arguments with "?" (all additional arguments are non-mandatory).

The definitions of the first four types in Table 1 are otherwise unchanged from the ST'09 definitions except that GENE EXPRESSION and TRANSCRIPTION extend on the former definition in recognizing REGULON-OPERON as an alternative unit of expression. LOCALIZATION, taking only PROTEIN type arguments in the ST'09 definition, is allowed to take any core entity argument. This expanded definition remains consistent with the scope of the corresponding GO term (GO:0051179). BINDING is similarly extended, giving it a scope largely consistent with GO:0005488 (binding) but also encompassing GO:0007155 (cell adhesion) (e.g. a bacterium binding another) and protein-organism binding. The three regulation types (REGULATION, POSITIVE REGULATION, and NEGATIVE REGULATION) likewise allow the new core entity types as arguments, but their definitions are otherwise unchanged from those in ST'09, that is, the GENIA ontology definitions. As in these resources, regulation types are used not only for the biological sense but also to capture statements of general causality (Kim et al., 2008). As in ST'09, all events of types discussed above require a *Theme* argument: only events involving an explicitly stated theme (of an appropriate type) should be extracted. All other arguments are optional.

The PROCESS type, new to ID, is used to annotate high-level processes such as virulence, infection and resistance that involve infectious organisms. This type differs from the others in that it has no mandatory arguments: the targeted processes should be extracted even if they have no explicitly stated participants, reflecting that they are of interest even without the further specification. When stated, the involved participants are captured using the generic role type *Participant*. Figure 1 shows an illustration of some of the the ID task extraction targets.

We term the first five event types in Table 1 taking exactly one *Theme* argument as their core argument *simple events*. In analysis we further differentiate *non-regulation events* (the first seven) and *regulation* (the last three), which is known to represent particular challenges for extraction in involving events as arguments, thus creating nested event structures.

2.4 Event modifications

The ID task defines two *event modification* extraction targets, NEGATION and SPECULATION. These modifications mark events as being explicitly negated (e.g. *virB is not expressed*) or stated in a speculative context (e.g. *virB may be expressed*). Both may apply simultaneously. The modification definitions are identical to the ST'09 ones, including the representation in which modifications (unlike events) are not assigned text bindings.

3 Data

The ID task data were newly annotated for the BioNLP Shared Task and are not based on any previously released resource. Annotation was performed by two teams, one in Tsujii laboratory (University of Tokyo) and one in Virginia Bioinformatics Institute (Virginia Tech). The entity and event annotation

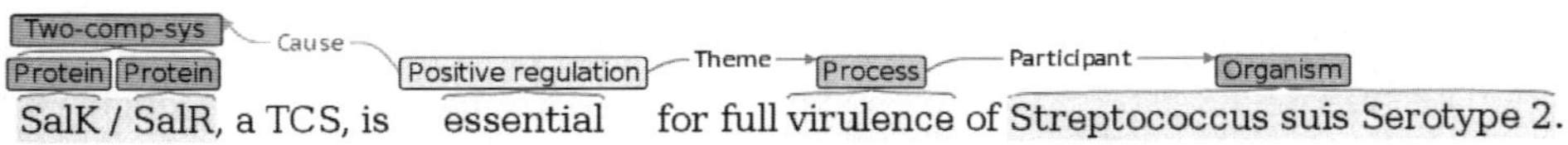

Figure 1: Example event annotation. The association of a TCS with an organism is captured through an event structure involving a PROCESS ("virulence") and POSITIVE REGULATION. Regulation types are used to capture also statements of general causality such as "is essential for" here. (Simplified from PMC ID 2358977)

Journal	#	Published
PLoS Pathogens	9	2006–2010
PLoS One	7	2008–2010
BMC Genomics	3	2008–2010
PLoS Genetics	2	2007–2010
Open Microbiology J.	2	2008–2010
BMC Microbiology	2	2008–2009
Other	5	2007–2008

Table 2: Corpus composition. Journals in which selected articles were published with number of articles (#) and publication years.

Entity type	prec.	rec.	F
PROTEIN	54.64	39.64	45.95
CHEMICAL	32.24	19.05	23.95
ORGANISM	90.38	47.70	62.44
TWO-COMPONENT-SYSTEM	87.69	47.24	61.40

Table 3: Automatic core entity tagging performance.

design was guided by previous studies on NER and event extraction in a closely related domain (Pyysalo et al., 2010; Ananiadou et al., 2011).

3.1 Document selection

The training and test data were drawn from the primary text content of recent full-text PMC open access documents selected by infectious diseases domain experts (Virginia Tech team) as representative publications on two-component regulatory systems. Table 2 presents some characteristics of the corpus composition. To focus efforts on natural language text likely to express novel information, we excluded tables, figures and their captions, as well as methods sections, acknowledgments, authors' contributions, and similar meta-content.

3.2 Annotation

Annotation was performed in two primary stages, one for marking core entities and the other for events and secondary entities. As a preliminary processing step, initial sentence segmentation was performed with the GENIA Sentence Splitter[2]. Segmentation errors were corrected during core entity annotation.

Core entity annotation was performed from the basis of an automatic annotation created using selected existing taggers for the target entities. The following tools and settings were adopted, with parameters tuned on initial annotation for two documents:

PROTEIN: NeMine (Sasaki et al., 2008) trained on the JNLPBA data (Kim et al., 2004) with threshold 0.05, filtered to only GENE and PROTEIN types.

ORGANISM: Linnaeus (Gerner et al., 2010) with "variant matching" for species names variants.

CHEMICAL: OSCAR3 (Corbett and Murray-Rust, 2006) with confidence 90%.

TWO-COMPONENT-SYSTEM: Custom regular expressions.

Initial automatic tagging was not applied for entities of the REGULON-OPERON type or the generic ENTITY type (for additional event arguments). All automatically generated annotations were at least confirmed through manual inspection, and the majority of the automatic annotations were revised in manual annotation. Table 3 summarizes the tagging performance of the automatic tools as measured against the final human-annotated training and development datasets.[3]

Annotation for the task extraction targets – events and event modifications – was created entirely manually without automatic annotation support to avoid any possible bias toward specific extraction methods or approaches. The Tsujii laboratory team orga-

[2] `http://www-tsujii.is.s.u-tokyo.ac.jp/~y-matsu/geniass/`

[3] It should be noted that these results are low in part due to differences in annotation criteria (see e.g. (Wang et al., 2009)) and to data tagged using the ID task annotation guidelines not being applied for training; training on the newly annotated data is expected to allow notably more accurate tagging.

Item	Train	Devel	Test	Total
Articles	15	5	10	30
Sentences	2,484	709	1,925	5118
Words	74,439	21,225	57,489	153,153
Core entities	6,525	1,976	4,239	12,740
Events	2,088	691	1,371	4150
Modifications	95	45	74	214

Table 4: Statistics of the ID corpus.

nized the annotation effort, with a coordinating annotator with extensive experience in event annotation (TO) leading annotator training and annotation scheme development. Detailed annotation guidelines (Pyysalo et al., 2011) extending on the GENIA annotation guidelines were developed jointly with all annotators and refined throughout the annotation effort. Based on measurements of inter-annotator consistency between annotations independently created by the two teams, made throughout annotator training and primary annotation (excluding final corpus cleanup), we estimate the consistency of the final entity annotation to be no lower than 90% F-score and that of the event annotation to be no lower than 75% F-score for the primary evaluation criteria (see Section 4).

3.3 Datasets and statistics

Initial annotation was produced for the selected sections (see Section 3.1) in 33 full-text articles, of which 30 were selected for the final dataset as representative of the extraction targets. These documents were split into training, development and test sets of 15, 5 and 10 documents, respectively. Participants were provided with all training and development set annotations and test set core entity annotations. The overall statistics of the datasets are given in Table 4.

As the corpus consists of full-text articles, it contains a somewhat limited number of articles, but in other terms it is of broadly comparable size to the largest of the BioNLP ST corpora: the corpus word count, for example, corresponds to that of a corpus of approximately 800 PubMed abstracts, and the core entity count is comparable to that in the ST'09 data. However, for reasons that may relate in part to the domain, the event count is approximately a third of that for the ST'09 data. In addition to having less training data, the entity/event ratio is thus considerably higher (i.e. there are more candidates for each true target), suggesting that the ID data could be expected to provide a more challenging extraction task.

4 Evaluation

The performance of participating systems was evaluated in terms of events using the standard precision/recall/F-score metrics. For the primary evaluation, we adopted the standard criteria defined in the BioNLP'09 shared task. In brief, for determining whether a reference annotation and a predicted annotation match, these criteria relax exact matching for event triggers and arguments in two ways: matching of text-bound annotation (event triggers and ENTITY type entities) allows limited boundary variation, and only core arguments need to match in nested event arguments for events to match. For details of the matching criteria, please refer to Kim et al. (2009).

The primary evaluation for the task requires the extraction of all event arguments (both core and additional; see Table 1) as well as event modifications (NEGATION and SPECULATION). This is termed the *full task*. We additionally report extraction results for evaluation where both the gold standard reference data and the submission events are reduced to only core arguments, event modifications are removed, and resulting duplicate events removed. We term this the *core task*. In terms of the subtask division applied in the BioNLP'09 Shared Task and the GE task of 2011, the core task is analogous to subtask 1 and the full task analogous to the combination of subtasks 1–3.

5 Results

5.1 Participation

Final results to the task were successfully submitted by seven participants. Table 5 summarizes the information provided by the participating teams. We note that full parsing is applied in all systems, with the specific choice of the parser of Charniak and Johnson (2005) with the biomedical domain model of McClosky (2009) and conversion into the Stanford Dependency representation (de Marneffe et al., 2006) being adopted by five participants. Further, five of the seven systems are predominantly machine learning-based. These can be seen as extensions of trends that were noted in analysis of the BioNLP

Rank	Team	Org	NLP		Events				Other resources	
			Word	Parse	Trig.	Arg.	Group.	Modif.	Corpora	Other
1	FAUST	3NLP	CoreNLP, SnowBall	McCCJ + SD	(UMass+Stanford as features)				GE	word clusters
2	UMass	1NLP	CoreNLP, SnowBall	McCCJ + SD	Joint, dual dec.+MIRA 1-best			-	GE	-
3	Stanford	3NLP	CoreNLP	McCCJ + SD	MaxEnt	Joint, MSTParser		-	GE	word clusters
4	ConcordU	2NLP	-	McCCJ + SD	dict	rules	rules	rules	-	triggers and hedge words
5	UTurku	1BI	Porter	McCCJ + SD	SVM	SVM	SVM	SVM	-	hedge words
6	PNNL	1CS, 1NLP, 2BI	Porter	Stanford	SVM	SVM	rules	-	GE	UMLS, triggers
7	PredX	1CS, 1NLP	LGP	LGP	dict	rules	rules	-	-	UMLS, triggers

Table 5: Participants and summary of system descriptions. Abbreviations: Trig./Arg./Group./Modif.=event trigger detection/argument detection/argument grouping/modification detection, BI=Bioinformatician, NLP=Natural Language Processing researcher, CS=Computer scientist, CoreNLP=Stanford CoreNLP, Porter=Porter stemmer, Snowball=Snowball stemmer McCCJ=McClosky-Charniak-Johnson parser, LGP=Link Grammar Parser, SD=Stanford Dependency conversion, UMLS=UMLS resources (e.g. lexicon, metamap)

ST'09 participation. In system design choices, we note an indication of increased use of joint models as opposed to pure pipeline designs, with the three highest-ranking systems involving a joint model.

Several participants compiled dictionaries of event trigger words and two dictionaries of hedge words from the data. Four teams, including the three top-ranking, used the GE task corpus as supplementary material, indicating that the GE annotations are largely compatible with ID ones (see detailed results below). This is encouraging for future applications of the event extraction approach: as manual annotation requires considerable effort and time, the ability to use existing annotations is important for the feasibility of adaptation of the approach to new domains.

While several participants made use of supporting syntactic analyses provided by the organizers (Stenetorp et al., 2011), none applied the analyses for supporting tasks, such as coreference or entity relation extraction results – at least in cases due to time constraints (Kilicoglu and Bergler, 2011).

5.2 Evaluation results

Table 6 presents the primary results by event type, and Table 7 summarizes these results. The full task requires the extraction of additional arguments and event modifications and involves multiple novel challenges from previously addressed domain tasks including a new subdomain, full-text documents, several new entity types and a new event category.

Team	recall	prec.	F-score
FAUST	48.03	65.97	55.59
UMass	46.92	62.02	53.42
Stanford	46.30	55.86	50.63
ConcordU	49.00	40.27	44.21
UTurku	37.85	48.62	42.57
PNNL	27.75	52.36	36.27
PredX	22.56	35.18	27.49

Table 7: Primary evaluation results.

Nevertheless, extraction performance for the top systems is comparable to the state-of-the-art results for the established BioNLP ST'09 task (Miwa et al., 2010) as well as its repetition as the 2011 GE task (Kim et al., 2011b), where the highest overall result for the primary evaluation criteria was also 56% F-score for the FAUST system (Riedel et al., 2011). This result is encouraging regarding the ability of the extraction approach and methods to generalize to new domains as well as their applicability specifically to texts on the molecular mechanisms of infectious diseases.

We note that there is substantial variation in the relative performance of systems for different entity types. For example, Stanford (McClosky et al., 2011) has relatively low performance for simple events but achieves the highest result for PROCESS, while UTurku (Björne and Salakoski, 2011) results show roughly the reverse. This suggests further potential for improvement from system combinations.

	FAUST	UMass	Stanford	ConcordU	UTurku	PNNL	PredX	Size
GENE EXPRESSION	**70.68**	66.43	54.00	56.57	64.88	53.33	0.00	512
TRANSCRIPTION	69.66	68.24	60.00	**70.89**	57.14	0.00	53.85	77
PROTEIN CATABOLISM	**75.00**	72.73	20.00	66.67	33.33	11.76	0.00	33
PHOSPHORYLATION	64.00	**66.67**	40.00	54.55	60.61	64.29	40.00	69
LOCALIZATION	33.33	14.29	31.58	20.00	**66.67**	20.69	0.00	49
Simple event total	***68.47***	*63.55*	*52.72*	*56.78*	*62.67*	*43.87*	*18.18*	*740*
BINDING	31.30	34.62	23.44	**40.00**	22.22	20.00	28.28	156
PROCESS	65.69	62.26	**73.57**	67.17	41.57	51.04	53.27	901
Non-regulation total	***63.78***	*60.68*	*63.59*	*62.43*	*46.39*	*47.34*	*43.65*	*1797*
REGULATION	**35.44**	30.49	17.67	19.43	22.96	0.00	2.16	267
POSITIVE REGULATION	47.50	**49.49**	34.78	23.41	41.28	24.60	21.02	455
NEGATIVE REGULATION	58.86	**60.45**	44.44	47.96	52.11	25.70	9.49	260
Regulation total	***47.07***	*46.65*	*33.02*	*28.87*	*39.49*	*18.45*	*9.71*	*982*
Subtotal	***57.28***	*55.03*	*52.09*	*46.60*	*43.33*	*37.53*	*28.38*	*2779*
NEGATION	0.00	0.00	0.00	22.92	**32.91**	0.00	0.00	96
SPECULATION	0.00	0.00	0.00	3.23	**15.00**	0.00	0.00	44
Modification total	*0.00*	*0.00*	*0.00*	*11.82*	***26.89***	*0.00*	*0.00*	*140*
Total	***55.59***	*53.42*	*50.63*	*44.21*	*42.57*	*36.27*	*27.49*	*2919*

Table 6: Primary evaluation F-scores by event type. The "size" column gives the number of annotations of each type in the given data (training+development). Best result for each type shown in bold.

The best performance for simple events and for PROCESS approaches or exceeds 70% F-score, arguably approaching a sufficient level for user-facing applications of the extraction technology. By contrast, BINDING and regulation events, found challenging in ST'09 and GE, remain problematic also in the ID task, with best overall performance below 50% F-score. Only two teams, UTurku and ConcordU (Kilicoglu and Bergler, 2011), attempted to extract event modifications, with somewhat limited performance. The difficulty of correct extraction of event modifications is related in part to the recursive nature of the problem (similarly as for nested regulation events): to extract a modification correctly, the modified event must also be extracted correctly. Further, only UTurku predicted any instances of secondary arguments. Thus, teams other than UTurku and ConcordU addressed only the core task extraction targets. With the exception of ConcordU, all systems clearly favor precision over recall (Table 7), in many cases having over 15% point higher precision than recall. This a a somewhat unexpected inversion, as the ConcordU system is one of the two rule-based in the task, an approach typically associated with high precision.

The five top-ranking systems participated also in the GE task (Kim et al., 2011b), which involves a subset of the ID extraction targets. This allows additional perspective into the relative performance of the systems. While there is a 13% point spread in overall results for the top five systems here, in GE all these systems achieved F-scores ranging between 50–56%. The results for FAUST, UMass and Stanford were similar in both tasks, while the ConcordU result was 6% points higher for GE and the UTurku result over 10% points higher for GE, ranking third after FAUST and UMass. These results suggest that while the FAUST and UMass systems in particular have some systematic (e.g. architectural) advantage at both tasks, much of the performance difference observed here between the top three systems and those of ConcordU and UTurku is due to strengths or weaknesses specific to ID. Possible weaknesses may relate to the treatment of multiple core entity types (vs. only PROTEIN in GE) or challenges related to nested entity annotations (not appearing in GE). A possible ID-specific strength of the three top-ranking systems is the use of GE data for training: Riedel and McCallum (2011) report an estimated 7% point improvement and McClosky et al. (2011) a 3% point improvement from use of this data; McGrath et al. (2011) estimate a 1% point improvement from direct corpus combination. The integration strategies applied in training these systems

Team	recall	prec.	F-score	Δ
FAUST	50.62	66.06	57.32	1.73
UMass	49.45	62.11	55.06	1.64
Stanford	48.87	56.03	52.20	1.57
ConcordU	50.77	43.25	46.71	2.50
UTurku	38.79	49.35	43.44	0.87
PNNL	29.36	52.62	37.69	1.42
PredX	23.67	35.18	28.30	0.81

Table 8: Core task evaluation results. The Δ column gives the F-score difference to the corresponding full task (primary) result.

could potentially be applied also with other systems, an experiment that could further clarify the relative strengths of the various systems. The top-ranking five systems all participated also in the EPI task (Ohta et al., 2011), for which UTurku ranked first with FAUST having comparable performance for the core task. While this supports the conclusion that ID performance differences do not reflect a simple universal ranking of the systems, due to many substantial differences between the ID and EPI setups it is not straightforward to identify specific reasons for relative differences to performance at EPI.

Table 8 summarizes the core task results. There are only modest and largely consistent differences to the corresponding full task results, reflecting in part the relative sparseness of additional arguments: in the training data, for example, only approximately 3% of instances of event types that can potentially take additional arguments had at least one additional argument. While event modifications represent a further 4% of full task extraction targets not required for the core task, the overall low extraction performance for additional arguments and modifications limits the practical effect of these annotation categories on the performance difference between systems addressing only the core targets and those addressing the full task.

6 Discussion and Conclusions

We have presented the preparation, resources, results and analysis of the Infectious Diseases (ID) task of the BioNLP Shared Task 2011. A corpus of 30 full-text publications on the two-component systems subdomain of infectious diseases was created for the task in a collaboration of event annotation and domain experts, adapting and extending the BioNLP'09 Shared Task (ST'09) event representation to the domain.

Seven teams submitted final results to the ID task. Despite the novel challenges of full papers, four new entity types, extension of event scopes and the introduction of a new event category for high-level processes, the highest results for the full ID task were comparable to the state-of-the-art performance on the established ST'09 data, showing that the event extraction approach and present systems generalize well and demonstrating the feasibility of event extraction for the infectious diseases domain. Analysis of results suggested further opportunities for improving extraction performance by combining the strengths of various systems and the use of other event resources.

The task design takes into account the needs of supporting practical applications, and its results and findings will be adopted in future development of the Pathosystems Resource Integration Center[4] (PATRIC). Specifically, PATRIC will combine domain named entity recognition and event extraction to mine the virulence factor literature and integrate the results with literature search and retrieval services, protein feature analysis, and systems such as Disease View.[5] Present and future advances at the ID event extraction task can thus assist biologists in efforts of substantial public health interest.

The ID task will be continued as an open shared task challenge with data, supporting resources, and evaluation tools freely available from the shared task site, `http://sites.google.com/site/bionlpst/`.

Acknowledgments

This work was supported by Grant-in-Aid for Specially Promoted Research (MEXT, Japan). This project has been funded in whole or in part with Federal funds from the National Institute of Allergy and Infectious Diseases, National Institutes of Health, Department of Health and Human Services, under Contract No. HHSN272200900040C, awarded to BWS Sobral.

[4] `http://patricbrc.org`

[5] See for example `http://patricbrc.org/portal/portal/patric/DiseaseOverview?cType=taxon&cId=77643`

References

Sophia Ananiadou, Sampo Pyysalo, Jun'ichi Tsujii, and Douglas B. Kell. 2010. Event extraction for systems biology by text mining the literature. *Trends in Biotechnology*, 28(7):381–390.

Sophia Ananiadou, Dan Sullivan, William Black, Gina-Anne Levow, Joseph J. Gillespie, Chunhong Mao, Sampo Pyysalo, BalaKrishna Kolluru, Junichi Tsujii, and Bruno Sobral. 2011. Named entity recognition for bacterial type IV secretion systems. *PLoS ONE*, 6(3):e14780.

M Ashburner, CA Ball, JA Blake, D Botstein, H Butler, JM Cherry, AP Davis, K Dolinski, SS Dwight, JT Eppig, MA Harris, DP Hill, L Issel-Tarver, A Kasarskis, S Lewis, JC Matese, JE Richardson, M Ringwald, GM Rubin, and G Sherlock. 2000. Gene ontology: tool for the unification of biology. *Nature genetics*, 25:25–29.

Jari Björne and Tapio Salakoski. 2011. Generalizing biomedical event extraction. In *Proceedings of the BioNLP 2011 Workshop Companion Volume for Shared Task*, Portland, Oregon, June. Association for Computational Linguistics.

Eugene Charniak and Mark Johnson. 2005. Coarse-to-Fine n-Best Parsing and MaxEnt Discriminative Reranking. In *Proceedings of the 43rd Annual Meeting of the Association for Computational Linguistics (ACL'05)*, pages 173–180.

Peter Corbett and Peter Murray-Rust. 2006. High-throughput identification of chemistry in life science texts. *Computational Life Sciences II*, pages 107–118.

Marie-Catherine de Marneffe, Bill MacCartney, and Christopher D. Manning. 2006. Generating Typed Dependency Parses from Phrase Structure Parses. In *Proceedings of the Fifth International Conference on Language Resources and Evaluation (LREC'06)*, pages 449–454.

Martin Gerner, Goran Nenadic, and Casey M. Bergman. 2010. LINNAEUS: a species name identification system for biomedical literature. *BMC bioinformatics*, 11(1):85+, February.

Yasuhiro Gotoh, Yoko Eguchi, Takafumi Watanabe, Sho Okamoto, Akihiro Doi, and Ryutaro Utsumi. 2010. Two-component signal transduction as potential drug targets in pathogenic bacteria. *Current Opinion in Microbiology*, 13(2):232–239. Cell regulation.

Halil Kilicoglu and Sabine Bergler. 2011. Adapting a general semantic interpretation approach to biological event extraction. In *Proceedings of the BioNLP 2011 Workshop Companion Volume for Shared Task*, Portland, Oregon, June. Association for Computational Linguistics.

Jin-Dong Kim, Tomoko Ohta, Yoshimasa Tsuruoka, Yuka Tateisi, and Nigel Collier, editors. 2004. *Introduction to the bio-entity recognition task at JNLPBA*, Geneva, Switzerland.

Jin-Dong Kim, Tomoko Ohta, and Jun'ichi Tsujii. 2008. Corpus annotation for mining biomedical events from lterature. *BMC Bioinformatics*, 9(1):10.

Jin-Dong Kim, Tomoko Ohta, Sampo Pyysalo, Yoshinobu Kano, and Jun'ichi Tsujii. 2009. Overview of BioNLP'09 Shared Task on Event Extraction. In *Proceedings of Natural Language Processing in Biomedicine (BioNLP) NAACL 2009 Workshop*, pages 1–9.

Jin-Dong Kim, Sampo Pyysalo, Tomoko Ohta, Robert Bossy, and Jun'ichi Tsujii. 2011a. Overview of BioNLP Shared Task 2011. In *Proceedings of the BioNLP 2011 Workshop Companion Volume for Shared Task*, Portland, Oregon, June. Association for Computational Linguistics.

Jin-Dong Kim, Yue Wang, Toshihisa Takagi, and Akinori Yonezawa. 2011b. Overview of the Genia Event task in BioNLP Shared Task 2011. In *Proceedings of the BioNLP 2011 Workshop Companion Volume for Shared Task*, Portland, Oregon, June. Association for Computational Linguistics.

M. Krallinger, A. Morgan, L. Smith, F. Leitner, L. Tanabe, J. Wilbur, L. Hirschman, and A. Valencia. 2008. Evaluation of text-mining systems for biology: overview of the Second BioCreative community challenge. *Genome biology*, 9(Suppl 2):S1.

Tino Krell, Jess Lacal, Andreas Busch, Hortencia Silva-Jimnez, Mara-Eugenia Guazzaroni, and Juan Luis Ramos. 2010. Bacterial sensor kinases: Diversity in the recognition of environmental signals. *Annual Review of Microbiology*, 64(1):539–559.

Thorsten Mascher, John D. Helmann, and Gottfried Unden. 2006. Stimulus perception in bacterial signal-transducing histidine kinases. *Microbiol. Mol. Biol. Rev.*, 70(4):910–938.

David McClosky, Mihai Surdeanu, and Christopher Manning. 2011. Event extraction as dependency parsing for bionlp 2011. In *Proceedings of the BioNLP 2011 Workshop Companion Volume for Shared Task*, Portland, Oregon, June. Association for Computational Linguistics.

David McClosky. 2009. *Any Domain Parsing: Automatic Domain Adaptation for Natural Language Parsing*. Ph.D. thesis, Department of Computer Science, Brown University.

Liam McGrath, Kelly Domico, Courtney Corley, and Bobbie-Jo Webb-Robertson. 2011. Complex biological event extraction from full text using signatures of linguistic and semantic features. In *Proceedings of the BioNLP 2011 Workshop Companion Volume for Shared Task*, Portland, Oregon, June. Association for Computational Linguistics.

Makoto Miwa, Sampo Pyysalo, Tadayoshi Hara, and Jun'ichi Tsujii. 2010. Evaluating dependency representation for event extraction. In *Proceedings of COLING'10*, pages 779–787.

Tomoko Ohta, Jin-Dong Kim, Sampo Pyysalo, Yue Wang, and Jun'ichi Tsujii. 2009. Incorporating GENETAG-style annotation to GENIA corpus. In *Proceedings of BioNLP'09*, pages 106–107.

Tomoko Ohta, Sampo Pyysalo, and Jun'ichi Tsujii. 2011. Overview of the Epigenetics and Post-translational Modifications (EPI) task of BioNLP Shared Task 2011. In *Proceedings of the BioNLP 2011 Workshop Companion Volume for Shared Task*, Portland, Oregon, June. Association for Computational Linguistics.

Sampo Pyysalo, Tomoko Ohta, Han-Cheol Cho, Dan Sullivan, Chunhong Mao, Bruno Sobral, Jun'ichi Tsujii, and Sophia Ananiadou. 2010. Towards event extraction from full texts on infectious diseases. In *Proceedings of BioNLP'10*, pages 132–140.

Sampo Pyysalo, Tomoko Ohta, Rafal Rak, Dan Sullivan, Chunhong Mao, Chunxia Wang, Bruno Sobral, Jun'ichi Tsujii, and Sophia Ananiadou. 2011. Annotation guidelines for infectious diseases event corpus. Technical report, Tsujii Laboratory, University of Tokyo. To appear.

Sebastian Riedel and Andrew McCallum. 2011. Robust biomedical event extraction with dual decomposition and minimal domain adaptation. In *Proceedings of the BioNLP 2011 Workshop Companion Volume for Shared Task*, Portland, Oregon, June. Association for Computational Linguistics.

Sebastian Riedel, David McClosky, Mihai Surdeanu, Andrew McCallum, and Chris Manning. 2011. Model combination for event extraction in bionlp 2011. In *Proceedings of the BioNLP 2011 Workshop Companion Volume for Shared Task*, Portland, Oregon, June. Association for Computational Linguistics.

Yutaka Sasaki, Yoshimasa Tsuruoka, John McNaught, and Sophia Ananiadou. 2008. How to make the most of NE dictionaries in statistical NER. *BMC bioinformatics*, 9 Suppl 11.

Pontus Stenetorp, Goran Topić, Sampo Pyysalo, Tomoko Ohta, Jin-Dong Kim, and Jun'ichi Tsujii. 2011. BioNLP Shared Task 2011: Supporting Resources. In *Proceedings of the BioNLP 2011 Workshop Companion Volume for Shared Task*, Portland, Oregon, June. Association for Computational Linguistics.

Peter Thomason and Rob Kay. 2000. Eukaryotic signal transduction via histidine-aspartate phosphorelay. *J Cell Sci*, 113(18):3141–3150.

Yue Wang, Jin-Dong Kim, Rune Sætre, Sampo Pyysalo, and Jun'ichi Tsujii. 2009. Investigating heterogeneous protein annotations toward cross-corpora utilization. *BMC Bioinformatics*, 10(403).

Chunxia Wang, Jocelyn Kemp, Isabel O. Da Fonseca, Raymie C. Equi, Xiaoyan Sheng, Trevor C. Charles, and Bruno W. S. Sobral. 2010. Sinorhizobium meliloti 1021 loss-of-function deletion mutation in chvi and its phenotypic characteristics. *Molecular Plant-Microbe Interactions*, 23(2):153–160.

Biomedical Event Extraction from Abstracts and Full Papers using Search-based Structured Prediction

Andreas Vlachos and **Mark Craven**
Department of Biostatistics and Medical Informatics
University of Wisconsin-Madison
{vlachos,craven}@biostat.wisc.edu

Abstract

In this paper we describe our approach to the BioNLP 2011 shared task on biomedical event extraction from abstracts and full papers. We employ a joint inference system developed using the search-based structured prediction framework and show that it improves on a pipeline using the same features and it is better able to handle the domain shift from abstracts to full papers. In addition, we report on experiments using a simple domain adaptation method.

1 Introduction

The term *biomedical event extraction* is used to refer to the task of extracting descriptions of actions and relations among one or more entities from the biomedical literature. The BioNLP 2011 shared task GENIA Task1 (BioNLP11ST-GE1) (Kim et al., 2011) focuses on extracting events from abstracts and full papers. The inclusion of full papers in the datasets is the only difference from Task1 of the BioNLP 2009 shared task (BioNLP09ST1) (Kim et al., 2009), which used the same task definition and abstracts dataset. Each event consists of a *trigger* and one or more *arguments*, the latter being proteins or other events. The protein names are annotated in advance and any token in a sentence can be a trigger for one of the nine event types. In an example demonstrating the complexity of the task, given the passage "... SQ 22536 suppressed **gp41**-induced **IL-10** production in monocytes", systems should extract the three nested events shown in Fig. 1d.

In our submission, we use the event extraction system of Vlachos and Craven (2011) which employs the search-based structured prediction framework (SEARN) (Daumé III et al., 2009). SEARN converts the problem of learning a model for structured prediction into learning a set of models for cost-sensitive classification (CSC). In CSC, each training instance has a vector of misclassification costs associated with it, thus rendering some mistakes in some instances to be more expensive than others. Compared to other structured prediction frameworks such as Markov Logic Networks (Poon and Vanderwende, 2010), SEARN provides high modeling flexibility but it does not requiring task-dependent approximate inference.

In this work, we show that SEARN is more accurate than a pipeline using the same features and it is better able to handle the domain shift from abstracts to full papers. Furthermore, we report on experiments with the simple domain adaptation method proposed by Daumé III (2007), which creates a version of each feature for each domain. While the results were mixed, this method improves our performance on full papers of the test set, for which little training data is available.

2 Event extraction decomposition

Figure 1 describes the event extraction decomposition that is used throughout the paper. Each stage has its own module to perform the classification needed.

In trigger recognition the system decides whether a token acts as a trigger for one of the nine event types or not. We only consider tokens that are tagged as nouns, verbs or adjectives by the parser, as they

Proceedings of BioNLP Shared Task 2011 Workshop, pages 36–40,

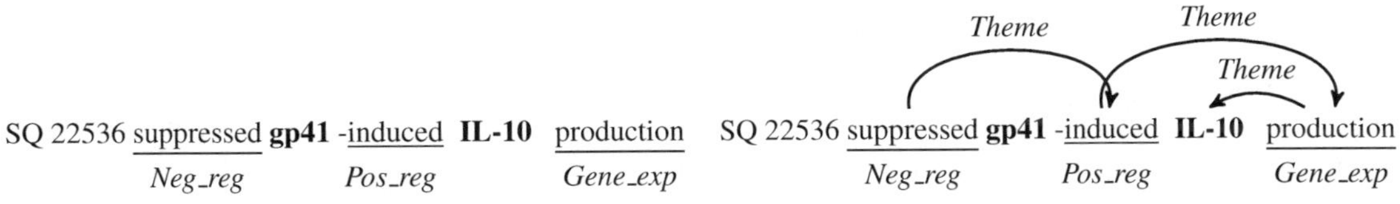

(a) Trigger recognition

(b) *Theme* assignment

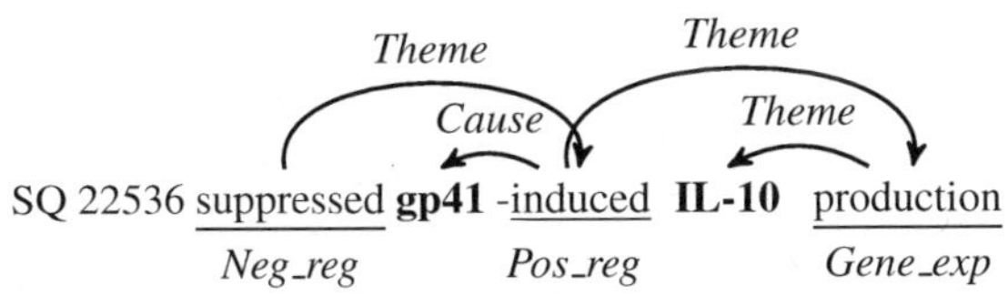

(c) *Cause* assignment

ID	type	*Trigger*	*Theme*	*Cause*
E1	*Neg_reg*	suppressed	*E2*	
E2	*Pos_reg*	induced	*E3*	gp41
E3	*Gene_exp*	production	IL-10	

(d) Event construction

Figure 1: The stages of our biomedical event extraction system.

cover the majority of the triggers in the data. The main features used in the classifier represent the lemma of the token which is sufficient to predict the event type correctly in most cases. In addition, we include features that conjoin each lemma with its part-of-speech tag and its immediate lexical and syntactic context, which allows us to handle words that can represent different event types, e.g. "activity" often denotes a *Regulation* event but in "binding activity" it denotes a *Binding* event instead.

In *Theme* assignment, we form an agenda of candidate trigger-argument pairs for all trigger-protein combinations in the sentence and classify them as *Themes* or not. Whenever a trigger is predicted to be associated with a *Theme*, we form candidate pairs between all the *Regulation* triggers in the sentence and that trigger as the argument, thus allowing the prediction of nested events. Also, we remove candidate pairs that could result in directed cycles, as they are not allowed by the task. In *Cause* assignment, we form an agenda of candidate trigger-argument pairs and classify them as *Causes* or not. We form pairs between *Regulation* class triggers that were assigned at least one *Theme*, and protein names and other triggers that were assigned at least one *Theme*.

The features used in these two stages are extracted from the syntactic dependency path and the textual string between the trigger and the argument. We extract the shortest unlexicalized dependency path connecting each trigger-argument pair using Dijkstra's algorithm, allowing the paths to follow either dependency direction. One set of features represents the shortest unlexicalized path between the pair and in addition we have sets of features representing each path conjoined with the lemma, the PoS tag and the event type of the trigger, the type of the argument and the first and last lemmas in the dependency path.

In the event construction stage, we convert the predictions of the previous stages into events. If a *Binding* trigger is assigned multiple *Themes*, we choose to form either one event per *Theme* or one event with multiple *Themes*. For this purpose, we group the arguments of each nominal *Binding* trigger according to the first label in their dependency path and generate events using the cross-product of these groups. For example, assuming the parse was correct and all the *Themes* recognized, "interactions of **A** and **B** with **C**" results in two *Binding* events with two *Themes* each, **A** with **C**, and **B** with **C** respectively. We add the exceptions that if two *Themes* are part of the same token (e.g. "**A/B** interactions"), or the trigger and one of the *Themes* are part of the same token, or the lemma of the trigger is "bind" then they form one *Binding* event with two *Themes*.

3 Structured prediction with SEARN

SEARN (Daumé III et al., 2009) forms the structured output prediction of an instance s as a sequence of T multiclass predictions $\hat{y}_{1:T}$ made by a hypothesis h. The latter is a weighted ensemble of classifiers that are learned jointly. Each prediction $\hat{y}_t$ can use features from s as well as from all the previous predictions $\hat{y}_{1:t-1}$, thus taking structure into account. These predictions are referred to as *actions* and we adopt this term in order to distinguish them from the structured output predictions.

The SEARN algorithm is presented in Alg. 1. In each iteration, SEARN uses the current hypothesis h to generate a CSC example for each action $\hat{y}_t$ chosen to form the prediction for each labeled instance s (steps 6-12). The cost associated with each action is estimated using the gold standard according to a loss function l which corresponds to the task evaluation metric (step 11). Using a CSC learning algorithm, a new hypothesis h_{new} is learned (step 13) which is combined with the current one according to the interpolation parameter β (step 14). h is initialized to the *optimal policy* (step 2) which is derived from the gold standard. In each iteration SEARN "corrupts" the optimal policy with the learned hypotheses. Thus, each h_{new} is adapted to the actions chosen by h instead of the optimal policy. The algorithm terminates when the dependence on the latter becomes insignificant.

Algorithm 1 SEARN

1: **Input:** labeled instances $\mathcal{S}$, *optimal policy* π, CSC learning algorithm *CSCL*, loss function ℓ
2: current policy $h = \pi$
3: **while** h depends significantly on π **do**
4:   Examples $E = \emptyset$
5:   **for** s **in** S **do**
6:     Predict $h(s) = \hat{y}_1 \ldots \hat{y}_T$
7:     **for** $\hat{y}_t$ **in** $h(s)$ **do**
8:       Extract features $\Phi_t = f(s, \hat{y}_{1:t-1})$
9:       **for each** possible action y_t^i **do**
10:         Predict $y'_{t+1:T} = h(s|\hat{y}_{1:t-1}, y_t^i)$
11:         Estimate $c_t^i = \ell(\hat{y}_{1:t-1}, y_t^i, y'_{t+1:T})$
12:       Add (Φ_t, c_t) to E
13:   Learn a classifier $h_{new} = CSCL(E)$
14:   $h = \beta h_{new} + (1 - \beta)h$
15: **Output:** hypothesis h without π

4 Biomedical event extraction with SEARN

In this section we describe how we learn the event extraction decomposition described in Sec. 2 under SEARN. Each instance is a sentence and the hypothesis learned in each iteration consists of a classifier for each stage of the pipeline, excluding event construction which is rule-based.

SEARN allows us to extract structural features for each action from the previous ones. During trigger recognition, we add as features the combination of the lemma of the current token combined with the event type (if any) assigned to the previous and the next token, as well as to the tokens that have syntactic dependencies with it. During *Theme* assignment, when considering a trigger-argument pair, we add features based on whether the pair forms an undirected cycle with previously predicted *Themes*, whether the trigger has been assigned a protein as a *Theme* and the candidate *Theme* is an event trigger (and the reverse), and whether the argument is the *Theme* of a trigger with the same event type. We also add a feature indicating whether the trigger has three *Themes* predicted already. During *Cause* assignment, we add features representing whether the trigger has been assigned a protein as a *Cause* and the candidate *Cause* is an event trigger.

Since the features extracted for an action depend on previous ones, we need to define a prediction order for the actions. In trigger recognition, we process the tokens from left to right since modifiers appearing before nouns tend to affect the meaning of the latter, e.g. "binding activity". In *Theme* and *Cause* assignment, we predict trigger-argument pairs in order of increasing dependency path length, assuming that, since they are the main source of features in these stages and shorter paths are less sparse, pairs containing shorter ones should be predicted more reliably. The loss function sums the number of false positive and false negative events, which is the evaluation measure of the shared task. The optimal policy is derived from the gold standard and returns the action that minimizes the loss over the sentence given the previous actions and assuming that all future actions are optimal.

In step 11 of Alg. 1, the cost of each action is estimated over the whole sentence. While this allows us to take structure into account, it can result in costs being affected by a part of the output that is not related to that action. This is likely to occur in event extraction, as sentences can often be long and contain disconnected event components in their output graphs. For this reason we use *focused costing* (Vlachos and Craven, 2011), in which the cost estimation for an action takes into account only the part of the output graph connected with that action.

	pipeline (R/P/F)			SEARN (R/P/F)		
trigger	49.1	64.0	55.6	83.2	28.6	42.6
Theme	43.7	78.6	56.2	63.8	72.0	67.6
Cause	13.9	61.0	22.6	33.9	53.8	41.6
Event	31.7	70.1	43.6	45.8	60.51	52.1

Table 1: Results on the development dataset.

5 Experiments

In our experiments, we perform multiclass CSC learning using our implementation of the online passive-aggressive (PA) algorithm proposed by Crammer et al. (2006). The aggressiveness parameter and the number of rounds in parameter learning are set by tuning on 10% of the training data and we use the variant named PA-II with prediction-based updates. For SEARN, we set the interpolation parameter β to 0.3. For syntactic parsing, we use the output of the parser of Charniak and Johnson (2005) adapted to the biomedical domain by McClosky (2010), as provided by the shared task organizers in the Stanford collapsed dependencies with conjunct dependency propagation (Stenetorp et al., 2011). Lemmatization is performed using *morpha* (Minnen et al., 2001). No other knowledge sources or tools are used.

In order to assess the benefits of joint learning under SEARN, we compare it against a pipeline of independently learned classifiers using the same features and task decomposition. Table 1 reports the Recall/Precision/F-score achieved in each stage, as well as the overall performance. SEARN obtains better performance on the development set by 8.5 F-score points. This increase is larger than the 7.3 points reported in Vlachos and Craven (2011) on the BioNLP09ST1 datasets which contain only abstracts. This result suggests that the gains of joint inference under SEARN are greater when learning from the additional data from full papers. Note that while the classifier learned with SEARN overpredicts triggers, the *Theme* and *Cause* classifiers maintain relatively high precision with substantially higher recall as they are learned jointly with it. As triggers that do not form events are ignored by the evaluation, trigger overprediction without event overprediction does not result in performance loss.

The results of our submission on the test dataset using SEARN were 42.6/61.2/50.2 (Recall/Precision/F-score) which ranked sixth in the shared task. In the *Regulation* events which are considered harder due to nesting, our submission was ranked fourth. This demonstrates the potential of SEARN for structured prediction, as the performance on regulation events depends partly on the performance on the simple ones on which our submission was ranked eighth.

After the end of the shared task, we experimented with the domain adaptation method proposed by Daumé III (2007), which creates multiple versions for each feature by conjoining it with the domain label of the instance it is extracted from (abstracts or full papers). While this improved the performance of the pipeline baseline by 0.3 F-score points, the performance under SEARN dropped by 0.4 points on the development data. Using the online service provided by the organizers, we evaluated the performance of the domain adapted SEARN-based system on the test set and the overall performance improved to 50.72 in F-score (would have ranked 5th). In particular, domain adaptation improved the performance on full papers by 1.22 points, thus reaching 51.22 in F-score. This version of the system would have ranked 3rd overall and 1st in the *Regulation* events in this part of the corpus. We hypothesize that these mixed results are due to the sparse features used in the stages of the event extraction decomposition, which become even sparser using this domain adaptation method, thus rendering the learning of appropriate weights for them harder.

6 Conclusions

We presented a joint inference approach to the BioNLP11ST-GE1 task using SEARN which converts a structured prediction task into a set of CSC tasks whose models are learned jointly. Our results demonstrate that SEARN achieves substantial performance gains over a standard pipeline using the same features.

Acknowledgments

The authors were funded by NIH/NLM grant R01 LM07050.

References

Eugene Charniak and Mark Johnson. 2005. Coarse-to-fine n-best parsing and MaxEnt discriminative reranking. In *Proceedings of the 43rd Annual Meeting on Association for Computational Linguistics*, pages 173–180.

Koby Crammer, Ofer Dekel, Joseph Keshet, Shai Shalev-Shwartz, and Yoram Singer. 2006. Online passive-aggressive algorithms. *Journal of Machine Learning Research*, 7:551–585.

Hal Daumé III, John Langford, and Daniel Marcu. 2009. Search-based structured prediction. *Machine Learning*, 75:297–325.

Hal Daumé III. 2007. Frustratingly easy domain adaptation. In *Proceedings of the 45th Annual Meeting of the Association for Computational Linguistics*, pages 256–263.

Jin-Dong Kim, Tomoko Ohta, Sampo Pyysalo, Yoshinobu Kano, and Jun'ichi Tsujii. 2009. Overview of BioNLP'09 shared task on event extraction. In *Proceedings of the BioNLP 2009 Workshop Companion Volume for Shared Task*, pages 1–9.

Jin-Dong Kim, Yue Wang, Toshihisa Takagi, and Akinori Yonezawa. 2011. Overview of the Genia Event task in BioNLP Shared Task 2011. In *Proceedings of the BioNLP 2011 Workshop Companion Volume for Shared Task*.

David McClosky. 2010. *Any domain parsing: Automatic domain adaptation for natural language parsing*. Ph.D. thesis, Department of Computer Science, Brown University.

Guido Minnen, John Carroll, and Darren Pearce. 2001. Applied morphological processing of English. *Natural Language Engineering*, 7(3):207–223.

Hoifung Poon and Lucy Vanderwende. 2010. Joint inference for knowledge extraction from biomedical literature. In *Proceedings of the Human Language Technologies: The 2010 Annual Conference of the North American Chapter of the Association for Computational Linguistics*, pages 813–821.

Pontus Stenetorp, Goran Topić, Sampo Pyysalo, Tomoko Ohta, Jin-Dong Kim, and Jun'ichi Tsujii. 2011. BioNLP Shared Task 2011: Supporting Resources. In *Proceedings of the BioNLP 2011 Workshop Companion Volume for Shared Task*.

Andreas Vlachos and Mark Craven. 2011. Search-based structured prediction applied to biomedical event extraction. In *Proceedings of the Fifteenth Conference on Computational Natural Language Learning*.

Event Extraction as Dependency Parsing for BioNLP 2011

David McClosky, Mihai Surdeanu, and Christopher D. Manning
Department of Computer Science
Stanford University
Stanford, CA 94305
{mcclosky,mihais,manning}@stanford.edu

Abstract

We describe the Stanford entry to the BioNLP 2011 shared task on biomolecular event extraction (Kim et al., 2011a). Our framework is based on the observation that event structures bear a close relation to dependency graphs. We show that if biomolecular events are cast as these pseudosyntactic structures, standard parsing tools (maximum-spanning tree parsers and parse rerankers) can be applied to perform event extraction with minimum domain-specific tuning. The vast majority of our domain-specific knowledge comes from the conversion to and from dependency graphs. Our system performed competitively, obtaining 3rd place in the Infectious Diseases track (50.6% *f*-score), 5th place in Epigenetics and Post-translational Modifications (31.2%), and 7th place in Genia (50.0%). Additionally, this system was part of the combined system in Riedel et al. (2011) to produce the highest scoring system in three out of the four event extraction tasks.

1 Introduction

The distinguishing aspect of our approach is that by casting event extraction as a dependency parsing, we take advantage of standard parsing tools and techniques rather than creating special purpose frameworks. In this paper, we show that with minimal domain-specific tuning, we are able to achieve competitive performance across the three event extraction domains in the BioNLP 2011 shared task.

At the heart of our system[1] is an off-the-shelf dependency parser, MSTParser[2] (McDonald et al., 2005; McDonald and Pereira, 2006), extended with event extraction-specific features and bookended by conversions to and from dependency trees. While features in MSTParser must be edge-factored and thus fairly local (e.g., only able to examine a portion of each event at once), decoding is performed globally allowing the parser to consider trade-offs. Furthermore, as MSTParser can use n-best decoders, we are able to leverage a reranker to capture global features to improve accuracy.

In §2, we provide a brief overview of our framework. We describe specific improvements for the BioNLP 2011 shared task in §3. In §4, we present detailed results of our system. Finally, in §5 we give some directions for future work.

2 Event Parsing

Our system includes three components: (1) anchor detection to identify and label event anchors, (2) event parsing to form candidate event structures by linking entities and event anchors, and (3) event reranking to select the best candidate event structure. As the full details on our approach are described in McClosky et al. (2011), we will only provide an outline of our methods here along with additional implementation notes.

Before running our system, we perform basic preprocessing on the corpora. Sentences need to be segmented, tokenized, and parsed syntactically. We use custom versions of these (except for Infectious Diseases where we use those from Stenetorp et al. (2011)). To ease event parsing, our

[1] nlp.stanford.edu/software/eventparser.shtml

[2] http://sourceforge.net/projects/mstparser/

Proceedings of BioNLP Shared Task 2011 Workshop, pages 41–45,
Portland, Oregon, USA, 24 June, 2011.

tokenizations are designed to split off suffixes which are often event anchors. For example, we split the token *RelA-induced* into the two tokens *RelA* and *induced*[3] since *RelA* is a protein and *induced* an event anchor. If this was a single token, our event parser would be unable to link them since it cannot predict self-loops in the dependency graph. For syntactic parsing, we use the self-trained biomedical parsing model from McClosky (2010) with the Charniak and Johnson (2005) reranking parser. We use its actual constituency tree, the dependency graph created by applying head percolation rules, and the Stanford Dependencies (de Marneffe and Manning, 2008) extracted from the tree (collapsed and uncollapsed).

Anchor detection uses techniques inspired from named entity recognition to label each token with an event type or *none*. The features for this stage are primarily drawn from Björne et al. (2009). We reduce multiword event anchors to their syntactic head.[4] We classify each token independently using a logistic regression classifier with L_2 regularization. By adjusting a threshold parameter, we can adjust the balance between precision and recall. We choose to heavily favor recall (i.e., overgenerate event anchors) as the event parser can drop extraneous anchors by not attaching any arguments to them.

The event anchors from anchor detection and the included entities (`.t1` files) form a "reduced" sentence, which becomes the input to event parsing. Thus, the only words in the reduced sentence are tokens believed to directly take part in events. Note, though, that we use the original "full" sentence (including the various representations of its syntactic parse) for feature generation. For full details on this process, see McClosky et al. (2011). As stated before, this stage consists of MSTParser with additional event parsing features. There are four decoding options for MSTParser, depending on (a) whether features are first- or second-order and (b) whether graphs produced are projective or non-projective. The projective decoders have complete n-best implementations whereas their non-projective counterparts are approximate. Nevertheless, these four decoders constitute slightly different views of the same data and can be combined inside the reranking framework. After decoding, we convert parses back to event structures. Details on this critical step are given in McClosky et al. (2011).

Event reranking, the final stage of our system, receives an n-best list of event structures from each decoder in the event parsing step. The reranker can use any global features of an event structure to rescore it and outputs the highest scoring structure. This is based on parse reranking (Ratnaparkhi, 1999; Collins, 2000) but uses features on event structures instead of syntactic constituency structures. We used Mark Johnson's `cvlm` estimator[5] (Charniak and Johnson, 2005) when learning weights for the reranking model. Since the reranker can incorporate the outputs from multiple decoders, we use it as an ensemble technique as in Johnson and Ural (2010).

3 Extensions for BioNLP 2011

This section outlines the changes between our BioNLP 2011 shared task submission and the system described in McClosky et al. (2011). The main differences are that all dataset-specific portions of the model have been factored out to handle the expanded Genia (GE) dataset (Kim et al., 2011b) and the new Epigenetics and Post-translational Modifications (EPI) and Infectious Diseases (ID) datasets (Ohta et al., 2011; Pyysalo et al., 2011, respectively). Other changes are relatively minor but documented here as implementation notes.

Several improvements were made to anchor detection, improving its accuracy on all three domains. The first is the use of distributional similarity features. Using a large corpus of abstracts from PubMed (30,963,886 word tokens of 335,811 word types), we cluster words by their syntactic contexts and morphological contents (Clark, 2003). We used the Ney-Essen clustering model with morphology to produce 45 clusters. Using these clusters, we extended the feature set for anchor detection from McClosky et al. (2011) as follows: for each lexicalized feature we create an equivalent feature where the corresponding word is replaced by its cluster ID. This yielded consistent improvements of at least 1 percentage point in both anchor detection and event

[3]The dash is removed since a lone dash would further confuse the syntactic parser.

[4]This does not affect performance if the approximate scorer is used, but it does impact scores if exact matching of anchor boundaries is imposed.

[5]`http://github.com/BLLIP/bllip-parser`

extraction in the development partition of the GE dataset.

Additionally, we improved the head percolation rules for selecting the head of each multiword event anchor. The new rules prohibit determiners and prepositions from being heads, instead preferring verbs, then nouns, then adjectives. There is also a small stop list to prohibit the selection of certain verbs ("has", "have", "is", "be", and "was").

In event parsing, we used the *morpha* lemmatizer (Minnen et al., 2001) to stem words instead of simply lowercasing them. This generally led to a small but significant improvement in event extraction across the three domains. Additionally, we do not use the feature selection mechanism described in McClosky et al. (2011) due to time restrictions. It requires running all parsers twice which is especially cumbersome when operating in a round-robin frame (as is required to train the reranker).

Also, note that our systems were only trained to do Task 1 (or "core") roles for each dataset. This was due to time restrictions and not system limitations.

3.1 Adapting to the Epigenetics track

For the EPI dataset, we adjusted our postprocessing rules to handle the CATALYSIS event type. Similar to REGULATION events in GE, CATALYSIS events do not accept multiple CAUSE arguments. We handle this by replicating such CATALYSIS events and assigning each new event a different CAUSE argument. To adapt the ontology features in the parser (McClosky et al., 2011, §3.3), we created a supertype for all non-CATALYSIS events since they behave similarly in many respects.

There are several possible areas for improvement in handling this dataset. First, our internal implementation of the evaluation criteria differed from the online scorer, sometimes by up to 6% f-score. As a result, the reranker optimized a noisy version of the evaluation criteria and potentially could have performed better. It is unclear why our evaluator scored EPI structures differently (it replicated the scores for GE) but it is worthy of investigation. Second, due to time constraints, we did not transfer the parser or reranker consistency features (e.g., non-REGULATION events should not take events as arguments) or the type ontology in the reranker to the EPI dataset. As a result, our results describe our system with incomplete domain-specific knowledge.

3.2 Adapting to the Infectious Diseases track

Looking only at event types and their arguments, ID is similar to GE. As a result, much of our domain-specific processing code for this dataset is based on code for GE. The key difference is that the GE post-processing code removes event anchors with zero arguments. Since ID allows PROCESS events to have zero or one anchors, we added this as an exception. Additionally, the ID dataset includes many nested entities, e.g., two-component system entities contain two other entities within their span. In almost all of these cases, only the outermost entity takes part in an event. To simplify processing, we removed all nested entities. Any events attaching to a nested entity were reattached to its outermost entity.

Given the similarities with GE, we explored simple domain adaptation by including the gold data from GE along with our ID training data. To ensure that the GE data did not overwhelm the ID data, we tried adding multiple copies of the ID data (see Table 1 and the next section).

As in EPI, we adjusted the type ontology in the parser for this dataset. This included "core entities" (as defined by the task) and a "PROTEIN-or-REGULON-OPERON" type (the type of arguments for GENE EXPRESSION and TRANSCRIPTION events). Also as in EPI, the reranker did not use the updated type ontology.

4 Results

For ID, we present experiments on merging GE with ID data (Table 1). Since GE is much larger than ID, we experimented with replicating the ID training partition. Our best performance came from training on three copies of the ID data and the training and development sections of GE. However, as the table shows, performance is stable for more than two copies of the ID data. Note that for this shared task we simply merged the two domains. We did not implement any domain adaptation techniques (e.g., labeling features based on the domain they come from (Daumé III, 2007)).

Table 2 shows the performance of the various parser decoders and their corresponding rerankers. The last line in each domain block lists the score of the reranker that uses candidates produced by all decoders. This reranking model always outperforms the best individual parser. Furthermore, the reranking models on top of individual decoders help in all but one situation (ID – 2N decoder). To our knowledge, our approach is the first to show that reranking with features generated from global event structure helps event extraction. Note that due to approximate 2N decoding in MSTParser, this decoder does not produce true n-best candidates and generally outputs only a handful of unique parses. Because of this, the corresponding rerankers suffer from insufficient training data and hurt performance in ID.

Finally, in Table 3, we give our results and ranking on the official test sets. Our results are 6 f points lower than the best submission in GE and EPI and 5 points lower in ID. Considering that the we used generic parsing tools with minimal customization (e.g., our parsing models cannot extract directed acyclic graph structures, which are common in this data), we believe these results are respectable.

5 Conclusion

Our participation in the BioNLP shared task proves that standard parsing tools (i.e., maximum-spanning tree parsers, parse rerankers) can be successfully used for event extraction. We achieved this by converting the original event structures to a pseudo-syntactic representation, where event arguments appear as modifiers to event anchors. Our analysis indicates that reranking always helps, which proves that there is merit in modeling non-local information in biomolecular events. To our knowledge, our approach is the first to use parsing models for biomedical event extraction.

During the shared task, we adapted our system previously developed for the 2009 version of the Genia dataset. This process required minimal effort: we did not add any new features to the parsing model; we added only two domain-specific post-processing steps (i.e., we allowed events without arguments in ID and we replicated CATALYSIS events with multiple CAUSE arguments in EPI). Our system's robust performance in all domains proves that our approach is portable.

A desired side effect of our effort is that we can easily incorporate any improvements to parsing models (e.g., parsing of directed acyclic graphs, dual decomposition, etc.) in our event extractor.

Model	Prec	Rec	f-score
ID	**59.3**	38.0	46.3
(ID×1) + GE	52.0	40.2	45.3
(ID×2) + GE	52.4	41.7	46.4
(ID×3) + GE	54.8	**45.0**	**49.4**
(ID×4) + GE	55.2	43.8	48.9
(ID×5) + GE	55.1	44.7	**49.4**

Table 1: Impact of merging several copies of ID training with GE training and development. Scores on ID development data (2N parser only).

Decoder(s)	Parser	Reranker
1P	49.0	49.4
2P	49.5	50.5
1N	**49.9**	50.2
2N	46.5	47.9
All	—	**50.7** ∗

(a) Genia results (task 1)

Decoder(s)	Parser	Reranker
1P	62.3	63.3
2P	62.2	63.3
1N	**62.9**	**64.6** ∗
2N	60.8	63.8
All	—	64.1

(b) Epigenetics results (core task)

Decoder(s)	Parser	Reranker
1P	46.0	48.5
2P	47.8	49.8
1N	48.5	49.4
2N	**49.4**	48.8
All	—	**50.2** ∗

(c) Infectious Diseases results (core task)

Table 2: Results on development sections in BioNLP f-scores. "∗" indicates the submission model for each domain.

Domain (task)	Prec	Rec	f-score	Ranking
GE (task 1)	61.1	42.4	50.0	7th
EPI (core)	70.2	56.9	62.8	5th
ID (core)	55.9	46.3	50.6	3rd

Table 3: BioNLP f-scores on the final test set.

Acknowledgments

We would like to thank the BioNLP shared task organizers for an enjoyable and interesting task and their quick responses to questions. We would also like to thank Sebastian Riedel for many interesting discussions. We gratefully acknowledge the support of the Defense Advanced Research Projects Agency (DARPA) Machine Reading Program under Air Force Research Laboratory (AFRL) prime contract no. FA8750-09-C-0181.

References

Jari Björne, Juho Heimonen, Filip Ginter, Antti Airola, Tapio Pahikkala, and Tapio Salakoski. 2009. Extracting complex biological events with rich graph-based feature sets. In *Proceedings of the BioNLP 2009 Workshop Companion Volume for Shared Task*, pages 10–18, Boulder, Colorado, June. Association for Computational Linguistics.

Eugene Charniak and Mark Johnson. 2005. Coarse-to-Fine n-Best Parsing and MaxEnt Discriminative Reranking. In *ACL*. The Association for Computer Linguistics.

Alexander Clark. 2003. Combining distributional and morphological information for part of speech induction. In *Proceedings of the tenth Annual Meeting of the European Association for Computational Linguistics (EACL)*, pages 59–66.

Michael Collins. 2000. Discriminative reranking for natural language parsing. In *Machine Learning: Proceedings of the Seventeenth International Conference (ICML 2000)*, pages 175–182, Stanford, California.

Hal Daumé III. 2007. Frustratingly easy domain adaptation. In *Conference of the Association for Computational Linguistics (ACL)*, Prague, Czech Republic.

Marie-Catherine de Marneffe and Christopher D. Manning. 2008. Stanford typed hierarchies representation. In *Proceedings of the COLING Workshop on Cross-Framework and Cross-Domain Parser Evaluation*.

Mark Johnson and Ahmet Engin Ural. 2010. Reranking the berkeley and brown parsers. In *Proceedings of the HLT: North American Chapter of the ACL (HLT-NAACL)*, pages 665–668. Association for Computational Linguistics, June.

Jin-Dong Kim, Sampo Pyysalo, Tomoko Ohta, Robert Bossy, and Jun'ichi Tsujii. 2011a. Overview of BioNLP Shared Task 2011. In *Proceedings of the BioNLP 2011 Workshop Companion Volume for Shared Task*, Portland, Oregon, June. Association for Computational Linguistics.

Jin-Dong Kim, Yue Wang, Toshihisa Takagi, and Akinori Yonezawa. 2011b. Overview of the Genia Event task in BioNLP Shared Task 2011. In *Proceedings of the BioNLP 2011 Workshop Companion Volume for Shared Task*, Portland, Oregon, June. Association for Computational Linguistics.

David McClosky, Mihai Surdeanu, and Chris Manning. 2011. Event extraction as dependency parsing. In *Proceedings of the Association for Computational Linguistics: Human Language Technologies 2011 Conference (ACL-HLT'11), Main Conference*, Portland, Oregon, June.

David McClosky. 2010. *Any Domain Parsing: Automatic Domain Adaptation for Parsing*. Ph.D. thesis, Computer Science Department, Brown University.

Ryan T. McDonald and Fernando C. N. Pereira. 2006. Online learning of approximate dependency parsing algorithms. In *Proceedings of EACL*. The Association for Computer Linguistics.

Ryan T. McDonald, Fernando Pereira, Kiril Ribarov, and Jan Hajic. 2005. Non-projective dependency parsing using spanning tree algorithms. In *Proceedings of HLT/EMNLP*. The Association for Computational Linguistics.

Guido Minnen, John Carroll, and Darren Pearce. 2001. Applied morphological processing of English. *Natural Language Engineering*, 7(03):207–223.

Tomoko Ohta, Sampo Pyysalo, and Jun'ichi Tsujii. 2011. Overview of the Epigenetics and Post-translational Modifications (EPI) task of BioNLP Shared Task 2011. In *Proceedings of the BioNLP 2011 Workshop Companion Volume for Shared Task*, Portland, Oregon, June. Association for Computational Linguistics.

Sampo Pyysalo, Tomoko Ohta, Rafal Rak, Dan Sullivan, Chunhong Mao, Chunxia Wang, Bruno Sobral, Jun'ichi Tsujii, and Sophia Ananiadou. 2011. Overview of the Infectious Diseases (ID) task of BioNLP Shared Task 2011. In *Proceedings of the BioNLP 2011 Workshop Companion Volume for Shared Task*, Portland, Oregon, June. Association for Computational Linguistics.

Adwait Ratnaparkhi. 1999. Learning to parse natural language with maximum entropy models. *Machine Learning*, 34(1-3):151–175.

Sebastian Riedel, David McClosky, Mihai Surdeanu, Andrew McCallum, and Christopher D. Manning. 2011. Model Combination for Event Extraction in BioNLP 2011. In *BioNLP 2011 Shared Task*.

Pontus Stenetorp, Goran Topić, Sampo Pyysalo, Tomoko Ohta, Jin-Dong Kim, and Jun'ichi Tsujii. 2011. BioNLP Shared Task 2011: Supporting Resources. In *Proceedings of the BioNLP 2011 Workshop Companion Volume for Shared Task*, Portland, Oregon, June. Association for Computational Linguistics.

Robust Biomedical Event Extraction with Dual Decomposition and Minimal Domain Adaptation

Sebastian Riedel **Andrew McCallum**
Department of Computer Science
University of Massachusetts, Amherst
{riedel,mccallum}@cs.umass.edu

Abstract

We present a joint model for biomedical event extraction and apply it to four tracks of the BioNLP 2011 Shared Task. Our model decomposes into three sub-models that concern (a) event triggers and outgoing arguments, (b) event triggers and incoming arguments and (c) protein-protein bindings. For efficient decoding we employ dual decomposition. Our results are very competitive: With minimal adaptation of our model we come in second for two of the tasks—right behind a version of the system presented here that includes predictions of the Stanford event extractor as features. We also show that for the Infectious Diseases task using data from the Genia track is a very effective way to improve accuracy.

1 Introduction

This paper presents the UMass entry to the BioNLP 2011 shared task (Kim et al., 2011a). We introduce a simple joint model for the extraction of biomedical events, and show competitive results for four tracks of the competition. Our model subsumes three tractable sub-models, one for extracting event triggers and outgoing edges, one for event triggers and incoming edges and one for protein-protein bindings. Fast and accurate joint inference is provided by combining optimizing methods for these three sub-models via dual decomposition (Komodakis et al., 2007; Rush et al., 2010). Notably, our model constitutes the first joint approach that explicitly predicts which protein should share the same binding event. So far this has either been done through post-processing heuristics (Björne et al., 2009; Riedel et al., 2009; Poon and Vanderwende, 2010), or through a local classifier at the end of a pipeline (Miwa et al., 2010).

Our model is very competitive. For Genia (GE) Task 1 (Kim et al., 2011b) we achieve the second-best results. In addition, the best-performing FAUST system (Riedel et al., 2011) is a variant of the model presented here. Its advantage stems from the fact that it uses predictions of the Stanford system (McClosky et al., 2011a; McClosky et al., 2011b), and hence performs model combination. The same holds for the Infectious Diseases (ID) track (Pyysalo et al., 2011), where we come in as second right behind the FAUST system. For the Epigenetics and Post-translational Modifications (EPI) track (Ohta et al., 2011) we achieve the 4th rank, partly because we did not aim to extract speculations, negations or cellular locations. Finally, for Genia Task 2 we rank 3rd—with the 1st rank achieved by the FAUST system.

In the following we will briefly describe our model and inference algorithm, as far as this is possible in limited space. Then we show our results on the three tasks and conclude. Note we will assume familiarity with the task, and refer the reader to the shared task overview paper for more details.

2 Biomedical Event Extraction

Our goal is to extract biomedical events as shown in figure 1a). To formulate the search for such structures as an optimization problem, we represent structures through a set of binary variables. Our representation is inspired by previous work (Riedel et al., 2009; Björne et al., 2009) and based on a projection of events to a labelled graph over tokens in the

Proceedings of BioNLP Shared Task 2011 Workshop, pages 46–50,
Portland, Oregon, USA, 24 June, 2011. ©2011 Association for Computational Linguistics

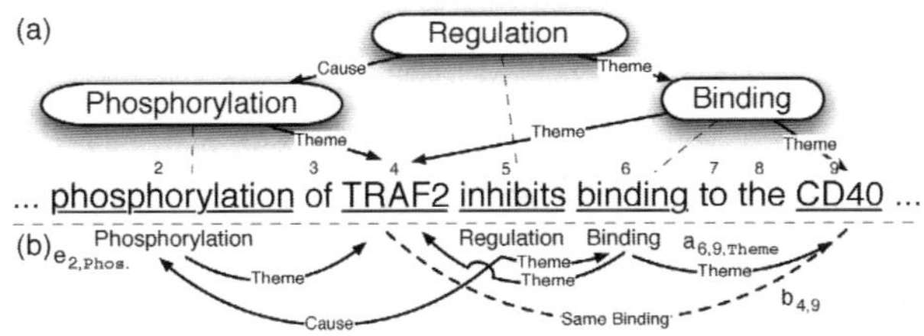

Figure 1: (a) sentence with target event structure; (b) projection to labelled graph.

sentence, as seen figure 1b).

We will first present some basic notation to simplify our exposition. For each sentence $\mathbf{x}$ we have a set candidate trigger words $\text{Trig}(\mathbf{x})$, and a set of candidate proteins $\text{Prot}(\mathbf{x})$. We will generally use the indices i and l to denote members of $\text{Trig}(\mathbf{x})$, the indices p, q for members of $\text{Prot}(\mathbf{x})$ and the index j for members of $\text{Cand}(\mathbf{x}) \stackrel{\text{def}}{=} \text{Trig}(\mathbf{x}) \cup \text{Prot}(\mathbf{x})$.

We label each candidate trigger i with an event Type $t \in \mathcal{T}$ (with $\texttt{None} \in \mathcal{T}$), and use the binary variable $e_{i,t}$ to indicate this labeling. We use binary variables $a_{i,l,r}$ to indicate that between i and l there is an edge labelled $r \in \mathcal{R}$ (with $\texttt{None} \in \mathcal{R}$).

The representation so far has been used in previous work (Riedel et al., 2009; Björne et al., 2009). Its shortcoming is that it does not capture whether two proteins are arguments of the same binding event, or arguments of two binding events with the same trigger. To overcome this problem, we introduce binary "same Binding" variables $b_{p,q}$ that are active whenever there is a binding event that has both p and q as arguments. Our inference algorithm will also need, for each trigger i and protein pair p, q, a binary variable $t_{i,p,q}$ that indicates that at i there is a binding event with arguments p and q. All $t_{i,p,q}$ are summarized in $\mathbf{t}$.

Constructing events from solutions $(\mathbf{e}, \mathbf{a}, \mathbf{b})$ can be done almost exactly as described by Björne et al. (2009). However, while Björne et al. (2009) group arguments according to ad-hoc rules based on dependency paths from trigger to argument, we simply query the variables $b_{p,q}$.

3 Model

We use the following objective to score the structures we like to extract:

$$s(\mathbf{e}, \mathbf{a}, \mathbf{b}) \stackrel{\text{def}}{=} \sum_{e_{i,t}=1} s_{\text{T}}(i,t) + \sum_{a_{i,j,r}=1} s_{\text{R}}(i,j,r) + \sum_{b_{p,q}=1} s_{\text{B}}(p,q)$$

with local scoring functions $s_{\text{T}}(i,t) \stackrel{\text{def}}{=} \langle \mathbf{w}_{\text{T}}, \mathbf{f}_{\text{T}}(i,t) \rangle$, $s_{\text{R}}(i,j,r) \stackrel{\text{def}}{=} \langle \mathbf{w}_{\text{R}}, \mathbf{f}_{\text{R}}(i,j,r) \rangle$ and $s_{\text{B}}(p,q) \stackrel{\text{def}}{=} \langle \mathbf{w}_{\text{B}}, \mathbf{f}_{\text{B}}(p,q) \rangle$.

Our model scores all parts of the structure in isolation. It is a joint model due to the three types of constraints we enforce. The first type acts on trigger labels and their *outgoing* edges. It includes constraints such as "an active label at trigger i requires at least one active outgoing `Theme` argument". The second type enforces consistency between trigger labels and their *incoming* edges. That is, if an incoming edge has a label that is not `None`, the trigger must not be labelled `None` either. The third type of constraints ensures that when two proteins p and q are part of the same binding (as indicated by $b_{p,q} = 1$), there needs to be a binding event at some trigger i that has p and q as arguments. We will denote the set of structures $(\mathbf{e}, \mathbf{a}, \mathbf{b})$ that satisfy all above constraints as $\mathcal{Y}$.

To learn $\mathbf{w}$ we choose the passive-aggressive online learning algorithm (Crammer and Singer, 2003). As loss function we apply a weighted sum of false positives and false negative labels and edges. The weighting scheme penalizes false negatives 3.8 times more than false positives.

3.1 Features

For feature vector $\mathbf{f}_{\text{T}}(i,t)$ we use a collection of representations for the token i: word-form, lemma, POS tag, syntactic heads, syntactic children; membership in two dictionaries used by Riedel et al. (2009).For $\mathbf{f}_{\text{R}}(a; i, j, r)$ we use representations of the token pair (i, j) inspired by Miwa et al. (2010) . They contain: labelled and unlabeled n-gram dependency paths; edge and vertex walk features (Miwa et al., 2010), argument and trigger modifiers and heads, words in between (for close distance i and j). For $\mathbf{f}_{\text{B}}(b; p, q)$ we use a small subset of the token pair representations in $\mathbf{f}_{\text{R}}$.

Algorithm 1 Dual Decomposition.

require:

R: *max. iteration,* α_t: *stepsizes*

$t \leftarrow 0 \; \boldsymbol{\lambda} \leftarrow 0 \; \boldsymbol{\mu} \leftarrow 0$

repeat

$(\bar{\mathbf{e}}, \bar{\mathbf{a}}) \leftarrow \text{bestIncoming}\left(-\boldsymbol{\lambda}\right)$

$(\mathbf{e}, \mathbf{a}) \leftarrow \text{bestOutgoing}\left(\mathbf{c}^{\text{out}}\left(\boldsymbol{\lambda}, \boldsymbol{\mu}\right)\right)$

$(\mathbf{b}, \mathbf{t}) \leftarrow \text{bestBinding}\left(\mathbf{c}^{\text{bind}}\left(\boldsymbol{\mu}\right)\right)$

$\lambda_{i,t} \leftarrow \lambda_{i,t} - \alpha_t \left(e_{i,t} - \bar{e}_{i,t}\right)$

$\lambda_{i,j,r} \leftarrow \lambda_{i,j,r} - \alpha_t \left(a_{i,j,r} - \bar{a}_{i,j,r}\right)$

$\mu^{\text{trig}}_{i,j,k} \leftarrow \left[\mu^{\text{trig}}_{i,j,k} - \alpha_t \left(e_{i,\texttt{Bind}} - t_{i,j,k}\right)\right]_+$

$\mu^{\text{arg1}}_{i,j,k} \leftarrow \left[\mu^{\text{arg1}}_{i,j,k} - \alpha_t \left(a_{i,j,\texttt{Theme}} - t_{i,j,k}\right)\right]_+$

$\mu^{\text{arg2}}_{i,j,k} \leftarrow \left[\mu^{\text{arg2}}_{i,j,k} - \alpha_t \left(a_{i,k,\texttt{Theme}} - t_{i,j,k}\right)\right]_+$

$t \leftarrow t + 1$

until no $\boldsymbol{\lambda}$, $\boldsymbol{\mu}$ changed or $t > R$

return$(\mathbf{e}, \mathbf{a}, \mathbf{b})$

3.2 Inference

Inference in our model amounts to solving

$$\arg \max_{(\mathbf{e},\mathbf{a},\mathbf{b}) \in \mathcal{Y}} s\left(\mathbf{e}, \mathbf{a}, \mathbf{b}\right). \quad (1)$$

Our approach to finding the maximizer is dual decomposition (Komodakis et al., 2007; Rush et al., 2010), a technique that allows us to exploit efficient search algorithms for tractable substructures of our problem. We divide the problem into three sub-problems: (1) finding the highest-scoring trigger labels and edges $(\mathbf{e}, \mathbf{a})$ such that constraints on triggers and their outgoing edges are fulfilled; (2) finding the highest-scoring trigger labels and edges $(\bar{\mathbf{e}}, \bar{\mathbf{a}})$ such that constraints on triggers and their incoming edges are fulfilled; (3) finding the highest-scoring pairs of proteins $\mathbf{b}$ to appear in the same binding, and make binding event trigger decisions $\mathbf{t}$ for these. Due to space constraints we only state that the first two problems can be solved exactly in $O\left(n^2 + nm\right)$ time while the last needs $O\left(m^2 n\right)$. Here n is the number of trigger candidates and m the number of proteins.

The subroutines to solve these three sub-problems are combined in algorithm 1—an instantiation of subgradient descent on the dual of an LP relaxation of problem 1. In the first three steps in the main loop of this algorithm, the individual sub-problems are solved. Note that to each subroutine a parameter is passed. For example, when finding the structure $(\bar{\mathbf{e}}, \bar{\mathbf{a}})$ that maximizes the objective under the incoming edge constraints, we pass the parameter $-\boldsymbol{\lambda}$. This parameter represents a set of *penalties* to be added to the objective used for the subproblem. In this case we have penalties $-\lambda_{i,e}$ to be added to the scores of trigger-label pairs (i, e), and penalties $-\lambda_{i,j,r}$ to be added for labelled edges $i \xrightarrow{r} j$.

One way to understand dual decomposition is as iterative tuning of the penalties such that eventually all individual solutions are consistent with each other. In our case this would mean, among other things, that the solutions $(\mathbf{e}, \mathbf{a})$ and $(\bar{\mathbf{e}}, \bar{\mathbf{a}})$ are identical. This tuning happens in the second part of the main loop which updates the *dual variables* $\boldsymbol{\lambda}$ and $\boldsymbol{\mu}$. We see, for example, how the penalties $\lambda_{i,e}$ are decreased by $e_{i,e} - \bar{e}_{i,e}$ scaled by a step-size α_t. Effectively this change to $\lambda_{i,e}$ will decrease the score of $\bar{e}_{i,e}$ within bestIn $(-\boldsymbol{\lambda})$ by α_t if $\bar{e}_{i,e}$ was true while $e_{i,e}$ was false in the current solutions.[1] If $\bar{e}_{i,e}$ was false but $e_{i,e}$ was true, the score is increased by α_t. If both agree, no change is needed.

Consistency between solutions also means that the binding decisions in $\mathbf{b}$ and $\mathbf{t}$ are consistent with the rest of the solution. This is achieved in algorithm 1 through tuning of the dual variables $\boldsymbol{\mu}$ but we omit details for brevity. For completeness we state how the penalties used for solving the other subproblems are set based on the dual variables $\boldsymbol{\lambda}$ and $\boldsymbol{\mu}$. We set $\mathbf{c}^{\text{out}}_{i,t}\left(\boldsymbol{\lambda}, \boldsymbol{\mu}\right) \stackrel{\text{def}}{=} \lambda_{i,t} + \delta_{t,\texttt{Bind}} \sum_{p,q} \mu^{\text{trig}}_{i,p,q}$; for the case that $j \in \text{Prot}\left(\mathbf{x}\right)$ we get $\mathbf{c}^{\text{out}}_{i,j,r}\left(\boldsymbol{\lambda}, \boldsymbol{\mu}\right) \stackrel{\text{def}}{=} \lambda_{i,j,r} + \sum_p \mu^{\text{arg1}}_{i,j,p} + \sum_q \mu^{\text{arg2}}_{i,q,j}$, otherwise $\mathbf{c}^{\text{out}}_{i,j,r}\left(\boldsymbol{\lambda}, \boldsymbol{\mu}\right) \stackrel{\text{def}}{=} \lambda_{i,j,r}$. For bestBind $(\mathbf{c})$ we set $\mathbf{c}^{\text{bind}}_{i,p,q}\left(\boldsymbol{\mu}\right) = -\mu^{\text{trig}}_{i,p,q} - \mu^{\text{arg1}}_{i,,p,q} - \mu^{\text{arg2}}_{i,,p,q}$.

3.3 Preprocessing

After basic tokenization and sentence segmentation, we generate a set of protein head tokens $\text{Prot}\left(\mathbf{x}\right)$ for each sentence $\mathbf{x}$ based on protein span definitions from the shared task. To ensure tokens contain not more than one protein we split them at protein boundaries. Parsing is performed using the Charniak-Johnson parser (Charniak and Johnson, 2005) with the self-trained biomedical parsing

[1] We refer to Koo et al. (2010) for details on how to set α_t.

	SVT	BIND	REG	TOT
Task 1	73.5	48.8	43.8	55.2
Task 1 (abst.)	71.5	50.8	45.5	56.1
Task 1 (full)	79.2	44.4	40.1	53.1
Task 2	71.4	38.6	39.1	51.0

Table 1: Results for the GE track, task 1 and 2; abst.=abstract; full=full text.

	I/G	BIND	REG	PRO	TOT
DEV	1/0	18.6	27.1	34.3	41.5
DEV	0/1	18.2	26.8	0.00	35.5
DEV	1/1	20.0	33.1	49.3	47.2
DEV	2/1	20.0	34.5	52.0	48.5
TEST	2/1	**34.6**	**46.4**	**62.3**	**53.4**

Table 2: ID results for different amounts of ID (I) and (G) training data.

model of McClosky and Charniak (2008). Finally, based on the set of trigger words in the training data, we generate a set of candidate triggers Trig $(\mathbf{x})$.

4 Results

We apply the same model to the GE, ID and EPI tracks, with minor modifications in order to deal with the different event type sets $\mathcal{T}$ and role sets $\mathcal{R}$ of each track. Training and testing together took between 30 (EPI) to 120 (GE) minutes using a single-core implementation.

4.1 Genia

Our results for GE task 1 and 2 can be seen in table 1. We also show results for abstracts only (abst.), and for full text only (full). Note that binding events (BIND) and general regulation events (REG) seem to be harder to extract in full text. Somewhat surprisingly, for simple events (SVT) the opposite holds. We also like to point out that for full text extraction we rank first—the second best FAUST system achieves an F1 score of 52.67.

4.2 Infectious Diseases

The Infectious Diseases track differs from the Genia track in two important ways. First, it introduces the event type `Process` that is allowed to have no arguments at all. Second, it comes with significantly less training data (152 vs 908 documents). We can accommodate the first difference by making simple changes in our inference algorithms. For example, for `Process` events we do not force the algorithm to pick a `Theme` argument.

To compensate for the lack of training data we simply add data from the GE track. This is reasonable because annotations overlap quite significantly. In table 2 we show the impact of mixing different amounts of ID data (I) and GE data (G) into the training set. We point out that adding the ID training set twice, and the GENIA set once, leads to the best performance (I/G=2/1). Remarkably, the F1 score for `Process` increases by including data, although this data does not include any such events. This may stem from a shared model of `None` arguments that is improved with more data.

4.3 Epigenetics and Post-translational Modifications

For this track a different set of events is to be predicted. However, it is straightforward to adapt our model and algorithms to this setting. For brevity we only report our total results here and omit a table with details. The first metric (ALL) includes negation, speculation and cellular location targets. We omitted these in our model and hence our result of 33.52 F1 is relatively weak. For the metric that neglects these aspects (CORE), we achieve 64.15 F1 and come in 4th. Note that in this metric the FAUST system, based on the model presented here, comes in as very close second.

5 Conclusion

We have presented a robust joint model for event extraction from biomedical text that performs well across all tasks. Remarkably, no feature set or parameter tuning was necessary to achieve this. We also show substantial improvements for the ID task by adding GENIA data into the training set.

Acknowledgements

This work was supported in part by the Center for Intelligent Information Retrieval. The University of Massachusetts gratefully acknowledges the support of Defense Advanced Research Projects Agency (DARPA) Machine Reading Program under Air Force Research Laboratory (AFRL) prime contract no. FA8750-09-C-0181. Any opinions, findings, and conclusion or recommendations expressed in this material are those of the authors and do not necessarily reflect the view of the DARPA, AFRL, or the US government.

References

Jari Björne, Juho Heimonen, Filip Ginter, Antti Airola, Tapio Pahikkala, and Tapio Salakoski. 2009. Extracting complex biological events with rich graph-based feature sets. In *Proceedings of the Natural Language Processing in Biomedicine NAACL 2009 Workshop (BioNLP '09)*, pages 10–18, Morristown, NJ, USA. Association for Computational Linguistics.

Eugene Charniak and Mark Johnson. 2005. Coarse-to-fine n-best parsing and maxent discriminative reranking. In *Proceedings of the 43rd Annual Meeting of the Association for Computational Linguistics (ACL '05)*, pages 173–180.

Koby Crammer and Yoram Singer. 2003. Ultraconservative online algorithms for multiclass problems. *Journal of Machine Learning Research*, 3:951–991.

Jin-Dong Kim, Sampo Pyysalo, Tomoko Ohta, Robert Bossy, and Jun'ichi Tsujii. 2011a. Overview of BioNLP Shared Task 2011. In *Proceedings of the BioNLP 2011 Workshop Companion Volume for Shared Task*, Portland, Oregon, June. Association for Computational Linguistics.

Jin-Dong Kim, Yue Wang, Toshihisa Takagi, and Akinori Yonezawa. 2011b. Overview of the Genia Event task in BioNLP Shared Task 2011. In *Proceedings of the BioNLP 2011 Workshop Companion Volume for Shared Task*, Portland, Oregon, June. Association for Computational Linguistics.

Nikos Komodakis, Nikos Paragios, and Georgios Tziritas. 2007. Mrf optimization via dual decomposition: Message-passing revisited. In *In ICCV*.

Terry Koo, Alexander M. Rush, Michael Collins, Tommi Jaakkola, and David Sontag. 2010. Dual decomposition for parsing with non-projective head automata. In *Proceedings of the 2010 Conference on Empirical Methods in Natural Language Processing (EMNLP)*.

David McClosky and Eugene Charniak. 2008. Self-training for biomedical parsing. In *Proceedings of the 46rd Annual Meeting of the Association for Computational Linguistics (ACL '08)*.

David McClosky, Mihai Surdeanu, and Chris Manning. 2011a. Event extraction as dependency parsing. In *Proceedings of the Association for Computational Linguistics: Human Language Technologies 2011 Conference (ACL-HLT'11), Main Conference* (to appear), Portland, Oregon, June.

David McClosky, Mihai Surdeanu, and Christopher D. Manning. 2011b. Event extraction as dependency parsing in BioNLP 2011. In *BioNLP 2011 Shared Task*.

Makoto Miwa, Rune Saetre, Jin-Dong D. Kim, and Jun'ichi Tsujii. 2010. Event extraction with complex event classification using rich features. *Journal of bioinformatics and computational biology*, 8(1):131–146, February.

Tomoko Ohta, Sampo Pyysalo, and Jun'ichi Tsujii. 2011. Overview of the Epigenetics and Post-translational Modifications (EPI) task of BioNLP Shared Task 2011. In *Proceedings of the BioNLP 2011 Workshop Companion Volume for Shared Task*, Portland, Oregon, June. Association for Computational Linguistics.

Hoifung Poon and Lucy Vanderwende. 2010. Joint Inference for Knowledge Extraction from Biomedical Literature. In *Human Language Technologies: The 2010 Annual Conference of the North American Chapter of the Association for Computational Linguistics*, pages 813–821, Los Angeles, California, June. Association for Computational Linguistics.

Sampo Pyysalo, Tomoko Ohta, Rafal Rak, Dan Sullivan, Chunhong Mao, Chunxia Wang, Bruno Sobral, Jun'ichi Tsujii, and Sophia Ananiadou. 2011. Overview of the Infectious Diseases (ID) task of BioNLP Shared Task 2011. In *Proceedings of the BioNLP 2011 Workshop Companion Volume for Shared Task*, Portland, Oregon, June. Association for Computational Linguistics.

Sebastian Riedel, Hong-Woo Chun, Toshihisa Takagi, and Jun'ichi Tsujii. 2009. A markov logic approach to bio-molecular event extraction. In *Proceedings of the Natural Language Processing in Biomedicine NAACL 2009 Workshop (BioNLP '09)*, pages 41–49.

Sebastian Riedel, David McClosky, Mihai Surdeanu, Christopher D. Manning, and Andrew McCallum. 2011. Model combination for event extraction in BioNLP 2011. In *BioNLP 2011 Shared Task*.

Alexander M. Rush, David Sontag, Michael Collins, and Tommi Jaakkola. 2010. On dual decomposition and linear programming relaxations for natural language processing. In *In Proc. EMNLP*.

Model Combination for Event Extraction in BioNLP 2011

Sebastian Riedel[a], **David McClosky**[b], **Mihai Surdeanu**[b],
Andrew McCallum[a], and **Christopher D. Manning**[b]
[a] Department of Computer Science, University of Massachusetts at Amherst
[b] Department of Computer Science, Stanford University
{riedel,mccallum}@cs.umass.edu
{mcclosky,mihais,manning}@stanford.edu

Abstract

We describe the FAUST entry to the BioNLP 2011 shared task on biomolecular event extraction. The FAUST system explores several stacking models for combination using as base models the UMass dual decomposition (Riedel and McCallum, 2011) and Stanford event parsing (McClosky et al., 2011b) approaches. We show that using stacking is a straightforward way to improving performance for event extraction and find that it is most effective when using a small set of stacking features and the base models use slightly different representations of the input data. The FAUST system obtained 1st place in three out of four tasks: 1st place in Genia Task 1 (56.0% f-score) and Task 2 (53.9%), 2nd place in the Epigenetics and Post-translational Modifications track (35.0%), and 1st place in the Infectious Diseases track (55.6%).

1 Introduction

To date, most approaches to the BioNLP event extraction task (Kim et al., 2011a) use a single model to produce their output. However, model combination techniques such as voting, stacking, and reranking have been shown to consistently produce higher performing systems by taking advantage of multiple views of the same data. The Netflix Prize (Bennett et al., 2007) is a prime example of this. System combination essentially allows systems to regularize each other, smoothing over the artifacts of each (c.f. Nivre and McDonald (2008), Surdeanu and Manning (2010)). To our knowledge, the only previous example of model combination for the BioNLP shared task was performed by Kim et al. (2009). Using a weighted voting scheme to combine the outputs from the top six systems, they obtained a 4% absolute f-score improvement over the best individual system.

This paper shows that using a straightforward model combination strategy on two competitive systems produces a new system with substantially higher accuracy. This is achieved with the framework of stacking: a *stacking* model uses the output of a *stacked* model as additional features.

While we initially considered voting and reranking model combination strategies, it seemed that given the performance gap between the UMass and Stanford systems that the best option was to include the predictions from the Stanford system into the UMass system (e.g., as in Nivre and McDonald (2008)). This has the advantage that one model (Umass) determines how to integrate the outputs of the other model (Stanford) into its own structure, whereas in reranking, for example, the combined model is required to output a complete structure produced by only one of the input models.

2 Approach

In the following we briefly present both the stacking and the stacked model and some possible ways of integrating the stacked information.

2.1 Stacking Model

As our stacking model, we employ the UMass extractor (Riedel and McCallum, 2011). It is based on a discriminatively trained model that jointly predicts trigger labels, event arguments and protein pairs in

Proceedings of BioNLP Shared Task 2011 Workshop, pages 51–55,
Portland, Oregon, USA, 24 June, 2011. ©2011 Association for Computational Linguistics

binding. We will briefly describe this model but first introduce three types of binary variables that will represent events in a given sentence. Variables $e_{i,t}$ are active if and only if the token at position i has the label t. Variables $a_{i,j,r}$ are active if and only if there is an event with trigger i that has an argument with role r grounded at token j. In the case of an entity mention this means that the mention's head is j. In the case of an event j is the position of its trigger. Finally, variables $b_{p,q}$ indicate whether or not two entity mentions at p and q appear as arguments in the same binding event.

Two parts form our model: a scoring function, and a set of constraints. The scoring function over the trigger variables $\mathbf{e}$, argument variables $\mathbf{a}$ and binding pair variables $\mathbf{b}$ is

$$s(\mathbf{e}, \mathbf{a}, \mathbf{b}) \stackrel{\text{def}}{=} \sum_{e_{i,t}=1} s_{\text{T}}(i,t) + \sum_{a_{i,j,r}=1} s_{\text{R}}(i,j,r) + \sum_{b_{p,q}=1} s_{\text{B}}(p,q)$$

with local scoring functions $s_{\text{T}}(i,t) \stackrel{\text{def}}{=} \langle \mathbf{w}_{\text{T}}, \mathbf{f}_{\text{T}}(i,t)\rangle$, $s_{\text{R}}(i,j,r) \stackrel{\text{def}}{=} \langle \mathbf{w}_{\text{R}}, \mathbf{f}_{\text{R}}(i,j,r)\rangle$ and $s_{\text{B}}(p,q) \stackrel{\text{def}}{=} \langle \mathbf{w}_{\text{B}}, \mathbf{f}_{\text{B}}(p,q)\rangle$.

Our model scores all parts of the structure in isolation. It is a joint model due to the nature of the constraints we enforce: First, we require that each active event trigger must have at least one `Theme` argument; second, only regulation events (or Catalysis events for the EPI track) are allowed to have `Cause` arguments; third, any trigger that is itself an argument of another event has to be labelled active, too; finally, if we decide that two entities p and q are part of the same binding (as indicated by $b_{p,q} = 1$), there needs to be a binding event at some trigger i that has p and q as arguments. We will denote the set of structures $(\mathbf{e}, \mathbf{a}, \mathbf{b})$ that satisfy these constraints as $\mathcal{Y}$.

Stacking with this model is simple: we only need to augment the local feature functions $\mathbf{f}_{\text{T}}(i,t)$, $\mathbf{f}_{\text{R}}(i,j,r)$ and $\mathbf{f}_{\text{B}}(p,q)$ to include predictions from the systems to be stacked. For example, for every system S to be stacked and every pair of event types (t', t_S) we add the features

$$f_{S,t',t_S}(i,t) = \begin{cases} 1 & h_S(i) = t_S \wedge t' = t \\ 0 & \text{otherwise} \end{cases}$$

to $\mathbf{f}_{\text{T}}(i,t)$. Here $h_S(i)$ is the event label given to token i according to S. These features allow different weights to be given to each possible combination of type t' that we want to assign, and type t_S that S predicts.

Inference in this model amounts to maximizing $s(\mathbf{e}, \mathbf{a}, \mathbf{b})$ over $\mathcal{Y}$. Our approach to solving this problem is dual decomposition (Komodakis et al., 2007; Rush et al., 2010). We divide the problem into three subproblems: (1) finding the best trigger label and set of outgoing edges for each candidate trigger; (2) finding the best trigger label and set of incoming edges for each candidate trigger; (3) finding the best pairs of entities to appear in the same binding. Due to space limitations we refer the reader to Riedel and McCallum (2011) for further details.

2.2 Stacked Model

For the stacked model, we use a system based on an event parsing framework (McClosky et al., 2011a) referred to as the Stanford model in this paper. This model converts event structures to dependency trees which are parsed using MSTParser (McDonald et al., 2005).[1] Once parsed, the resulting dependency tree is converted back to event structures. Using the Stanford model as the stacked model is helpful since it captures tree structure which is not the focus in the UMass model. Of course, this is also a limitation since actual BioNLP event graphs are DAGs, but the model does well considering these restrictions. Additionally, this constraint encourages the Stanford model to provide different (and thus more useful for stacking) results.

Of particular interest to this paper are the four possible decoders in MSTParser. These four decoders come from combinations of feature order (first or second) and whether the resulting dependency tree is required to be projective.[2] Each decoder presents a slightly different view of the data and thus has different model combination properties. Projectivity constraints are not captured in the UMass model so these decoders incorporate novel information.

To produce stacking output from the Stanford system, we need its predictions on the training, devel-

[1] http://sourceforge.net/projects/mstparser/

[2] For brevity, the second-order non-projective decoder is abbreviated as 2N, first-order projective as 1P, etc.

	UMass			FAUST+All		
	R	P	F1	R	P	F1
GE T1	48.5	64.1	55.2	49.4	64.8	56.0
GE T2	43.9	60.9	51.0	46.7	63.8	53.9
EPI (F)	28.1	41.6	33.5	28.9	44.5	35.0
EPI (C)	57.0	73.3	64.2	59.9	80.3	68.6
ID (F)	46.9	62.0	53.4	48.0	66.0	55.6
ID (C)	49.5	62.1	55.1	50.6	66.1	57.3

Table 1: Results on test sets of all tasks we submitted to. T1 and T2 stand for task 1 and 2, respectively. C stands for CORE metric, F for FULL metric.

opment and test sets. For predictions on test and development sets we used models learned from the the complete training set. Predictions over training data were produced using crossvalidation. This helps to avoid a scenario where the stacking model learns to rely on high accuracy at training time that cannot be matched at test time.

Note that, unlike Stanford's individual submission in this shared task, the stacked models in this paper do not include the Stanford reranker. This is because it would have required making a reranker model for each crossvalidation fold.

We made 19 crossvalidation training folds for Genia (GE) (Kim et al., 2011b), 12 for Epigenetics (EPI), and 17 for Infectious Diseases (ID) (Kim et al., 2011b; Ohta et al., 2011; Pyysalo et al., 2011, respectively). Note that while ID is the smallest and would seem like it would have the fewest folds, we combined the training data of ID with the training and development data from GE. To produce predictions over the test data, we combined the training folds with 6 development folds for GE, 4 for EPI, and 1 for ID.

3 Experiments

Table 1 gives an overview of our results on the test sets for all four tasks we submitted to. Note that for the EPI and ID tasks we show the CORE metric next to the official FULL metric. The former is suitable for our purposes because it does not measure performance for negations, speculations and cellular locations—all of these we did not attempt to predict.

We compare the UMass standalone system to the FAUST+All system which stacks the Stanford 1N, 1P, 2N and 2P predictions. For all four tasks we observe substantial improvements due to stacking. The increase is particular striking for the EPI track, where stacking improves *f*-score by more than 4.0 points on the CORE metric.

System	SVT	BIND	REG	TOTAL
UMass	74.7	**47.7**	42.8	54.8
Stanford 1N	71.4	38.6	32.8	47.8
Stanford 1P	70.8	35.9	31.1	46.5
Stanford 2N	69.1	35.0	27.8	44.3
Stanford 2P	72.0	36.2	32.2	47.4
FAUST+All	**76.9**	43.5	44.0	**55.9**
FAUST+1N	76.4	45.1	43.8	55.6
FAUST+1P	75.8	43.1	**44.6**	55.7
FAUST+2N	74.9	42.8	43.8	54.9
FAUST+2P	75.7	46.0	44.1	55.7
FAUST+All (triggers)	76.4	41.2	43.1	54.9
FAUST+All (arguments)	76.1	41.7	43.6	55.1

Table 2: BioNLP *f*-scores on the development section of the Genia track (task 1) for several event categories.

To analyze the impact of stacking further, Table 2 shows a breakdown of our results on the Genia development set. Presented are *f*-scores for simple events (SVT), binding events (BIND), regulation events (REG) and the set of all event types (TOTAL). We compare the UMass standalone system, various Stanford-standalone models and stacked versions of these (FAUST+X).

Remarkably, while there is a 7 point gap between the best individual Stanford system and the standalone UMass systems, integrating the Stanford prediction still leads to an *f*-score improvement of 1. This can be seen when comparing the UMass, Stanford 1N and FAUST+All results, where the latter stacks 1N, 1P, 2N and 2P. We also note that stacking the projective 1P and 2P systems helps almost as much as stacking all Stanford systems. Notably, both 1P and 2P do not do as well in isolation when compared to the 1N system. When stacked, however, they do slightly better. This suggests that projectivity is a missing aspect in the UMass standalone system.

The FAUST+All (triggers) and FAUST+All (arguments) lines represent experiments to determine whether it is useful to incorporate only portions of the stacking information from the Stanford system. Given the small gains over the original UMass system, it is clear that stacking information is only useful when attached to triggers and arguments. Our theory is that most of our gains come from when the UMass and Stanford systems disagree on triggers and the Stanford system provides not only its triggers but also their attached arguments to the UMass system. This is supported by a pilot experiment where we trained the Stanford model to use the UMass triggers and saw no benefit from stacking (even when both triggers and arguments were used).

Table 3 shows our results on the development set of the ID task, this time in terms of recall, precision and *f*-score. Here the gap between Stanford-only results, and the UMass results, is much smaller. This seems to lead to more substantial improvements for stacking: FAUST+All obtains a *f*-score 2.2 points larger than the standalone UMass system. Also note that, similarly to the previous table, the projective systems do worse on their own, but are more useful when stacked.

Another possible approach to stacking *conjoins* all the original features of the stacking model with the predicted features of the stacked model. The hope is that this allows the learner to give different weights to the stacked predictions in different contexts. However, incorporating Stanford predictions by conjoining them with all features of the UMass standalone system (FAUST+2P-Conj in Table 3) does not help here.

We note that for our results on the ID task we augment the training data with events from the GE training set. Merging both training sets is reasonable since there is a significant overlap between both in terms of events as well as lexical and syntactic patterns to express these. When building our training set we add each training document from GE once, and each ID training document twice—this lead to substantially better results than including ID data only once.

4 Discussion

Generally stacking has led to substantial improvements across the board. There are, however, some exceptions. One is binding events for the GE task. Here the UMass model still outperforms the best stacked system (see Table 2). Likewise, for full papers in the Genia test set, the UMass model still does slightly better with 53.1 *f*-score compared to 52.7 *f*-score. This suggests that a more informed combination of our systems (e.g., metaclassifiers) could lead to better performance.

System	Rec	Prec	F1
UMass	46.2	51.1	48.5
Stanford 1N	43.1	49.1	45.9
Stanford 1P	40.8	46.7	43.5
Stanford 2N	41.6	53.9	46.9
Stanford 2P	42.8	48.1	45.3
FAUST+All	47.6	**54.3**	**50.7**
FAUST+1N	45.8	51.6	48.5
FAUST+1P	47.6	52.8	50.0
FAUST+2N	45.4	52.4	48.6
FAUST+2P	**49.1**	52.6	**50.7**
FAUST+2P-Conj	48.0	53.2	50.4

Table 3: Results on the development set for the ID track.

5 Conclusion

We have presented the FAUST entry to the BioNLP 2011 shared task on biomolecular event extraction. It is based on stacking, a simple approach for model combination. By using the predictions of the Stanford entry as features of the UMass model, we substantially improved upon both systems in isolation. This helped us to rank 1st in three of the four tasks we submitted results to. Remarkably, in some cases we observed improvements despite a 7.0 *f*-score margin between the models we combined.

In the future we would like to investigate alternative means for model combination such as reranking, union, intersection, and other voting techniques. We also plan to use dual decomposition to encourage models to agree. In particular, we will seek to incorporate an MST component into the dual decomposition algorithm used by the UMass system.

Acknowledgments

We thank the BioNLP shared task organizers for setting this up and their quick responses to questions. This work was supported in part by the Center for Intelligent Information Retrieval. We gratefully acknowledge the support of the Defense Advanced Research Projects Agency (DARPA) Machine Reading Program under Air Force Research Laboratory (AFRL) prime contract no. FA8750-09-C-0181.

References

James Bennett, Stan Lanning, and Netflix. 2007. The netflix prize. In *KDD Cup and Workshop in conjunction with KDD*.

Jin-Dong Kim, Tomoko Ohta, Sampo Pyysalo, Yoshinobu Kano, and Jun'ichi Tsujii. 2009. Overview of BioNLP'09 shared task on event extraction. In *Proceedings of the Workshop on BioNLP: Shared Task*, pages 1–9. Association for Computational Linguistics.

Jin-Dong Kim, Sampo Pyysalo, Tomoko Ohta, Robert Bossy, and Jun'ichi Tsujii. 2011a. Overview of BioNLP Shared Task 2011. In *Proceedings of the BioNLP 2011 Workshop Companion Volume for Shared Task*, Portland, Oregon, June. Association for Computational Linguistics.

Jin-Dong Kim, Yue Wang, Toshihisa Takagi, and Akinori Yonezawa. 2011b. Overview of the Genia Event task in BioNLP Shared Task 2011. In *Proceedings of the BioNLP 2011 Workshop Companion Volume for Shared Task*, Portland, Oregon, June. Association for Computational Linguistics.

Nikos Komodakis, Nikos Paragios, and Georgios Tziritas. 2007. MRF optimization via dual decomposition: Message-passing revisited. In *ICCV*.

David McClosky, Mihai Surdeanu, and Chris Manning. 2011a. Event extraction as dependency parsing. In *Proceedings of the Association for Computational Linguistics: Human Language Technologies 2011 Conference (ACL-HLT'11), Main Conference*, Portland, Oregon, June.

David McClosky, Mihai Surdeanu, and Christopher D. Manning. 2011b. Event extraction as dependency parsing in BioNLP 2011. In *BioNLP 2011 Shared Task*.

Ryan T. McDonald, Fernando Pereira, Kiril Ribarov, and Jan Hajic. 2005. Non-projective dependency parsing using spanning tree algorithms. In *Proceedings of HLT/EMNLP*. The Association for Computational Linguistics.

Joakim Nivre and Ryan McDonald. 2008. Integrating graph-based and transition-based dependency parsers. In *Proceedings of ACL-08: HLT*, pages 950–958, Columbus, Ohio, June. Association for Computational Linguistics.

Tomoko Ohta, Sampo Pyysalo, and Jun'ichi Tsujii. 2011. Overview of the Epigenetics and Post-translational Modifications (EPI) task of BioNLP Shared Task 2011. In *Proceedings of the BioNLP 2011 Workshop Companion Volume for Shared Task*, Portland, Oregon, June. Association for Computational Linguistics.

Sampo Pyysalo, Tomoko Ohta, Rafal Rak, Dan Sullivan, Chunhong Mao, Chunxia Wang, Bruno Sobral, Jun'ichi Tsujii, and Sophia Ananiadou. 2011. Overview of the Infectious Diseases (ID) task of BioNLP Shared Task 2011. In *Proceedings of the BioNLP 2011 Workshop Companion Volume for Shared Task*, Portland, Oregon, June. Association for Computational Linguistics.

Sebastian Riedel and Andrew McCallum. 2011. Robust biomedical event extraction with dual decomposition and minimal domain adaptation. In *BioNLP 2011 Shared Task*.

Alexander M. Rush, David Sontag, Michael Collins, and Tommi Jaakkola. 2010. On dual decomposition and linear programming relaxations for natural language processing. In *Proc. EMNLP*.

Mihai Surdeanu and Christopher D. Manning. 2010. Ensemble models for dependency parsing: Cheap and good? In *Proceedings of the North American Chapter of the Association for Computational Linguistics Conference (NAACL-2010)*, Los Angeles, CA, June.

BioNLP shared Task 2011 - Bacteria Biotope

Robert Bossy[1], Julien Jourde[1], Philippe Bessières[1], Maarten van de Guchte[2], Claire Nédellec[1]

[1]MIG UR1077 [2]Micalis UMR 1319
INRA, Domaine de Vilvert
78352 Jouy-en-Josas, France
forename.name@jouy.inra.fr

Abstract

This paper presents the Bacteria Biotope task as part of the BioNLP Shared Tasks 2011. The Bacteria Biotope task aims at extracting the location of bacteria from scientific Web pages. Bacteria location is a crucial knowledge in biology for phenotype studies. The paper details the corpus specification, the evaluation metrics, summarizes and discusses the participant results.

1 Introduction

The Bacteria Biotope (BB) task is one of the five main tasks of the BioNLP Shared Tasks 2011. The BB task consists of extracting bacteria location events from Web pages, in other words, citations of places where a given species lives. Bacteria locations range from plant or animal hosts for pathogenic or symbiotic bacteria, to natural environments like soil or water. Challenges for Information Extraction (IE) of relations in Biology are mostly devoted to the identification of bio-molecular events in scientific papers where the events are described by relations between named entities, *e.g.* genic interactions (Nédellec, 2005), protein-protein interactions (Pyysalo et al., 2008), and more complex molecular events (Kim et al., 2011). However, this far from reflects the diversity of the potential applications of text mining to biology. The objective of previous challenges has mostly been focused on modeling biological functions and processes using the information on elementary molecular events extracted from text.

The BB task is the first step towards linking information on bacteria at the molecular level to ecological information. The information on bacterial habitats and properties of these habitats is very abundant in literature, in particular in Systematics literature (*e.g. International Journal of Systematic and Evolutionary Microbiology*), however it is rarely available in a structured way (Hirschman et al., 2008; Tamames and de Lorenzo, 2009). The NCBI GenBank nucleotide *isolation source* field (GenBank) and the JGI Genome OnLine Database (GOLD) *isolation site* field are incomplete with respect to the microbial diversity and are expressed in natural language. The two critical missing steps in terms of biotope knowledge modeling are (1) the automatic population of databases with organism/location pairs that are extracted from text, and (2) the normalization of the habitat name with respect to biotope ontologies. The BB task mainly aims at solving the first information extraction issue. The second classification issue is handled through the categorization of locations into eight types.

2 Context

According to NCBI statistics there are nearly 900 bacteria with complete genomes, which account for more than 87% of total complete genomes. Consequently, molecular studies in bacteriology are shifting from species-centered to full diversity investigation. The current trend in high-throughput experiments targets diversity related fields, typically phylogeny or ecology. In this context, adaptation properties, biotopes and biotope properties become critical information. Illustrative questions are:

Proceedings of BioNLP Shared Task 2011 Workshop, pages 56–64,
Portland, Oregon, USA, 24 June, 2011.

- Is there a phylogenetic correlation between species that share the same biotope?
- What are common metabolic pathways of species that live in given conditions, especially species that survive in extreme conditions?
- What are the molecular signaling patterns in host relationships or population relationships (*e.g.* in biofilms)?

Recent metagenomic experiments produce molecular data associated with a habitat rather than a single species. This raises new challenges in computational biology and data integration, such as identifying known and new species that belong to a metagenome.

Not only will these studies require comprehensive databases that associate bacterial species to their habitat, but they also require a formal description of habitats for property inference. The bacteria biotope description is potentially very rich since any physical object, from a cell to a continent, can be a bacterial habitat. However these relations are much simpler to model than with general formal spatial ontologies. A given place is a bacterial habitat if the bacteria and the habitat are physically in contact, while the relative position of the bacteria and its dissemination are not part of the BB task model.

The BB Task requires the locations to be assigned different types (*e.g.* soil, water). We view location typing as a preliminary step of more fine-grained modeling in location ontologies. Some classifications for bacteria biotopes have been proposed by some groups (Floyd et al., 2005; Hirschman et al., 2008; Field et al., 2008; Pignatelli et al., 2009). The Environment Ontology project (EnvO) is developing an ambitious detailed environment ontology for supporting standard manual annotation of environments of all types of organisms and biological samples (Field et al., 2008). In a similar way, the GOLD group at JGI defined a standard classification for bacteria population metagenome projects. Developing methods for the association of such biotope classes to organisms remains an open question. EnvDB (Pignatelli et al., 2009) is an attempt to inventory isolation sources of bacteria as recorded in GenBank and to map them to a three level hierarchy of 71 biotope classes. The assignment of bacterial samples in one of the EnvDB classes is supported by a text-mining tool based on a Naïve Bayes (NB) classifier applied to a bag of words representing the associated reference title and abstract. Unfortunately, the low number of paper references associated with the isolation source field (46 %) limits the scope of the method.

The BB task has a similar goal, but directly applies to natural language texts thus avoiding the issue of database incompleteness. As opposed to database-based approaches, biotope information density is higher but the task has to include bacteria and location identification, as well as information extraction to relate them.

The eight types of locations in the BB task capture high-level information for further ontology mappings. The location types are *Host*, *HostPart*, *Geographical* and *Environmental*. *Environmental* is broadly defined to qualify locations that are not associated to hosts, in a similar way to what was described by Floyd et al. (Floyd et al., 2005). In addition, the BB task types exclude artificially constructed biotopes (*e.g.* bacteria growing in labs on a specific medium) and laboratory mutant bacteria. The *Environmental* class is divided into *Food*, *Medical*, *Soil* and *Water*. Locations that are none of these subtypes are classified as *Environmental*.

The exact geographical location (*e.g.* latitude and longitude coordinates) has less importance here than in eukaryote ecology because most of the biotope properties vary along distances smaller than the precision of the current positioning technologies. Geographical names are only useful in bacteria biotope studies when the physico-chemical properties of the location can be inferred. For the sake of simplicity, the locations of bacteria host (*e.g.* the stall of the infected cow) are not taken into account despite their richness (Floyd et al., 2005).

The important information conveyed by the locations, especially of *Environment* type, is the function of the bacterium in its ecosystem rather than the substance of the habitat. Indeed the final goal is to extract habitat properties and bacteria phenotypes. Beyond the identification of locations, their properties (*e.g.* temperature, pH, salinity, oxygen) are of high interest for phenotypes (*e.g.* thermophily, acidophily, halophily) and trophism studies. This information is difficult to extract, and is often incomplete or even not available in papers (Tamames and de Lorenzo., 2009). Hopefully, some properties can be automatically retrieved with the help of specialized databases, which give the physico-chemical properties of locations, such as hosts (plant, animal, human organs), soils (see WebSoilSurvey, Corine Land Cover), water, or chemical pollutants.

From a linguistic point of view, the BB task differs from other IE molecular biology tasks while it raises some issues common to biomedicine and more general IE tasks. The documents are scientific Web pages intended for non-experts such as encyclopedia notices. The information is dense compared to scientific papers. Documents are structured as encyclopedia pages, with the main focus on a single species or a few species of the same genus or family. The frequency of anaphora and coreferences is unusually high. The location entities are denoted by complex expressions with semantic boundaries instead of rigid designators.

3 Task description

The goal of the BB task is illustrated in Figure 1.

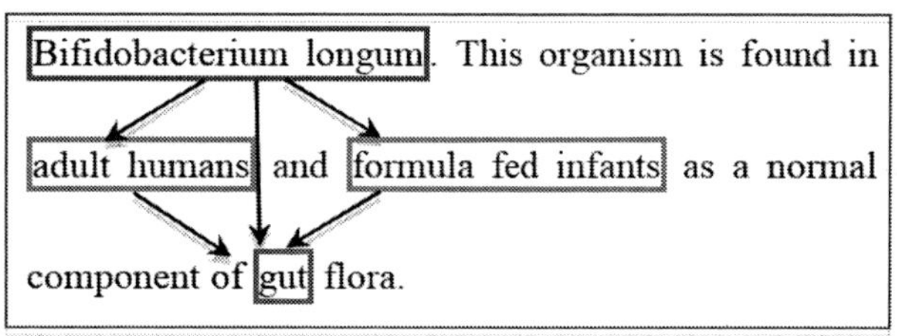

Figure 1. Example of information to be extracted in the BB Task.

The entities to be extracted are of two main types: bacteria and locations. They are text-bound and their position has to be predicted. Relations are of type *Localization* between bacteria and locations, and *PartOf* between hosts and host parts. In the example in Figure 1, *Bifidobacterium longum* is a bacterium. *adult humans* and *formula fed infants* denote host locations for the bacteria. *gut* is also a bacteria location, part of the two hosts and thus of type host part.

Coreference relations between entities denoting the same information represent valid alternatives for the relation arguments. For example, the three taxon names in Figure 2 are equivalent.

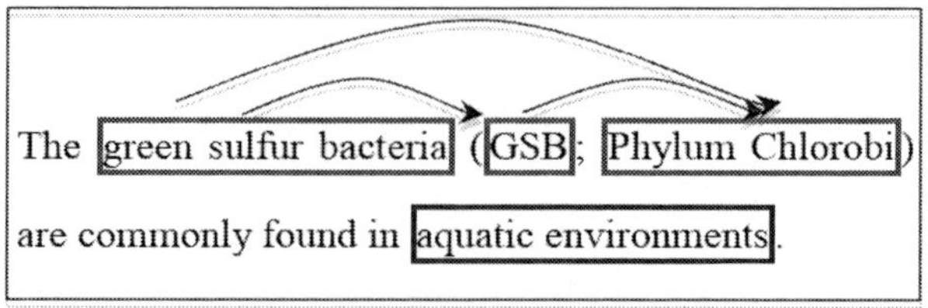

Figure 2. Coreference example.

The coreference relation between pairs of entities is binary, symmetric and transitive. Coreference sets are equivalence sets defined as the transitive closure of the binary coreference relation. Their annotation is provided in the training and development sets, but it does *not* have to be predicted in the test set.

4 Corpus description

The corpus sources are the following bacteria sequencing project Web pages:

- Genome Projects referenced at NCBI;
- Microbial Genomics Program at JGI;
- Bacteria Genomes at EBI;
- Microorganisms sequenced at Genoscope;
- Encyclopedia pages from MicrobeWiki.

The documents are publicly available and quite easy to understand by non-experts compared to scientific papers on similar topics. From the 2,086 downloaded documents, 105 were randomly selected for the BB task. A quarter of the corpus was retained for test evaluation. The rest was split into train and development sets. Table 1 gives the distribution of the entities and relations per corpus. The distribution of the five document sources in the test corpus reflects the distribution of the training set and no other criteria. Food is therefore underrepresented.

	Training+Dev	**Test**
Document	78 (65 + 13)	27 (26 %)
Bacteria	538	121 (18 %)
Environment	62	16 (21 %)
Host	486	101 (17 %)
HostPart	217	84 (28 %)
Geographical	111	25 (18 %)
Water	70	21 (23 %)
Food	46	0 (0 %)
Medical	24	2 (8 %)
Soil	26	20 (43 %)
Coreference	484	100 (17 %)
Total entities	1,580	390

	Training+Dev	Test
Localization	998	250 (20 %)
Part of Host	204	78 (28 %)
Total relations	1,202	328

Table 1. Corpus Figures.

5 Annotation methodology

HTML tags and irrelevant metadata were stripped from the corpus. The Alvis pipeline (Nédellec et al., 2009) pre-annotated the species names that are potential bacteria and host names. A team of 7 scientists manually annotated the entities, coreferences and relations using the Cadixe XML editor (Cadixe). Each document was processed by two independent annotators in a double-blind manner. Conflicts were automatically detected, resolved by annotator negotiation and irrelevant documents (*e.g.* without bacterial location) were removed. The remaining inconsistencies among documents were resolved by the two annotators assisted by a third person acting as an arbitrator.

The annotator group designed the detailed annotation guidelines in two phases. First, they annotated a set of 10 documents, discussed the options and wrote detailed guidelines with representative and illustrative examples. During the annotation of the rest of the documents, new cases were discussed by email and the guidelines amended accordingly.

Location types. The main issues under debate were the definition of location types, boundaries of annotations and coreferences. Additional annotation specifications concerned the exclusion of overly general locations (*e.g. environment*, *zone*), artificially constructed biotopes and indirect effects of bacteria on distant places. For instance, a disease symptom occurring in a given host part does not imply the presence of the bacteria in this place, whereas infection does. Boundaries of types were also an important point of discussion since the definite formalization of habitat categories was at stake. For instance we decided to exclude land environment citations (*fields*, *deserts*, *savannah*, *etc.*) from the type *Soil*, and thus enforced a strict definition of soil bacteria. The most controversial type was host parts. We decided to include fluids, secretions and excretions (which are not strictly organs). Therefore, the host parts category required specifications to determine at which point of dissociation from the original host is a habitat not a host part anymore (e.g. *mother's milk* vs. *industrial milk*, *rhizosphere* as host part instead of soil).

Boundaries. The bacteria name boundaries do not include any external modifiers (*e.g. two A. baumannii strains*). Irrelevant modifiers of locations are considered outside the annotation boundaries (*e.g. responsible for a hospital epidemic*). All annotations are contiguous and span on a single fragment in the same way as the other BioNLP Shared Tasks. This constraint led us to consider cases where several annotations occur side by side. The preferred approach was to have one distinct annotation for each different location (*e.g. contact with infected animal products or through the air*). In the case of head or modifier factorization, the annotation depends on the information conveyed by the factorized part. If the head is not relevant to determine the location type, then each term is annotated separately (*e.g. tropical and temperate zones*). Conversely, if the head is the most informative with regards to the location type, a single annotation spans the whole fragment (*fresh and salt water*).

Coreferences. Two expressions are considered as coreferential and thus valid solution alternatives, if they convey the same information. For instance, complete taxon names and non-ambiguous abbreviations are valid alternatives (*e.g. Borrelia garinii vs. B. garinii*), while ambiguous anaphora ellipses are not (*e.g.* as in "*[..] infected with Borrelia duttonii. Borrelia then multiplies [..]*"). The ellipsis of the omitted specific name (*dutotonii*) leaves the ambiguous generic name (*Borrelia*).

The full guidelines document is available for download on the BioNLP Shared Task Bacteria Biotope page[1].

6 Evaluation procedure

6.1 Campaign organization

The training and development corpora with the reference annotations were made available to the participants by December 1st 2010 on the BioNLP Shared Tasks pages together with the evaluation software. The test corpus, which does not contain

[1] https://sites.google.com/site/bionlpst/home/bacteria-biotopes/ BioNLP-ST_2011_Bacteria_Biotopes_Guidelines.pdf

any annotation, was made available by March, 1st 2011. The participants sent the predicted annotations to the BioNLP Shared Task organizers by March 10th. Each participant submitted a single final prediction set. The detailed evaluation results were computed, provided to the participants and published on the BioNLP website by March, 11th.

6.2 Evaluation metrics

The evaluation metrics are based on precision, recall and the F-measure. In the following section, the *PartOf* and *Localization* relations will both be referred to as events. The metrics measure the accuracy of the participant prediction of events with respect to the reference annotation of the test corpus. Predicted entities that are not event arguments are ignored and they do not penalize the score. Each event E_r in the reference set is matched to the predicted event E_p that maximizes the event similarity function S. The recall is the sum of the S results divided by the number of events in the reference set. Each event E_p in the predicted set is matched to the reference event E_r that maximizes S. The precision is the sum of the S results divided by the number of events in the predicted set. Participants were ranked by the F-score defined as the harmonic mean between precision and recall.

E_{ab}, the event similarity between a reference *Localization* event a and a predicted *Localization* event b, is defined as:

$$E_{ab} = B_{ab} \,.\, T_{ab} \,.\, J_{ab}$$

- B_{ab} is the bacteria boundary component defined as: if the *Bacterium* arguments of both the predicted and reference events have exactly the same boundaries, then $B_{ab} = 1$, otherwise $B_{ab} = 0$. Bacteria name boundary matching is strict since boundary mistakes usually yield a different taxon.
- T_{ab} is the location type prediction component defined as: if the *Location* arguments of both the predicted and reference events are of the same type, then $T_{ab} = 1$, otherwise $T_{ab} = 0.5$. Thus type errors divide the score by two.
- J_{ab} is the location boundary component defined as: if the *Location* arguments of the predicted and reference events overlap, then

$$J_{ab} = \frac{LEN_a + LEN_b}{OV_{ab}} - 1$$

where LEN_a and LEN_b are the length of the *Localization* arguments of predicted and reference events, and OV_{ab} is the length of the overlapping segment between the *Localization* arguments of the predicted and reference events. If the arguments do not overlap, then J_{ab} is 0. This formula is a Jaccard index applied to overlapping segments. Location boundary matching is relaxed, though the Jaccard index rewards predictions that approach the reference.

For *PartOf* events between *Hosts* and *HostParts*, the matching score P_{ab} is defined as: if the *Host* arguments of the reference and predicted events overlap and the *Part* arguments of the reference and predicted events overlap, then $P_{ab} = 1$, otherwise $P_{ab} = 0$. Boundary matching of *PartOf* arguments is relaxed, since boundary mistakes are already penalized in E_{ab}.

Arguments belonging to the same coreference set are strictly equivalent. In other words, the argument in the predicted event is correct if it is equal to the reference entity or to any item in the reference entity coreference set.

7 Results

7.1 Participating systems

Three teams submitted predictions to the BB task. The first team is from the University of Turku (UTurku); their system is generic and produced predictions for every BioNLP Shared Task. This system uses ML intensely, especially SVMs, for entity recognition, entity typing and event extraction. UTurku adapted their system for the BB task by using specific NER patterns and external resources (Björne and Salakoski, 2011).

The second team is from the Japan Advanced Institute of Science and Technology (JAIST); their system was specifically designed for this task. They used CRF for entity recognition and typing, and classifiers for coreference resolution and event extraction (Nguyen and Tsuruoka, 2011).

The third team is from Bibliome INRA; their system was specifically designed for this task (Ratkovik et al., 2011). This team has the same affiliation as the BB Task authors, however great care was taken to prevent communication on the subject between task participants and the test set annotators.

The results of the three submissions according to the official metrics are shown in Table 2. The scores are micro-averaged: *Localization* and *PartOf* relations have the same weight. Given the novelty and the complexity of the task, these first results are quite encouraging. Almost half of the relations are correctly predicted. The Bibliome team achieved the highest F-measure with a balanced recall and precision (45%).

	Recall	Precision	F-score
Bibliome	**45**	45	**45**
JAIST	27	42	33
UTurku	17	**52**	26

Table 2. Bacteria Biotope Task results.

7.2 Systems description and result analysis

All three systems perform the same distinct sub-tasks: bacteria name detection, detection and typing of locations, coreference resolution and event extraction. The following description of the approaches used by the three systems in each subtask will be supported by intermediate results.

Bacteria name detection. Interestingly the three participants used three different resources for the detection of bacteria names: the List of Prokaryotic Names with Standing in Nomenclature (LPNSN) by UTurku, names in the genomic BLAST page of NCBI by JAIST and the NCBI Taxonomy by Bibliome.

Bibliome	84
JAIST	55
UTurku	16

Table 3. Bacteria entity recall.

Table 3 shows a disparity in the bacteria entity recall of participants. The merits of each resource cannot be deduced directly from these figures since they have been exploited in different manners. UTurku and JAIST systems injected the resource as features in a ML algorithm, whereas Bibliome directly projected the resource on the corpus with additional rule-based abbreviation detection.

However there is some evidence that the resources have a major impact on the result. According to Sneath and Brenner (1992) LPNSN is necessarily incomplete. NCBI BLAST only contains names of species for which a complete genome has been published. The NCBI Taxonomy used by INRA only contains names of taxa for which some sequence was published. It appears that all the lists are incomplete. However, the bacteria referenced by the sequencing projects, which are mentioned in the corpus should all be recorded by the NCBI Taxonomy.

Location detection and typing. As stated before, locations are not necessarily denoted by rigid designators. This was an interesting challenge that called for the use of external resources and linguistic analysis with a broad scope.

UTurku and JAIST both used WordNet, a sensible choice since it encompasses a wide vocabulary and is also structured with synsets and hyperonymy relations. The WordNet entries were injected as features in the participant ML-based entity recognition and typing subsystems.

It is worth noting that JAIST also used word clustering based on MEMM for entity detection. This method has things in common with distributional semantics. JAIST experiments demonstrated a slight improvement using word clustering, but further exploration of this idea may prove to be valuable.

Alternatively, the Bibliome system extracted terms from the corpus using linguistic criteria classified them as locations and predicted their type, by comparing them to classes in a habitat-specific ontology. This prediction uses both linguistic analysis of terms and the hierarchical structure of the ontology. Bibliome also used additional resources for specific types: the NCBI Taxonomy for type *Host* and Agrovoc countries for type *Geographical*.

	Bibliome	JAIST	UTurku
Host	**82**	49	28
Host part	**72**	36	28
Geo.	29	**60**	53
Environment	**53**	10	11
Water	**83**	32	2
Soil	**86**	37	34

Table 4. Location entity recall by type. The number of entities of type *Food* and *Medical* in the test set is too low to be significant. The scores are computed using T_{ab} and J_{ab}.

The location entity recall in Table 4 shows that Bibliome consistently outperformed the other groups for all types except for *Geographical*. This demonstrates the strength of exploiting a resource with strong semantics (ontology vs. lexicon) and with mixed semantic and linguistic rules.

In order to evaluate the impact of *Location* entity boundaries and types, we computed the final score by relaxing T_{ab} and J_{ab} measures. We re-defined T_{ab} as always equal to 1, in other words the type of the localization was not evaluated. We also re-defined J_{ab} as: if the *Location* arguments overlap, then $J_{ab} = 1$, otherwise $J_{ab} = 0$. This means that boundaries were relaxed. The relaxed scores are shown in Table 5. While the difference is not significant for JAIST and UTurku, the Bibliome results exhibit a 9 point increase. This demonstrates that the Bibliome system is efficient at predicting which entities are locations, while the other participants predict more accurately the boundaries and types.

	Recall	Prec.	F-score	Diff.
Bibliome	54	54	54	**+9**
JAIST	29	45	35	+2
UTurku	19	56	28	+2

Table 5. Participants score using relaxed location boundaries and types.

Coreference resolution. The corpus exhibits an unusual number of anaphora, especially bacteria coreferences since a single bacterium species is generally the central topic of a document. The Bibliome submission is the only one that performed bacteria coreference resolution. Their system is rule-based and dealt with referential "it", bi-antecedent anaphora and more importantly sortal anaphora. The JAIST system has a bacteria coreference module based on ML. However the submission was done without coreference resolution since their experiments did not show any performance improvement.

Event extraction. Both UTurku and JAIST approached the event extraction as a classification task using ML (SVM). Bibliome exploited the co-occurrence of arguments and the presence of trigger words from a predefined list. Both UTurku and Bibliome generate events in the scope of a sentence, whereas JAIST generates events in the scope of a paragraph.

As shown in Table 6, UTurku achieved the best score for *PartOf* events. For all participants, the prediction is often correct (between 60 and 80%) while the recall is rather low (20 to 32%).

		Recall	Precis.	F-score
	Host	**61**	48	**53**
	Host part	**53**	42	**47**
	Geo.	13	38	19
B.	Env.	**29**	24	**26**
	Water	**60**	**55**	**57**
	Soil	**69**	**59**	**63**
	Part-of	23	79	36
	Host	30	43	36
	Host part	18	**68**	28
	Geo.	**52**	35	**42**
J.	Env.	5	0	0
	Water	19	27	23
	Soil	21	42	28
	Part-of	31	61	41
	Host	15	**51**	23
	Host part	9	40	15
	Geo.	32	**40**	36
U.	Env.	6	50	11
	Water	1	7	2
	Soil	12	21	15
	Part-of	**32**	**83**	**46**

Table 6. Event extraction results per type.

Conversely, the score of the *Localization* relation by UTurku has been penalized by its low recognition of bacteria names (16%). This strongly affects the score of *Localizations* since the bacterium is the only expected agent argument. The good results of Bibliome are partly explained by its high bacteria name recall of 84%.

The lack of coreference resolution might penalize the event extraction recall. To test this hypothesis, we computed the recall by taking only into account events where both arguments occur in the same sentence. The goal of this selection is to remove most events denoted through a coreference. The recall difference was not significant for Bibliome and JAIST, however UTurku recall raised by 12 points (29%). That experiment confirms that UTurku low recall is explained by coreferences rather than the quality of event extraction. The paragraph scope chosen by JAIST probably compensates the lack of coreference resolution.

As opposed to Bibliome, the precision of the *Localization* relation prediction by JAIST and UTurku, is high compared to the recall, with a noticeable exception of geographical locations. The difference between participants seems to be caused by the geographical entity recognition step more than the relation itself. This is shown by the difference between the entity and the event recall (Table 4 and 6 respectively).. The worst predicted type is *Environment,* which includes diverse locations, such as agricultural, natural and industrial sites and residues. This reveals significant room for improvement for *Water*, *Soil* and *Environment* entity recognition.

8 Discussion

The participant papers describe complementary methods for tackling BB Task's new goals. The novelty of the task prevents participants from deeply investing in all of the issues together. Depending on the participants, the effort was focused on different issues with various approaches: entity recognition and anaphora resolution based on extensive use of background knowledge, and relation prediction based on linguistic analysis of syntactic dependencies. Moreover, these different approaches revealed to be complementary with distinct strengths and limitations. In the future, one may expect that the integration of these promising approaches will improve the current score.

The corpus of BioNLP BB Task 2011 consists of a set of Web pages that were selected for their readability. However, some corpus traits make the IE task more difficult compared to scientific papers. For example, the relaxed style of some pages tolerates some typographic errors (*e.g. morrow* instead of *marrow*) and ambiguous anaphora. The genome sequencing project documents aim at justifying the sequencing of bacteria. This results in abundant descriptions of potential uses and locations that should not be predicted as actual locations. Their correct prediction requires complex analysis of modalities (possibility, probability, negation). Some pages describe the action of hosted bacteria at the molecular level, such as cellular infection. Terms related to the cell are ambiguous locations because they may refer to either bacteria or host cells.

Scientific papers form a much richer source of bacterial location information that is exempt from such flaws. However, as opposed to Web pages, most of them are not publicly available and they are in PDF format.

The typology of locations was designed according to the BB Task corpus with a strong bias towards natural environments since bioremediation and plant growth factor are important motivations for bacteria sequencing. It could be necessary to revise it according to a broader view of bacterial studies where pathogenicity and more generally human and animal health are central issues.

9 Conclusion

The Bacteria Biotope Task corpus and objectives differ from molecular biology text-mining of scientific papers. The annotation strategy and the analysis of the participant results contributed to the construction of a preliminary review of the nature and the richness of its linguistic specificities. The participant results are encouraging for the future of the Bacteria Biotope issue. The degree of sophistication of participating systems shows that the community has technologies, which are mature enough to address this crucial biology question. However, the results leave a large room for improvement.

The Bacteria Biotope Task was an opportunity to extend molecular biology text-mining goals towards the support of bacteria biodiversity studies such as metagenomics, ecology and phylogeny. The prediction of bacterial location information is the very first step in this direction. The abundance of scientific papers dealing with this issue and describing location properties form a potentially rich source for further extensions.

Acknowledgments

The authors thank Valentin Loux for his valuable contribution to the definition of the Bacteria Biotope task. This work was partially supported by the French Quaero project.

References

Jari Björne and Taio Salakoski. 2011. Generalizing Biomedical Event Extraction. *Proceedings of the BioNLP 2011 Workshop Companion Volume for Shared Task.*

Cadixe. http://caderige.imag.fr/Articles/CADIXE-XML-Annotation.pdf

Corine Land Cover. http://www.eea.europa.eu/themes/landuse/interactive/clc-download

EnvDB database. http://metagenomics.uv.es/envDB/

EnvO Project. http://gensc.org/gc_wiki/index.php/EnvO_Project

Dawn Field [et al]. 2008. Towards a richer description of our complete collection of genomes and metagenomes: the "Minimum Information about a Genome Sequence" (MIGS) specification. *Nature Biotechnology*. 26: 541-547.

Melissa M. Floyd, Jane Tang, Matthew Kane and David Emerson. 2005. Captured Diversity in a Culture Collection: Case Study of the Geographic and Habitat Distributions of Environmental Isolates Held at the American Type Culture Collection. Applied and Environmental Microbiology. 71(6):2813-23.

GenBank. http://www.ncbi.nlm.nih.gov/

GOLD. http://www.genomesonline.org/cgi-bin/GOLD/bin/gold.cgi

Lynette Hirschman, Cheryl Clark, K. Bretonnel Cohen, Scott Mardis, Joanne Luciano, Renzo Kottmann, James Cole, Victor Markowitz, Nikos Kyrpides, Norman Morrison, Lynn M. Schriml, Dawn Field. 2008. Habitat-Lite: a GSC case study based on free text terms for environmental metadata. *Omics*. 12(2):129-136.

Jin-Dong Kim, Tomoko Ohta, Sampo Pyysalo, Yoshinobu Kano, Jun'ichi Tsujii. 2010. Extracting bio-molecular events from literature - the BioNLP'09 shared task. *Special issue of the International Journal of Computational Intelligence.*

MicrobeWiki. http://microbewiki.kenyon.edu/index.php/MicrobeWiki

Microbial Genomics Program at JGI. http://genome.jgi-psf.org/programs/bacteria-archaea/index.jsf

Microorganisms sequenced at Genoscope. http://www.genoscope.cns.fr/spip/Microorganisms-sequenced-at.html

Claire Nédellec. 2005. Learning Language in Logic - Genic Interaction Extraction Challenge" in *Proceedings of the Learning Language in Logic (LLL05)* workshop joint to *ICML'05*. Cussens J. and Nédellec C. (eds). Bonn.

Claire Nédellec, Adeline Nazarenko, Robert Bossy. 2008..Information Extraction. *Ontology Handbook*. S. Staab, R. Studer (eds.), Springer Verlag, 2008.

Nhung T. H. Nguyen and Yoshimasa Tsuruoka. 2011. Extracting Bacteria Biotopes with Semi-supervised Named Entity Recognition and Coreference Resolution. *Proceedings of the BioNLP 2011 Workshop Companion Volume for Shared Task.*

Miguel Pignatelli, Andrés Moya, Javier Tamames. (2009). EnvDB, a database for describing the environmental distribution of prokaryotic taxa. *Environmental Microbiology Reports*. 1:198-207.

Prokaryote Genome Projects at NCBI. http://www.ncbi.nlm.nih.gov/genomes/lproks.cgi

Sampo Pyysalo, Antti Airola, Juho Heimonen, Jari Björne, Filip Ginter and Tapio Salakoski. 2008. Comparative analysis of five protein-protein interaction corpora. *BMC Bioinformatics*. vol 9. Suppl 3. S6.

Zorana Ratkovic, Wiktoria Golik, Pierre Warnier, Philippe Veber, Claire Nédellec. 2011. BioNLP 2011 Task Bacteria Biotope – The Alvis System. *Proceedings of the BioNLP 2011 Workshop Companion Volume for Shared Task.*

Peter H. A. Sneath and Don J. Brenner. 1992. "Official" Nomenclature Lists. *American Society for Microbioloy News*. 58, 175.

Javier Tamames and Victor de Lorenzo. 2010. EnvMine: A text-mining system for the automatic extraction of contextual information. *BMC Bioinformatics*. 11:294.

Web Soil Survey. http://websoilsurvey.nrcs.usda.gov/

BioNLP Shared Task 2011 – Bacteria Gene Interactions and Renaming

Julien Jourde[1], Alain-Pierre Manine[2], Philippe Veber[1], Karën Fort[3], Robert Bossy[1], Erick Alphonse[2], Philippe Bessières[1]

[1]Mathématique, Informatique et Génome – Institut National de la Recherche Agronomique
MIG INRA UR1077
F78352 Jouy-en-Josas, France
forename.lastname@jouy.inra.fr

[2]PredictiveDB
16, rue Alexandre Parodi
F75010 Paris, France
{apmanine,alphonse}@predictivedb.com

[3]LIPN – Université Paris-Nord/CNRS UMR7030 and
INIST CNRS UPS76 – F54514
Vandœuvre-lès-Nancy, France
karen.fort@inist.fr

Abstract

We present two related tasks of the BioNLP Shared Tasks 2011: Bacteria Gene Renaming (Rename) and Bacteria Gene Interactions (GI). We detail the objectives, the corpus specification, the evaluation metrics, and we summarize the participants' results. Both issued from PubMed scientific literature abstracts, the Rename task aims at extracting gene name synonyms, and the GI task aims at extracting genic interaction events, mainly about gene transcriptional regulations in bacteria.

1 Introduction

The extraction of biological events from scientific literature is the most popular task in Information Extraction (IE) challenges applied to molecular biology, such as in LLL (Nédellec, 2005), BioCreative Protein-Protein Interaction Task (Krallinger et al., 2008), or BioNLP (Demner-Fushman et al., 2008). Since the BioNLP 2009 shared task (Kim et al., 2009), this field has evolved from the extraction of a unique binary interaction relation between proteins and/or genes towards a broader acceptation of biological events including localization and transformation (Kim et al., 2008). In the same way, the tasks Bacteria Gene Interactions and Bacteria Gene Renaming deal with the extraction of various molecular events capturing the mechanisms relevant to gene regulation in prokaryotes. The study of bacteria has numerous applications for health, food and industry, and overall, they are considered as organisms of choice for the recent integrative approaches in systems biology, because of their relative simplicity. Compared to eukaryotes, they allow easier and more in-depth analysis of biological functions and of their related molecular mechanisms.

Processing literature on bacteria raises linguistic and semantic specificities that impact text analysis. First of all, gene renaming is a frequent phenomenon, especially for model bacteria. Hence, the abundance of gene synonyms that are not morphological variants is high compared to eukaryotes. The history of bacterial gene naming has led to drastic amounts of homonyms and synonyms which are often missing (or worse, erroneous) in gene databases. In particular, they often omit old gene names that are no longer used in new publications, but that are critical for exhaustive bibliography search. Polysemy makes the situation even worse, as old names frequently happen to be reused to denote different genes. A correct and complete gene synonym table is crucial to biology studies, for instance when integrating large scale experimental data using distinct nomenclatures. Indeed this information can save a lot of bibliographic research time. The Rename Task is a new task in text-mining for biology that aims at extracting explicit mentions of renaming relations. It is a critical step in gene name normalization that is needed for further extraction of biological events such as genic interactions.

Regarding stylistics, gene and protein interactions are not formulated in the same way for eukaryotes and prokaryotes. Descriptions of interactions and regulations in bacteria include more knowledge about their molecular actors and mechanisms, compared to the literature on eukaryotes. Typically in bacteria literature, the genic regulations are more

Proceedings of BioNLP Shared Task 2011 Workshop, pages 65–73,
Portland, Oregon, USA, 24 June, 2011. ©2011 Association for Computational Linguistics

likely expressed by direct binding of the protein, while in eukaryote literature, non-genic agents related to environmental conditions are much more frequent. The bacteria GI Task is based on (Manine et al., 2010) which is a semantic re-annotation of the LLL challenge corpus (Nédellec, 2005), where the description of the GI events in a fine-grained representation includes the distinction between expression, transcription and other action events, as well as different transcription controls (e.g. regulon membership, promoter binding). The entities are not only protein agent and gene target but extend to families, complexes and DNA sites (binding sites, promoters) in order to better capture the complexity of the regulation at a molecular level. The task consists in relating the entities with the relevant relations.

2 Rename Task Description

The goal of the Rename task is illustrated by Figure 1. It consists in predicting renaming relations between text-bound gene names given as input. The only type of event is *Renaming* where both arguments are of type *Gene*. The event is directed, the former and the new names are distinguished. Genes and proteins were not distinguished because of the high frequency of metonymy in renaming events. The relation to predict between genes is a *Renaming* of a former gene name into a new one. In the example of Figure 1, YtaA, YvdP and YnzH are the former names of three proteins renamed CotI, CotQ and CotU, respectively.

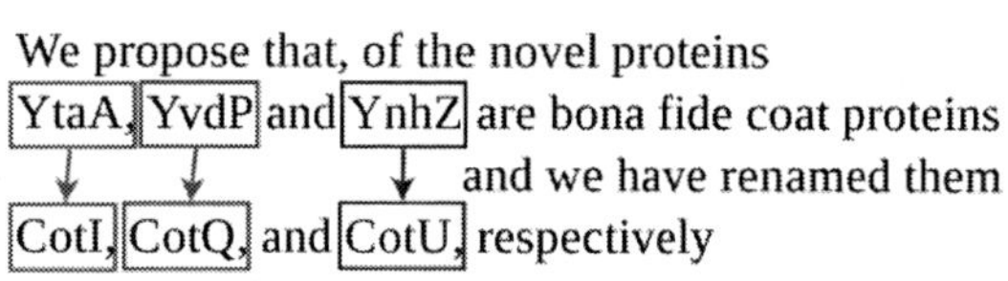

Figure 1: Examples of relations to be extracted.

2.1 Rename Task corpus

The Rename Task corpus is a set of 1,836 PubMed references of bacterial genetic and genomic studies, including title and abstract. A first set of 23,000 documents was retrieved, identifying the presence of the bacterium *Bacillus subtilis* in the text and/or in the MeSH terms. *B. subtilis* documents are particularly rich in renaming mentions. Many genes were renamed in the middle of the nineties, so that the new names matched those of the *Escherichia coli* homologues. The 1,843 documents the most susceptible to mention renaming were automatically filtered according to two non exclusive criteria:

1. Either the document mentions at least two gene synonyms as recorded in the fusion of seven *B. subtilis* gene nomenclatures. This led to a set of 703 documents.
2. Or the document contains a renaming expression from a list that we manually designed and tested (e.g. rename, also known as). It is an extension of a previous work by (Weissenbacher, 2004). A total of 1,140 new documents not included in the first set match this criteria.

About 70% of the documents (1,146) were kept in the training data set. The rest was split into the development and test sets, containing 246 and 252 documents respectively. Table 1 gives the distribution of genes and renaming relations per corpus. Gene names were automatically annotated in the documents with the nomenclature of *B. subtilis*. Gene names involved in renaming acts were manually curated. Among the 21,878 gene mentions in the three corpus, 680 unique names are involved in renaming relations which represents 891 occurrences of genes.

	Training + Dev.	*Test*
Documents	(1,146 + 246) 1,392	252 (15%)
Gene names	18,503	3,375 (15%)
Renamings	373	88 (24%)

Table 1: Rename Task corpus content.

2.2 Rename Task annotation and guidelines

Annotation procedure The corpus was annotated in a joint effort of MIG/INRA and INIST/CNRS. The reference annotation of the Rename Task corpus was done in two steps, a first annotation step by science information professionals of INIST with MIG initial specifications, a second checking step by people at MIG. Two annotators and a project manager were in charge of the task at INIST. The documents were annotated using the Cadixe editor[1]. We

[1]http://caderige.imag.fr/Articles/CADIXEXML-Annotation.pdf

provided to them detailed annotation guidelines that were largely modified in the process. A subset of 100 documents from the first set of 703 was annotated as a training session. This step was used to refine the guidelines according to the methodology described in (Bonneau-Maynard et al., 2005). Several inter-annotator agreements coefficients were computed to measure the discrepancy between annotators (Fort et al., 2009). With a *kappa* and *pi* scores (for more details on those, see (Artstein and Poesio, 2008)), the results can be considered satisfactory. The manual analysis of the 18 discrepancies led to enrich the annotation guidelines. The first hundreds of documents of the second set did not mention any renaming, leading to concentrate the annotation efforts on the first set. These documents actually contained renamings, but nearly exclusively concerning other kinds of biological entities (protein domains, molecules, cellular ultrastructures, etc.).

Guidelines In order to simplify the task, only short names of gene/protein/groups in *B. subtilis* were considered. Naming conventions set short names of four letters long with an upper case letter at the end for all genes (e.g. gerE) and the same names with the upper case of the initial letter (e.g. GerE) and long names for the proteins (e.g. Spore germination protein gerE). But many irregular gene names exist (e.g. tuf), which are considered as well. It also happens that gene or protein name lists are abbreviated by factorization to form a sequence. For instance queCDEF is the abbreviation of the list of gene names queC, queD, queE and queF. Such aggregations are acceptable gene names as well. In any case, these details were not needed by the task participants since the corpus was provided with tagged gene names.

Most renaming relations involve couples of the same type, genes, proteins or aggregations. Only 18 relations link mixed couples of genes and proteins. In case of ambiguity, annotators would consult international gene databases and an internal INRA database to help them determine whether a given couple of names were actually synonyms.

Multiple occurrences of the same renaming relation were annotated independently, and had to be predicted. The renaming pairs are directed, the former and the new forms have to be distinguished. When the renaming order was not explicit in the document, the rule was to annotate by default the first member of the couple as the new form, and the second one as the former form. Figure 2 presents the most common forms of renaming.

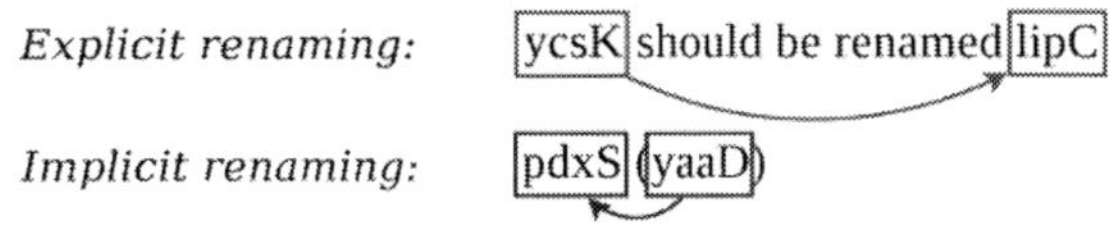

Figure 2: Common types of relations to be extracted.

Revised annotations INIST annotations were systematically checked by two experts in Bioinformatics from INRA. Mainly, encoding relations (e.g. the gene encoding sigma K (sigK)) that are not renaming cases were purged. Given the number of ambiguous annotations, we designed a detailed typology in order to justify acceptance or rejection decisions in seven different sub-cases hereafter presented. Three positive relations figure in Table 2, where the underlined names are the former names and the framed names are the new ones. Explicit renaming relations occur in 261 sentences, synonymy-like relations in 349 sentences, biological proof-based relations in 76 sentences.

Explicit renaming relation is the easiest positive case to identify. In the example, the aggregation of gene names ykvJKLM is clearly renamed by the authors as queCDEF. Although the four genes are con-

Explicit renaming
PMID 15767583: Genetic analysis of ykvJKLM mutants in Acinetobacter confirmed that each was essential for queuosine biosynthesis, and the genes were renamed queCDEF.
Implicit renaming
PMID 8002615: Analysis of a suppressor mutation ssb (kinC) of sur0B20 (spo0A) mutation in *Bacillus subtilis* reveals that kinC encodes a histidine protein kinase.
Biological proof
PMID 1744050: DNA sequencing established that spoIIIF and spoVB are a single monocistronic locus encoding a 518-amino-acid polypeptide with features of an integral membrane protein.

Table 2: Positive examples of the Rename Task.

catenated, there is no evidence mentioned of them acting as an operon. Furthermore, despite the context involving mutants of Acinetobacter, the aggregation belongs correctly to *B. subtilis*.

Implicit renaming is an asymmetric relation since one of the synonyms is intended to replace the other one in future uses. The example presents two renaming relations between former names ssb and spo0A, and new names kinC and sur0B20, respectively. The renaming relation between ssb and kinC has a different orientation due to additional information in the reference. Like in the preceding example, the renaming is a consequence of a genetic mutation experiment. Mutation names represent an important transversal issue that is discussed below.

Biological proof is a renaming relation induced by an explicit scientific conclusion while the renaming is not, as in the example where experiments reveal that two loci spoIIIF and spoVB are in fact the same one and then become synonyms. Terms such as "allelic to" or "identical to" usually qualify such conclusions. Predicting biological proof-based relations requires some biological modeling.

The next three cases are negative (Table 3). Underlined gene and protein names are involved in a relation which is not a renaming relation.

Protein encoding relation occurs between a gene and the protein it codes for. Some mentions may look like renaming relations. The example presents the gene yeaC coding for MoxR. No member of the couple is expected to replace the other one.

Homology measures the similarity between gene or protein sequences. Most of the homology mentions involve genes or proteins from different species

Protein encoding
PMID 8969499: The putative products of ORFs yeaB (Czd protein), yeaC (MoxR), yebA (CNG-channel and cGMP-channel proteins from eukaryotes),

Genetic homology
PMID 10619015: Dynamic movement of the ParA-like Soj protein of *B. subtilis* and its dual role in nucleoid organization and developmental regulation.

Operon | Regulon | Family
PMID 3127379: Three promoters direct transcription of the sigA (rpoD) operon in *Bacillus subtilis*.

Table 3: Negative examples of the Rename Task.

(orthologues). The others compare known gene or protein sequences of the same species (paralogues). This may be misleading since the similarity mention may look like biological proof-based relations, as between ParA and Soj in Table 3.

Operon, regulon or family renaming involves objects that may look like genes, proteins or simple aggregations of gene or protein names but that are perceptibly different. The objects represent more than one gene or protein and the renaming does not necessarily affect all of them. More problematic, their name may be the same as one of the genes or proteins they contain, as in the example where sigA and rpoD are operons but are also known as gene names. Here, sigA (and so rpoD) represents at least two different genes. For the sake of clarity, operons, regulons and families are rejected, even if all the genes are clearly named, as in an aggregation.

The last point concerns ***mutation*** which are frequent in Microbiology for revealing gene phenotypes. They carry information about the original gene names (e.g., rvtA11 is a mutant name created by adding 11 to rvtA). But partial names cannot be partially annotated, that is to say, the original part (rvtA) should not be annotated in the mutation name (rvtA11). Most of these names are local names, and should not be annotated because of their restricted scope. It may happen so that the mutation name is registered as a synonym in several international databases. To avoid inconsistencies, all renamings involving a mutation referenced in a database were accepted, and only biological proof-based and explicit renamings involving a strict non-null unreferenced mutation (a null mutation corresponds to a total suppression of a gene) were accepted.

2.3 Rename Task evaluation procedure

The evaluation of the Rename task is given in terms of recall, precision and F-score of renaming relations. Two set of scores are given: the first set is computed by enforcing strict direction of renaming relations, the second set is computed with relaxed direction. Since the relaxed score takes into account renaming relations even if the arguments are inverted, it will necessarily be greater or equal than the strict score. The participant score is the relaxed score, the strict score is given for information. Relaxed scores are informative with respect to the application goal. The motivation of the Rename task is to keep bacteria gene synonyms tables up to date. The choice of the canonical name among synonyms for denoting a gene is done by the bacteriology community, and it may be independent of the anteriority or novelty of the name. The annotation of the reference corpus showed that the direction was not always decidable, even for a human reader. Thus, it would have been unfair to evaluate systems on the basis of unsure information.

2.4 Results of the Rename Task participants

Final submissions were received from three teams, the University of Turku (Uturku), the University of Concordia (Concordia) and the Bibliome team from MIG/INRA. Their results are summarized in Table 4. The ranking order is given by the overall F-score for relations with relaxed argument order.

Team	*Prec.*	*Recall*	*F-score*
Univ. of Turku	**95.9**	**79.6**	**87.0**
Concordia Univ.	74.4	65.9	69.9
INRA	57.0	73.9	64.4

Table 4: Participant scores at the Rename Task.

Uturku achieved the best F-score with a very high precision and a high recall. Concordia achieved the second F-score with balanced precisions and recalls. Bibliome is five points behind with a better recall but much lower precision. Both UTurku and Concordia predictions rely on dependencies (Charniak-Johnson and Stanford respectively, using McClosky model), whereas Bibliome predictions rely on bag of words. This demonstrates the high value of dependency parsing for this task, in particular for the precision of predictions. We notice that UTurku system uses machine learning (SVM) and Concordia uses rules based on trigger words. The good results of UTurku confirms the hypothesis that gene renaming citations are highly regular in scientific literature. The most frequently missed renamings belong to the Biological Proof category (see Table 2). This is expected because the renaming is formulated as a reasoning where the conclusion is only implicit.

2.5 Discussion

The very high score of Uturku method leads us to conclude that the task can be considered as solved by a linguistic-based approach. Whereas Bibliome used an extensive nomenclature considered as exhaustive and sentence filtering using a SVM, Uturku used only two nomenclatures in synergy but with more sophisticated linguistic-based methods, in particular syntactic analyses. Bibliome methods showed that a too high dependence to nomenclatures may decrease scores if they contain compromised data. However, the use of an extensive nomenclature as done by Bibliome may complement Uturku approach and improve recall. It is also interesting that both systems do not manage renamings crossing sentence boundaries.

The good results of the renaming task will be exploited to keep synonym gene lists up to date with extensive bibliography mining. In particular this will contribute to enriching SubtiWiki, a collaborative annotation effort on *B. subtilis* (Flórez et al., 2009; Lammers et al., 2010).

3 Gene Interactions Task description

The goal of the Bacteria GI Task is illustrated by Figure 3. The genes cotB and cotC are related to their two promoters, not named here, by the relation *PromoterOf*. The protein GerE is related to these promoters by the relation *BindTo*. As a consequence, GerE is related to cotB and cotC by an *Interaction* relation. According to (Kim et al., 2008), the need to define specialized relations replacing one unique and general interaction relation was raised in (Manine et al., 2009) for extracting genic interactions from text. An ontology describes relations and entities (Manine et al., 2008) catching a model of gene transcription to which biologists implicitly refer in their publications. Therefore, the ontology is mainly oriented towards the description of a structural model of genes, with molecular mechanisms of their transcription and associated regulations.

The corpus roughly contains three kinds of genic

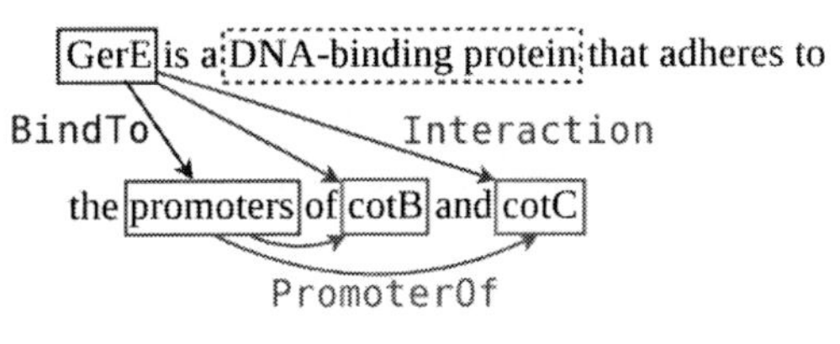

Figure 3: Examples of relations to be extracted.

interaction mentions, namely regulations, regulon membership and binding. The first case corresponds to interactions the mechanism of which is not explicitly given in the text. The mention only tells that the transcription of a given gene is influenced by a given protein, either positively (activation), negatively (inhibition) or in an unspecified way. The second kind of genic interaction mention (regulon membership) basically conveys the same information, using the regulon term/concept. The regulon of a gene is the set of genes that it controls. In that case, the interaction is expressed by saying that a gene is a member of some regulon. The third and last kind of mention provides with more mechanistic details on a regulation, since it describes the binding of a protein near the promoter of a target gene. This motivates the introduction of *Promoter* and *Site* entities, which correspond to DNA regions. It is thus possible to extract the architecture of a regulatory DNA region, linking a protein agent to its gene target (see Figure 3).

The set of entity types is divided into two main groups, namely 10 genic entities and 3 kinds of action (Table 5). Genic entities represent biological objects like a gene, a group of genes or a gene product. In particular, a *GeneComplex* annotation corresponds to an operon, which is a group of genes that are contiguous in the genome and under the control of the same promoter. The annotation *GeneFamily* is used to denote either genes involved in the same biological function or genes with sequence homologies. More importantly, *PolymeraseComplex* annotations correspond to the protein complex that is responsible for the transcription of genes. This complex includes several subunits (components), combined with a sigma factor, that recognizes specific promoters on the DNA sequence.

The second group of entities are phrases expressing either molecular processes (e.g. sequestration, dephosphorylation, etc.) or the molecular state of the bacteria (e.g. presence, activity or level of a protein). They represent some kind of action that can be performed on a genic entity. Note that transcription and expression events were tagged as specific actions, because they play a specific part in certain relations (see below).

The annotation of entities and actions was provided to the participants, and the task consisted in extracting the relations listed in Table 6.

Name	*Example*
Gene	cotA
GeneComplex	sigX-ypuN
GeneFamily	class III heat shock genes
GeneProduct	yvyD gene product
Protein	CotA
PolymeraseComplex	SigK RNA polymerase
ProteinFamily	DNA-binding protein
Site	upstream site
Promoter	promoter regions
Regulon	regulon
Action	activity | level | presence
Expression	expression
Transcription	transcription

Table 5: List of molecular entities and actions in GI.

Name	*Example*
ActionTarget	**expression** of *yvyD*
Interaction	**ComK** negatively regulates *degR* expression
RegulonDependence	*sigmaB* **regulon**
RegulonMember	*yvyD* is member of sigmaB **regulon**
BindTo	**GerE** adheres to the *promoter*
SiteOf	**-35 sequence** of the *promoter*
PromoterOf	the *araE* **promoter**
PromoterDependence	*GerE*-controlled **promoter**
TranscriptionFrom	**transcription** from the *upstream site*
TranscriptionBy	**transcription** of cotD by *sigmaK RNA polymerase*

Table 6: List of relations in GI.

The relations are binary and directed, and rely the entities defined above. The three kinds of interactions are represented with an *Interaction* annotation, linking an agent to its target. The other relations provide additional details on the regulation, like elementary components involved in the reaction (sites, promoters) and contextual information (mainly provided by the *ActionTarget* relations). A formal definition of relations and relation argument types can be found on the Bacteria GI Task Web page.

3.1 Bacteria Gene Interactions corpus

The source of the Bacteria GI Task corpus is a set of PubMed abstracts mainly dealing with the transcription of genes in *Bacillus subtilis*. The semantic annotation, derived from the ontology of (Manine et al., 2008), contains 10 molecular entities, 3 different actions, and 10 specialized relations. This is applied to 162 sentences from the LLL set (Nédellec, 2005), which are provided with manually checked linguistic annotations (segmentation, lemmatization, syntactic dependencies). The corpus was split into 105 sentences for training, 15 for development and 42 for test. Table 7 gives the distribution of the entities and actions per corpus and Table 8 gives the distribution of the relations per corpus.

3.2 Annotation procedures and guidelines

The semantic annotation scheme was developed by two annotators through a series of independent annotations of the corpus, followed by reconciliation steps, which could involve concerted modifications (Manine et al., 2010). As a third and final stage, the corpus was reviewed and the annotation simplified to make it more appropriate to the contest. The final annotation contains 748 relations distributed in nine categories, 146 of them belonging to the test set.

Entity or action	*Train. + Dev.*	*Test*
Documents	(105+15) 120	42
Protein	219	85
Gene	173	56
Transcription	53	21
Promoter	49	10
Action	45	22
PolymeraseComplex	43	14
Expression	29	6
Site	22	8
GeneComplex	19	4
ProteinFamily	12	3
Regulon	11	2
GeneProduct	10	3
GeneFamily	6	5

Table 7: Distribution of entities and actions in GI.

Relation	*Train. + Dev.*	*Test*
Interaction	208	64
ActionTarget	173	47
PromoterOf	44	8
BindTo	39	4
PromoterDependence	36	4
TranscriptionBy	36	8
SiteOf	23	6
RegulonMember	17	2
TranscriptionFrom	14	2
RegulonDependence	12	1

Table 8: Distribution of relations in GI.

The annotation scheme was generally well suited to accurately represent the meaning of the sentences in the corpus, with one notable exception. In the corpus, there is a common phrasing telling that a protein P regulates the transcription of a gene G by a given sigma factor S. In that case, the only annotated interactions are between the couples (P, G) and (S, G). This representation is not completely satisfactory, and a ternary relation involving P, S and G would have been more adequate.

Additional specific rules were needed to cope with linguistic issues. First, when the argument of a relation had coreferences, the relation was repeated for each maximally precise coreference of the argument. Second, in case of a conjunction like "sigmaA and sigmaX holoenzymes", there should ideally be two entities (namely "sigmaA holoenzyme" and "sigmaX holoenzyme"); however, this is not easy to represent using the BioNLP format. In this situation, we grouped the two entities into a single one. These cases were rare and unlikely affected the feasibility of the task, since entities were provided in the test set.

3.3 Gene Interactions evaluation procedure

The training and development corpora with the reference annotations were made available to participants by December, 1st on the BioNLP shared Task pages together with evaluation software. The test corpus with the entity annotations has been made available by March, 1st. The participants sent the predicted annotations to the BioNLP shared Task organizers by March, 10th. The evaluation results were computed and provided to the participants and on the Web site the same day. The participants are evaluated and ranked according to two scores: F-score for all event types together, and F-score for the *Interaction* event type. In order for a predicted event to count as a hit, both arguments must be the same as in the reference in the right order and the event type must be the same as in the reference.

3.4 Results of GI Task participants

There was only one participant, whose results are shown in Tables 9 and 10. Some relations were not significantly represented in the test set and thus the corresponding results should be considered with caution. This is the case for *RegulonMember* and *TranscriptionFrom*, only represented two times each in the test. The lowest recall, 17%, obtained for the *SiteOf* relation is explained by its low representation in the corpus: most of the test errors come from a difficult sentence with coreferences.

The recall of 56% for the *Interaction* relation certainly illustrates the heterogeneity of this category, gathering mentions of interactions at large, as well as precise descriptions of gene regulations. For instance, Figure 4 shows a complex instance where all of the interactions were missed. Surprisingly, we also found false negatives in rather trivial examples ("*ykuD was transcribed by SigK RNA polymerase from T4 of sporulation.*"). Uturku used an SVM-based approach for extraction, and it is thus delicate to account for the false negatives in a simple and concise way.

Event	*U. Turku scores*
Global Precision	85
Global Recall	71
Global F-score	77
Interaction Precision	75
Interaction Recall	56
Interaction F-score	64

Table 9: University of Turku global scores.

Event	*Prec.*	*Rec.*	*F-score*
Global	85	71	77
ActionTarget	94	92	93
BindTo	75	75	75
Interaction	75	56	64
PromoterDependence	100	100	100
PromoterOf	100	100	100
RegulonDependence	100	100	100
RegulonMember	100	50	67
SiteOf	100	17	29
TranscriptionBy	67	50	57
TranscriptionFrom	100	100	100

Table 10: University of Turku scores for each relation.

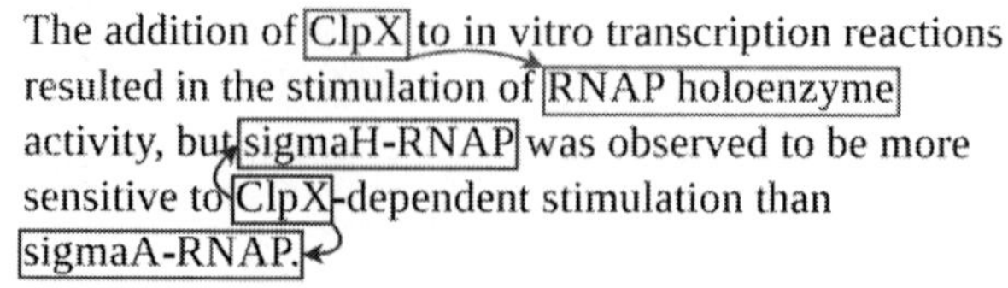

Figure 4: Examples of three missed interactions.

3.5 Discussion

The GI corpus was previously used in a relation extraction work (Manine et al, 2009) based on Inductive Logic Programming (Muggleton and Raedt, 1994). However a direct comparison of the results is not appropriate here since the annotations were partially revised, and the evaluation setting was different (leave-one-out in Manine's work, test set in the challenge).

Nevertheless, we note similar tendencies if we compare relative results between relations. In particular, it was also found in Manine's paper that *SiteOf*, *TranscriptionBy* and *Interaction* are the most difficult relations to extract. It is also worth to mention that both approaches rely on syntactic dependencies, and use the curated dependencies provided in the corpus. Interestingly, the approach by the University of Turku reports a slightly lower F-measure with dependencies calculated by the Charniak parser (about 1%, personal communication). This information is especially important in order to consider a production setting.

4 Conclusion

The quality of results for both challenges suggests that current methods are mature enough to be used in semi-automatic strategies for genome annotation, where they could efficiently assist biological experts involved in collaborative annotation efforts (Lammers et al., 2010). However, the false positive rate, notably for the *Interaction* relation, is still too high for the extraction results to be used as a reliable source of information without a curation step.

Acknowlegments

We thank Françoise Tisserand and Bernard Talercio (INIST) for their work on the Rename corpus, and the QUAERO Programme funded by OSEO (French agency for innovation) for its support.

References

Artstein R., Poesio M. (2008). Inter-coder agreement for Computational Linguistics. *Computational Linguistics*, 34(4):555-96.

Björne J., Heimonen J., Ginter F., Airola A., Pahikkala T., Salakoski T. (2009). Extracting complex biological events with rich graph-based feature sets. BioNLP'09 Proc. *Workshop Current Trends in Biomedical Natural Language Processing: Shared Task*, pp. 10-18.

Bonneau-Maynard H., Rosset S., Ayache C., Kuhn A., Mostefa D. (2005). Semantic annotation of the French Media Dialog Corpus. *Interspeech-2005*, pp. 3457-60.

Demner-Fushman D., Ananiadou S., Cohen K.B., Pestian J., Tsujii J., Webber B. (2008). Themes in biomedical natural language processing: BioNLP08. *BMC Bioinformatics*, 9(Suppl. 11):S1.

Flórez L.A., Roppel S.F., Schmeisky A.G., Lammers C.R., Stülke J. (2009). A community-curated consensual annotation that is continuously updated: The *Bacillus subtilis* centred wiki SubtiWiki. *Database*, 2009:bap012.

Fort K., François C., Ghribi M. (2010). Évaluer des annotations manuelles dispersées : les coefficients sont-ils suffisants pour estimer l'accord inter-annotateurs ? *17e Conf. Traitement Automatique des Langues Naturelles (TALN 2010).*

Kim J.D., Ohta T., Tsujii J. (2008) Corpus annotation for mining biomedical events from literature. *BMC Bioinformatics*, 9:10.

Kim J.D., Ohta T., Pyysalo S., Kano Y., Tsujii J. (2009). Overview of BioNLP'09 shared task on event extraction. *BioNLP'09 Proc. Workshop Current Trends in Biomedical Natural Language Processing: Shared Task*, pp. 1-9.

Krallinger M., Leitner F., Rodriguez-Penagos C., Valencia A. (2008). Overview of the protein-protein interaction annotation extraction task of BioCreative II. *Genome Biology*, 9(Suppl. 2):S4.

Lammers C.R., Flórez L.A., Schmeisky A.G., Roppel S.F., Mäder U., Hamoen L., Stülke J. (2010). Connecting parts with processes: SubtiWiki and SubtiPathways integrate gene and pathway annotation for *Bacillus subtilis. Microbiology*, 156(3):849-59.

Manine A.P., Alphonse E., Bessières P. (2008). Information extraction as an ontology population task and its application to genic interactions. *20th IEEE Int. Conf. Tools with Artificial Intelligence (ICTAI'08)*, pp. 74-81.

Manine A.P., Alphonse E., Bessières P. (2009). Learning ontological rules to extract multiple relations of genic interactions from text. *Int. J. Medical Informatics*, 78(12):e31-8.

Manine A.P., Alphonse E., Bessières P. (2010). Extraction of genic interactions with the recursive logical theory of an ontology. *Lecture Notes in Computer Sciences*, 6008:549-63.

Muggleton S., Raedt L.D. (1994) Inductive Logic Programming: Theory and methods. *J. Logic Programming*, 19-20:629-79.

Nédellec C. (2005). Learning Language in Logic – Genic Interaction Extraction Challenge. *Proc. 4th Learning Language in Logic Workshop (LLL'05)*, pp. 31-7.

Weissenbacher, D. (2004). La relation de synonymie en Génomique. *RECITAL 2004 Conference.*

Overview of the Protein Coreference task in BioNLP Shared Task 2011

Ngan Nguyen
University of Tokyo
Hongo 7-3-1, Bunkyoku, Tokyo
nltngan@is.s.u-tokyo.ac.jp

Jin-Dong Kim
Database Center for Life Science
Yayoi 2-11-16, Bunkyo-ku, Tokyo
jdkim@dbcls.rois.ac.jp

Jun'ichi Tsujii
Microsoft Research Asia
5 Dan Ling Street, Haiian District, Beijing
jtsujii@microsoft.com

Abstract

This paper summarizes the Protein Coreference Resolution task of BioNLP Shared Task 2011. After 7 weeks of system development period, the task received final submissions from 6 teams. Evaluation results show that state-of-the-art performance on the task can find 22.18% of protein coreferences with the precision of 73.26%. Analysis of the submissions shows that several types of anaphoric expressions including definite expressions, which occupies a significant part of the problem, have not yet been solved.

1 Introduction

While named entity recognition (NER) and relation or event extraction are regarded as standard tasks of information extraction (IE), coreference resolution (Ng, 2010; Bejan and Harabagiu, 2010) is more and more recognized as an important component of IE for a higher performance. Without coreference resolution, the performance of IE is often substantially limited due to an abundance of coreference structures in natural language text, i.e. information pieces written in text with involvement of a coreference structure are hard to be captured (Miwa et al., 2010). There have been several attempts for coreference resolution, particularly for newswire texts (Strassel et al., 2008; Chinchor, 1998). It is also one of the lessons from BioNLP Shared Task (BioNLP-ST, hereafter) 2009 that coreference structures in biomedical text substantially hinder the progress of fine-grained IE (Kim et al., 2009).

To address the problem of coreference resolution in molecular biology literature, the Protein Coreference (COREF) task is arranged in BioNLP-ST 2011 as a supporting task. While the task itself is not an IE task, it is expected to be a useful component in performing the main IE tasks more effectively. To establish a stable evaluation and to observe the effect of the results of the task to the main IE tasks, the COREF task particularly focuses on finding anaphoric protein references.

The benchmark data sets for developing and testing coreference resolution system were developed based on various manual annotations made to the Genia corpus (Ohta et al., 2002). After 7 weeks of system development phase, for which training and development data sets with coreference annotation were given, six teams submitted their prediction of coreferences for the test data. The best system according to our primary evaluation criteria is evaluated to find 22.18% of anaphoric protein references at the precision of 73.26%.

This paper presents overall explanation of the COREF task, which includes task definition (Section 2), data preparation (Section 4), evaluation methods (Section 5), results (Section 7), and thorough analyses (Section 8) to figure out what are remaining problems for coreference resolution in biomedical text.

2 Problem Definition

This section provides an explanation of the coreference resolution task in our focus, through examples.

Figure 1 shows an example text segmented into four sentences, S2 - S5, where anaphoric coreferences are illustrated with colored extends and arrows. In the figure, protein names are highlighted in purple, T4 - T10, and anaphoric protein references, e.g. pronouns and definite noun phrases, are highlighted in red, T27, T29, T30, T32, of which the an-

Proceedings of BioNLP Shared Task 2011 Workshop, pages 74–82,
Portland, Oregon, USA, 24 June, 2011. ©2011 Association for Computational Linguistics

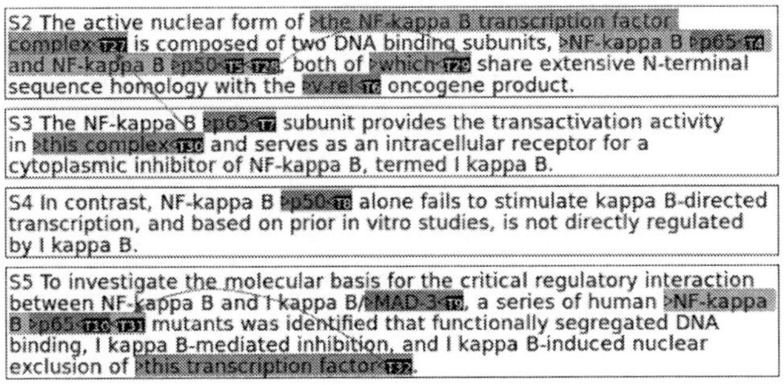

Figure 1: Protein coreference annotation

tecedents are indicated by arrows if found in the text. In the example, the definite noun phrase (NP), *this transcription factor* (T32), is a coreference to *p65* (T10). Without knowing the coreference structure, it becomes hard to capture the information written in the phrase, *nuclear exclusion of this transcription factor*, which is *localization of p65 (out of nucleus)* according to the framework of BioNLP-ST.

A standard approach would include a step to find candidate anaphoric expressions that may refer to proteins. In this task, pronouns, e.g. *it* or *they*, and definite NPs that may refer to proteins, e.g. *the transcription factor* or *the inhibitor* are regarded as candidates of anaphoric protein references. This step corresponds to *markable detection* and *anaphoricity determination* steps in the jargon of MUC. The next step would be to find the antecedents of the anaphoric expressions. This step corresponds to *anaphora resolution* in the jargon of MUC.

3 Task Setting

In the task, the training, development and test data sets are provided in three types of files: the text, the protein annotation, and the coreference annotation files. The *text* files contain plain texts which are target of annotation. The *protein annotation* files provide gold annotation for protein names in the texts, and the *coreference annotation* files provide gold annotation for anaphoric references to those protein names. The protein annotation files are given to the participants, together with all the training, development and test data sets. The coreference annotation files are not given with the test data set, and the task for the participants is to produce them automatically.

In protein annotation files, annotations for protein names are given in a stand-off style encoding. For example, those highlighted in purple in Figure 1 are protein names, which are given in protein annotation files as follows:

```
T4   Protein 275 278 p65
T5   Protein 294 297 p50
T6   Protein 367 372 v-rel
T7   Protein 406 409 p65
T8   Protein 597 600 p50
T9   Protein 843 848 MAD-3
T10  Protein 879 882 p65
```

The first line indicates *there is a protein reference in the span that begins at 275th character and ends before 278th character, of which the text is "p65", and the annotation is identified by the id, "T4"*

The coreference annotation files include three sort of annotations. First, annotations for anaphoric protein references are given. For example, those in red in Figure 1 are anaphoric protein references:

```
T27 Exp 179 222   the N.. 215 222 complex
T29 Exp 307 312   which
T30 Exp 459 471   this .. 464 471 complex
T32 Exp 1022 1047 this .. 1027 1047 tra..
```

The first line indicates that *there is an anaphoric protein reference in the specified span, of which the text is "the NF-kappa B transcription factor complex" (truncated due to limit of space)*, and that *its minimal expression is "complex"*. Second, noun phrases that are antecedents of the anaphoric references are also given in the coreference annotation file. For example, T28 and T31 (highlighted in blue) are antecedents of T29 and T32, respectively, and thus given in the file:

```
T28 Exp 264 297   NF-ka..
T31 Exp 868 882   NF-ka..
```

Third, the coreference relation between the anaphoric expressions and their antecedents are given in predicate-argument expressions[1]:

```
R1   Coref Ana:T29 Ant:T28 [T5, T4]
R2   Coref Ana:T30 Ant:T27
R3   Coref Ana:T32 Ant:T31 [T10]
```

The first line indicates *there is a coreference relation, R1, of which the anaphor is T29 and the antecedent is T28, and the relation involves two protein names, T5 and T4.*

Note that, sometimes, an anaphoric expression, e.g. *which* (T29), is connected to more than one protein names, e.g. *p65* (T4) and *p50* (T5). Sometimes, coreference structures do not involve any specific protein names, e.g. T30 and T27. In order

[1]Due to limitation of space, argument names are abbreviated, e.g. "Ana" for "Anaphora", and "Ant" for "Antecedent"

to establish a stable evaluation, our primary evaluation will focus only on coreference structures that involve specific protein names, e.g. T29 and T28, and T32 and T31. Among the three, only two, R1 and R3, involves specific protein references, T4 and T5, and T10. Thus, finding of R2 will be ignored in the primary evaluation. However, those not involving specific protein references are also provided in the training data to help system development, and will be considered in the secondary evaluation mode. See section 5 for more detail.

4 Data Preparation

The data sets for the COREF task are produced based on three resources: MedCO coreference annotation (Su et al., 2008), Genia event annotation (Kim et al., 2008), and Genia Treebank (Tateisi et al., 2005). Although the three have been developed independently from each other, they are annotations made to the same corpus, the Genia corpus (Kim et al., 2008). Since COREF was focused on finding anaphoric references to proteins (or genes), only relevant annotations were extracted from the MedCO corpus though the following process:

1. From MedCo annotation, coreference entities that were pronouns and definite base NPs were extracted, which became candidate anaphoric expressions. The base NPs were determined by consulting Genia Tree Bank.
2. Among the candidate anaphoric expressions, those that could not be protein references were filtered out. This process was done by checking the head noun of NPs. For example, definite NPs with "cell' as their head noun were filtered out. The remaining ones became candidate protein coreferences.
3. The candidate protein coreferences and their antecedents according to MedCo annotation were included in the data files for COREF task.
4. The protein name annotations from Genia event annotation were added to the data files to determine which coreference expressions involve protein name references.

Table 1 summarizes the coreference entities in the training, development, and test sets for COREF task. In the table, the anaphoric entities are classified into four types as follows:

RELAT indicates relative pronouns or relative adjectives, e.g. *that*, *which*, or *whose*.

PRON indicates pronouns, e.g. *it*.

Type		Train	Dev	Test
Anaphora	RELAT	1193	254	349
	PRON	738	149	269
	DNP	296	58	91
	APPOS	9	1	3
	N/C	11	1	2
Antecedent		2116	451	674
TOTAL		4363	914	1388

Table 1: Statistics of coreference entities in COREF data sets: N/C = not-classified.

DNP indicates definite NPs or demonstrative NPs, e.g. NPs that begin with *the*, *this*, etc.

APPOS indicates coreferences in apposition.

5 Evaluation

The coreference resolution performance is evaluated in two modes.

The *Surface coreference mode* evaluates the performance of finding anaphoric protein references and their antecedents, regardless whether the antecedents actually embed protein names or not. In other words, it evaluates the ability to predict the coreference relations as provided in the gold coreference annotation file, which we call *surface coreference links*.

The *protein coreference mode* evaluates the performance of finding anaphoric protein references with their links to actual protein names (*protein coreference links*). In the implementation of the evaluation, the chain of surface coreference linkes is traced until an antecedent embedding a protein name is found. If a protein-name-embedding antecedent is connected to an anaphora through only one surfs link, we call the antecedent a *direct protein antecedent*. If a protein-name-embedding anteceden is connected to an anaphora through more than one surface link, we call it an *indirect protein antecedent*, and the antecedents in the middle of the chain *intermediate antecedents*. The performance evaluated in this mode may be directly connected to the potential performance in main IE tasks: the more the (anaphoric) protein references are found, the more the protein-related events may be found. For this reason, the protein coreference mode is chosen as the primary evaluation mode.

Evaluation results for both evaluation modes are

given in traditional precision, recall and f-score, which are similar to (Baldwin, 1997).

5.1 Surface coreference

A response expression is matched with a gold expression following partial match criterion. In particular, a response expression is considered correct when it covers the minimal boundary, and is included in the maximal boundary of expression. Maximal boundary is the span of expression annotation, and minimal boundary is the head of expression, as defined in MUC annotation schemes (Chinchor, 1998). A response link is correct when its two argument expressions are correctly matched with those of a gold link.

5.2 Protein coreference

This is the primary evaluation perspective of the protein coreference task. In this mode, we ignore coreference links that do not reference to proteins. Intermediate antecedents are also ignored.

Protein coreference links are generated from the surface coreference links. A protein coreference link is composed of an anaphoric expression and a protein reference that appears in its direct or indirect antecedent. Below is an example.

```
Example:
R1   Coref Ana:T29 Ant:T28 [T5, T4]
R2   Coref Ana:T30 Ant:T27
R3   Coref Ana:T32 Ant:T31 [T10]
R4   Coref Ana:T33 Ant:T32
```

In this example, supposing that there are four surface links in the coreference annotation file (T29,T28), (T30,T27), (T32,T31), and (T33, T32), in which T28 contains two protein mentions T5, T4, and T31 contains one protein mention T10; thus, the protein coreference links generated from these surface links are (T29,T4), (T29,T5), (T32,T10), and (T33, T10). Notice that T33 is connected with T10 through the intermediate expression T32.

Response expressions and generated response result links are matched with gold expressions and links correspondingly in a way similar to the surface coreference evaluation mode.

6 Participation

We received submissions from six teams. Each team was requested to submit a brief description of their team, which was summarized in Table 2.

Team	Member	Approach & Tools
UU	1 NLP	ML (Yamcha SVM, Reconcile)
UZ	5 NLP	RB (-)
CU	2 NLP	RB (-)
UT	1 biochemist	ML (SVM-Light)
US	2 AI	ML (SVM-Light)
UC	3 NLP, 1 BioNLP	ML (Weka SVM)

Table 2: Participation. UU = UofU, UZ = UZH, CU=ConcordU, UT = UTurku, UZ = UZH, US = Uszeged, UC = UCD_SCI, RB = Rule-based, ML = Machine learning-based.

TEAM	RESP	C	P	R	F
UU	86	63	73.26	22.18	34.05
UZ	110	61	55.45	21.48	30.96
CU	87	55	63.22	19.37	29.65
UT	61	41	67.21	14.44	23.77
US	259	9	3.47	3.17	3.31
UC	794	2	0.25	0.70	0.37

Table 3: Protein coreference results. Total number of gold link = 284. RESP=response, C=correct, P=precision, R=recall, F=fscore

The *tool* column shows the external tools used in resolution processing. Among these tools, there is only one team used an external coreference resolution framework, *Reconcile*, which achieved the state-of-the-art performance for supervised learning-based coreference resolution (Stoyanov et al., 2010b).

7 Results

7.1 Protein coreference results

Evaluation results in the protein coreference mode are shown in Table 3. The UU team got the highest f-score 34.05%. The UZ and CU teams are the second- and third-best teams with 30.96% and 29.65% f-score correspondingly, which are comparable to each other. Unfortunately, two teams, US and UC could not produce meaningful results, and the other four teams show performance optimized for high precision. It was expected that the 22.18% of protein coreferences may contribute to improve the performance on main task, which was not observed this time, unfortunately.

The first ranked system by UU utilized Recon-

TEAM	RESP	C	P	R	F
UU	360	43	11.94	20.48	15.09
UZ	736	51	6.93	24.29	10.78
CU	365	36	9.86	17.14	12.52
UT	452	50	11.06	23.81	15.11
US	259	4	1.54	1.90	1.71
UC	797	1	0.13	0.48	0.20

Table 4: Surface coreference results. Total number of gold link = 210. RESP=response, C=correct, P=precision, R=recall, F=fscore

	UU	UT
S-correct & P-missing	8	29
S-missing & P-correct	16	5

Table 5: Count of anaphors that have different status in different evaluation modes. S = surface coreference evaluation mode, P = protein coreference evaluation mode

cile which was originally developed for newswire domain. It supports the hypothesis that machine learning-based coreference resolution tool trained on different domains can be helpful for the bio medical domain; however, it still requires some adaptations.

7.2 Surface coreference results

Table 4 shows the evaluation results in the surface link mode. The overall performances of all the systems are low, in which recalls are much higher than the precisions. One possible reason of the low results is because most of the teams focus on resolving pronominal coreference; however, they failed to solve some difficult types of pronoun such as "it", "its", "these", "them", and "which", which occupy the majority of anaphoric pronominal expressions (Table 1). Definite anaphoric expressions were ignored by almost all of the systems (except one submission).

The results show that the protein coreference resolution is not a trivial task; and many parts remains challenging. In next section, we analyze about potential reason of the low results, and discuss possible directions for further improvement.

Ex 1	GOLD
T5	DQalpha and DQbeta trans heterodimeric HLA-DQ molecules
T6	such *trans-dimers*
T7	which
R1	T6 T5 [T3, T4]
R2	T7 T6
	RESP
T5	such trans-dimers
T6	which
R1	T6 T5
Ex 2	GOLD
T18	Five *members* of this family (MYC, SCL, TAL-2, LYL-1 and E2A)
T20	their
R3	T20 T18 [T3, T2, T5, T4]
	RESP
T19	Five members
T20	their
R2	T20 T19

Table 6: Example of surface-correct & protein-missing cases. Protein names are underlined, and the min-values are in italic.

8 Analysis

8.1 Why the rankings based on the two evaluation methods are not the same?

Comparing with the protein coreference mode, we can see the rankings based on two evaluation methods are different. In order to find out what led to this interesting difference, we further analyzed the submissions from the two teams UT and UU. The UT team achieved the highest f-score in the surface evaluation mode, but was in the fourth rank in the protein evaluation mode. Meanwhile, the score of UU team was slightly less than the UT team in the former mode, but got the highest in the later (Table 3 and Table 4). In other words, there is no clear correlation between the two evaluation results.

Because the two precisions in surface evaluation mode are not much different, the recalls were the main contribution in the difference of f-score. Analyzing the correct and missing examples in both evaluation modes, we found that there are anaphors whose surface links are correct, while the protein links with the same anaphors are evaluated as missing; and vice versa with missing surface links and correct protein links. Counts of anaphors of each type are shown in Table 5. In this table, the cell at column *UT* and row *S-correct and P-missing* can be interpreted as following. There are 29 anaphors in the UT response whose surface links are correct but protein links are missing, which contributes positively to the recall in *surface coreference mode*, and negatively to that in *protein coreference mode*.

Table 6 shows two examples of *S-correct and P-missing*. In the first example, we can see that the gold antecedent proteins are contained in an indirect antecedent. Therefore, when the intermediate antecedent is correctly detected by the surface link *R1*, but the indirect antecedent is not detected, the anaphor is not linked to it antecedent proteins "DQalpha" and "DQbeta". Another reason is because response antecedents do not include antecedent proteins. This is actually the problem of expression boundary detection. An example of this is example 2 (Table 6), in which the response surface link *R2* is correct, but the protein links to the four proteins are not detected, because the response antecedent "five members" does not include the protein mentions "SCL, TAL-2, LYL-1 and E2A". However, the response antecedent expression is correct because it contains the minimal boundary "members".

For *S-missing and P-correct*, we found that anaphors are normally directly linked to antecedent proteins. In other words, expression boundary is same as protein boundary. Another case is that response antecedents contain the antecedent proteins, but are evaluated as incorrect because the expression boundary of the response expression is larger than the gold expression. An example is shown in Table 7 where the response expression "a second GCR, termed GCRbeta" includes the gold expression "GCRbeta". Therefore, although the surface link is incorrect because the response expression is evaluated as incorrect, the protein coreference link receives a full score .

The difference reflects the characteristics of the two evaluation methods. The analysis result also shows the affect of markable detection or expression detection on the resolution evaluation result.

8.2 Protein coreference analysis

We want to see how well each system performs on each type of anaphor. However, the type information

Ex 3	GOLD
T17	GCRbeta
T18	which
R2	T18 T17 [T4]
	RESP
T16	a second GCR, termed GCRbeta
T19	which
R2	T19 T16

Table 7: Examples of S-missing and P-correct

is not explicitly included in the response, so it has to be induced automatically. We done this by finding the first word of anaphoric expression; then, we combine it with *1* if the expression is a single-word expression, or *2* if the expression is multi-word, to create a sub type value for each anaphor of both gold and response anaphors. After that, subtypes are mapped with the anaphor types specified in Section 4 using the mapping in Table 10.

Protein coreference resolution results by sub type are given in Table 9 and 8. It can be easily seen in Table 9 which team performed well on which type of anaphor. In particular, the CU system was good at resolving the RELAT, APPOS and other types. The UU team performed well on the DNP type. And for the PRON type, UZ was the best team. In theory, knowing this, we can combine strengths of the teams to tackle all the types.

We analyzed false positive protein anaphora links to see what types of anaphora are solved by each system. The recalls in Table 11 are calculated based on the anaphor type information manually annotated in the gold data. Comparing with those in Table 9, there is a small difference due to the automatic induction of anaphoric types based on sub types. It can be seen in the table 11 that only 77.5 percent of RELAT-typed anaphora links were resolved (by CU team), although this type is supposed to be the easiest type. Examining the output data, we found that the system tends to choose the nearest expression as the antecedent of a relative pronoun; however, this is not always correct, as in the following examples from the UofU submission: "We also identified *functional Aiolos-binding sites*$_1a$ in the *Bcl-2 promoter*$_1b$, *which*$_1$ are able to activate the luciferase reporter gene.", and "Furthermore, the analysis of IkappaBalpha turnover demonstrated *an increased*

	PRON both-2	P-it-1	P-its-1	P-one-2	P-that-1	P-their-1	P-these-2	DNP this-2	D-those-1	RELAT which-1	R-whose-1	N/C
UU			36.4		64.4		2	13.3	18.2	62	5	30.8
UZ		46.2	35.7		53.3	7.1		12.5	5.4	59	66.7	15.4
CU					62					70.9	5	42.1
UT		9.5	36.8	10	34.6				9.5	5		30.8
US			13.9			22.9						
UC	28.6	9.1										

Table 8: Fine-grained results (f-score, %)

Team	PRON P	P-R	P-F	DNP P	D-R	D-F	RELAT P	R-R	R-F	Others P	O-R	O-F
UU	79.0	11.5	20.1	66.7	5.9	10.8	71.3	56.0	62.7	100.0	18.3	30.8
UZ	62.9	16.9	26.7	12.5	4.4	6.5	71.4	46.7	56.5	50.0	9.1	15.4
CU	-	-	-	-	-	-	64.6	68.0	66.2	50.0	36.4	42.1
UT	72.7	12.3	21.1	14.3	1.5	2.7	73.3	29.3	41.9	100.0	18.2	30.8
US	27.3	6.9	11.0	-	-	-	-	-	-	-	-	-
UC	9.1	1.5	2.6	-	-	-	-	-	-	-	-	-

Table 9: Protein coreference results by coreference type (fscore, %). P = precision, R = recall, F = f-score. O = Others.

TEAM	A	R	D	P	O
UU	0.0	62.0	5.7	11.1	0.0
UZ	0.0	49.3	4.3	17.0	0.0
CU	0.0	77.5	0.0	0.0	0.0
UT	0.0	32.4	1.4	11.9	14.3
US	0.0	0.0	0.0	6.7	0.0
UC	0.0	0.0	1.4	0.7	0.0

Table 11: Exact recalls by anaphor type, based on manual *type* annotation. A=APPOS, R=RELAT, D=DNP, P=PRON, O=OTHER

degradation of *IkappaBalpha*$_{2a}$ in *HIV-1-infected cells*$_{2b}$ *that*$_2$ may account for the constitutive DNA binding activity.". Expressions with the same index are coreferential expressions. The *a* subscript indicates correct antecedent, and *b* subscript indicates the wrong one. In these examples, the relative pronoun *that* and *which* are incorrectly linked with the nearest expression, which is actually part of post-modifier or the correct antecedent expression.

For the DNP type, recall of the best system is less than 6 percent (Table 11), although it is an important type which occupies almost one fifth of all protein links (Table 1). There is only one team, the UC team, attempted to tackle *the* anaphor; however, it resulted in many spurious links. The other teams did not make any prediction on this type. A possible reason of this is because there are much more non-anaphoric definite noun phrases than anaphoric ones, which making it difficult to train an effective classier for anaphoricity determination. We have to seek for a better method for solving the DNP links, in order to significantly improve protein coreference resolution system.

Concerning the PRON type, Table 8 shows that except for *that-1*, no other figures are higher than 50 percent f-score. This is an interesting observation because pronominal anaphora problem has been reported with much higher results on other domains(Raghunathan et al., 2010), and also on other bio data (hsiang Lin and Liang, 2004). One of the reasons for the low recall is because target anaphoric pronouns in the bio domain are neutral-gender and third-person pronouns(Nguyen and Kim, 2008), which are difficult to resolve than other types of pronouns(Stoyanov et al., 2010a).

8.3 Protein coreference analysis - Intermediate antecedent

As mentioned in the task setting, anaphors can directly link to their antecedent, or indirectly link via one or more intermediate antecedents. We counted the numbers of correct direct and indirect protein coreference links in each submission (Table 12).

Sub type	Type	Count	Sub type	Type	Count	Sub type	Type	Count
both_1	PRON	2	both_2	PRON	4	either_1	PRON	0
it_1	PRON	17	its_1	PRON	61	one_2	PRON	1
such_2	DNP	2	that_1	RELAT	37	the_2	DNP	20
their_1	PRON	27	them_1	PRON	1	these_1	PRON	1
these_2	DNP	26	they_1	PRON	5	this_1	PRON	1
this_2	DNP	20	those_1	PRON	9	which_1	RELAT	37
whose_1	RELAT	1	whose_2	RELAT	0	(others)	N/C	11

Table 10: Mapping from sub type to coreference type. Count = number of anaphors

TEAM	A	R	R	D	D	P	P	O
	Di	Di	In	Di	In	Di	In	Di
UU		44		4		15		
UZ		35		2	1	23		
CU		54	1					
UT		22	1	1		16		1
US						8	1	
UC					1	1		
Total	1	64	7	65	5	126	9	7

Table 12: Numbers of correct protein coreference links by anaphor type and by number of antecedents, based on manual *type* annotation. A=APPOS, R=RELAT, D=DNP, P=PRON, O=Others. Di=direct, In=indirect.

APPOS and Others types do not have any intermediate antecedent, thus there is only one column marked with *D* (direct protein coreference link). We can see in this table that very few indirect links were detected. Therefore, there is place to improve our resolution system by focusing on detection of such links.

8.4 Surface coreference results

Because inclusion of all expressions was not a requirement of shared task submission, the submitted results may not contain expressions that do not involve in any coreference links. Therefore, it is unfair to evaluate expression detection based on the response expressions.

Evaluation results for anaphoricity determination are shown in Table 13. The calculation is performed as following. Supposing that every anaphor has a response link, the number of anaphors is number of distinct anaphoric expressions inferred from the response links, which is given in the first column. The total number of gold anaphors are also calculated in similar way. Since response expressions are lined with gold expressions before evaluation, we provided the alignment rate for reference in the second column of the table. The third and forth columns show the precisions and recalls. In theory, low anaphoricity determination precision results in many spurious response links, while low recall becomes the bottle neck for the overall coreference resolution recall. Therefore, we can conclude that the low performance of anaphoricity determination contribute to the low coreference evaluation results (Table 4, Table 3).

Team	Resp	Align	P	R	F
UU	360	94.2	19.4	33.3	24.6
UZ	736	75.8	22.0	77.1	34.2
CU	365	89.6	15.3	26.7	19.5
UT	452	92.0	18.1	39.0	24.8
US	259	9.3	6.2	7.6	6.8
UC	797	6.8	1.1	4.3	1.8

Table 13: Anaphoricity determination results. Total number of gold anaphors = 210. Resp = number of response anchors, Align = alignment rate(%), P = precision (%), R = recall (%), F = f-score (%)

9 Conclusion

The coreference resolution supporting task of BioNLP Shared Task 2011 has drawn attention from researchers of different interests. Although the overall results are not good enough to be helpful for the main shared tasks as expected, the analysis results in this paper shows the coreference types which have and have not yet been successfully solved. Tackling the remained problems in expression boundary detection, anaphoricity determination and resolution algorithms for difficult types of anaphors such as definite noun phrases should be the future work. Then, it would be interesting to see how much coreference can contribute to event extraction.

References

B. Baldwin. 1997. Cogniac: High precision with limited knowledge and linguistic resources. In *Proceedings of the ACL'97/EACL'97 Workshop on Operational Factors in Practical, Robust Anaphora Resolution*, pages 38–45, Madrid, Spain.

Cosmin Bejan and Sanda Harabagiu. 2010. Unsupervised event coreference resolution with rich linguistic features. In *Proceedings of the 48th Annual Meeting of the Association for Computational Linguistics*, pages 1412–1422, Uppsala, Sweden, July. Association for Computational Linguistics.

Nancy Chinchor. 1998. Overview of MUC-7/MET-2. In *Message Understanding Conference (MUC-7) Proceedings*.

Yu hsiang Lin and Tyne Liang. 2004. Pronominal and sortal anaphora resolution for biomedical literature. In *In Proceedings of ROCLING XVI: Conference on Computational Linguistics and Speech Processing*.

Jin-Dong Kim, Tomoko Ohta, and Jun'ichi Tsujii. 2008. Corpus annotation for mining biomedical events from lterature. *BMC Bioinformatics*, 9(1):10.

Jin-Dong Kim, Tomoko Ohta, Sampo Pyysalo, Yoshinobu Kano, and Jun'ichi Tsujii. 2009. Overview of BioNLP'09 Shared Task on Event Extraction. In *Proceedings of Natural Language Processing in Biomedicine (BioNLP) NAACL 2009 Workshop*, pages 1–9.

Makoto Miwa, Rune Sætre, Jin-Dong Kim, and Jun'ichi Tsujii. 2010. Event extraction with complex event classification using rich features. *Journal of Bioinformatics and Computational Biology (JBCB)*, 8(1):131–146, February.

Vincent Ng. 2010. Supervised noun phrase coreference research: The first fifteen years. In *Proceedings of the ACL*, pages 1396–1411.

Ngan Nguyen and Jin-Dong Kim. 2008. Exploring domain differences for the design of a pronoun resolution system for biomedical texts. In *Proceedings of 22nd International Conference on Computational Linguistics (COLING-2008)*.

T Ohta, Y Tateisi, H Mima, and J Tsujii. 2002. Genia corpus: an annotated research abstract corpus in molecular biology domain. *Proceedings of the Human Language Technology Conference (HLT 2002), San Diego, California*, pages 73–77.

Karthik Raghunathan, Heeyoung Lee, Sudarshan Rangarajan, Nate Chambers, Mihai Surdeanu, Dan Jurafsky, and Christopher Manning. 2010. A multi-pass sieve for coreference resolution. In *Proceedings of the 2010 Conference on Empirical Methods in Natural Language Processing*, pages 492–501, October.

V. Stoyanov, C. Cardie, N. Gilbert, E. Riloff, D. Buttler, and D. Hysom. 2010a. Coreference resolution with reconcile. In *Proceedings of the Conference of the 48th Annual Meeting of the Association for Computational Linguistics (ACL 2010)*.

V. Stoyanov, C. Cardie, N. Gilbert, E. Riloff, D. Buttler, and D. Hysom. 2010b. Reconcile: A coreference resolution platform. In *Tech Report - Cornell University*.

Stephanie Strassel, Mark Przybocki, Kay Peterson, Zhiyi Song, and Kazuaki Maeda. 2008. Linguistic Resources and Evaluation Techniques for Evaluation of Cross-Document Automatic Content Extraction. In *Proceedings of the 6th International Conference on Language Resources and Evaluation (LREC 2008)*.

Jian Su, Xiaofeng Yang, Huaqing Hong, Yuka Tateisi, and Jun'ichi Tsujii. 2008. Coreference Resolution in Biomedical Texts: a Machine Learning Approach. In *Ontologies and Text Mining for Life Sciences'08*.

Yuka Tateisi, Akane Yakushiji, Tomoko Ohta, and Jun'ichi Tsujii. 2005. Syntax annotation for the genia corpus. In *International Joint Conference on Natural Language Processing*, pages 222–227, Jeju Island, Korea, October.

Overview of the Entity Relations (REL) supporting task of BioNLP Shared Task 2011

Sampo Pyysalo* **Tomoko Ohta*** **Jun'ichi Tsujii**†
*Department of Computer Science, University of Tokyo, Tokyo, Japan
†Microsoft Research Asia, Beijing, China
{smp,okap}@is.s.u-tokyo.ac.jp, jtsujii@microsoft.com

Abstract

This paper presents the Entity Relations (REL) task, a supporting task of the BioNLP Shared Task 2011. The task concerns the extraction of two types of part-of relations between a gene/protein and an associated entity. Four teams submitted final results for the REL task, with the highest-performing system achieving 57.7% F-score. While experiments suggest use of the data can help improve event extraction performance, the task data has so far received only limited use in support of event extraction. The REL task continues as an open challenge, with all resources available from the shared task website.

1 Introduction

The BioNLP Shared Task 2011 (BioNLP ST'11) (Kim et al., 2011a), the follow-up event to the BioNLP'09 Shared Task (Kim et al., 2009), was organized from August 2010 (sample data release) to March 2011. The shared task was divided into two stages, with supporting tasks carried out before the main tasks. The motivation for this task setup drew in part from analysis of the results of the previous shared task, which suggested that events that involve coreference or entity relations represent particular challenges for extraction. To help address these challenges and encourage modular extraction approaches, increased sharing of successful solutions, and an efficient division of labor, the two were separated into independent supporting tasks on Coreference (CO) (Nguyen et al., 2011) and Entity Relations in BioNLP ST'11. This paper presents the Entity Relations (REL) supporting task.

2 Task Setting

In the design of the REL task, we followed the general policy of the shared task in assuming named entity recognition (NER) as a given starting point: participants were provided with manually annotated gold standard annotations identifying gene/protein names in all of the training, development, and final test data. By limiting effects due to NER performance, the task remains more specifically focused on the key challenge studied.

Following the results and analysis from previous studies (Pyysalo et al., 2009; Ohta et al., 2010), we chose to limit the task specifically to relations involving a gene/protein named entity (NE) and one other entity. Fixing one entity involved in each relation to an NE helps assure that the relations are "anchored" to real-world entities, and the specific choice of the gene/protein NE class further provides a category with several existing systems and substantial ongoing efforts addressing the identification of those referents through named entity recognition and normalization (Leaman and Gonzalez, 2008; Hakenberg et al., 2008; Krallinger et al., 2008; Morgan et al., 2008; Wermter et al., 2009). The recognition of biologically relevant associations of gene/protein NEs is a key focus of the main event extraction tasks of the shared task. By contrast, in the REL task setting, only one participant in each binary relation is a gene/protein NE, while the other can be either a non-name reference such as *promoter* or the name of an entity not of the gene/protein type (e.g. a complex).[1] Motivated in part by the relatively limited number of existing methods for the detec-

[1]Pronominal references are excluded from annotation scope.

Proceedings of BioNLP Shared Task 2011 Workshop, pages 83–88,
Portland, Oregon, USA, 24 June, 2011. ©2011 Association for Computational Linguistics

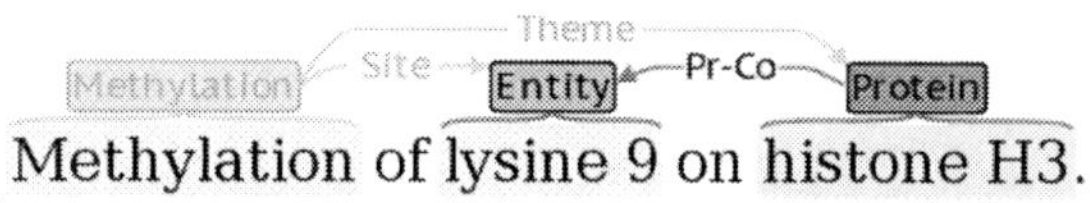

Figure 1: Simple REL annotation example showing a PROTEIN-COMPONENT (PR-CO) relation between "histone H3" and "lysine 9". An associated METHYLATION event and its arguments (shaded, not part of the REL task targets) shown for context.

Item	Training	Devel	Test
Abstract	800	150	260
Word	176,146	33,827	57,256
Protein	9,297	2,080	3,589
Relation	1,857	480	497
PROTEIN-COMPONENT	1,302	314	334
SUBUNIT-COMPLEX	555	166	163

Table 1: REL dataset statistics.

tion of such entity references, their detection is included in the task: participants must recognize these secondary entities in addition to extracting the relations they participate in. To limit the demands of this NER-type task, these entities are not assigned specific types but rather the generic type ENTITY, and exact matching of their boundaries is not required (see Section 4).

The general task setting encompasses a rich set of potential relation extraction targets. For the task, we aimed to select relations that minimize overlap between the targets of other tasks while maintaining relevance as a supporting goal. As the main tasks primarily target events ("things that happen") involving change in entities, we chose to focus in the REL task on what we have previously termed "static relations" (Pyysalo et al., 2009), that is, relations such as part-of that hold between entities without necessary implication of causality or change. A previous study by Van Landeghem et al. (2010) indicated that this class of relations may benefit event extraction. We based our choice of specific target relation on previous studies of entity relations domain texts (Pyysalo et al., 2009; Ohta et al., 2010), which indicated that part-whole relations are by far the most frequent class of relevant relations for the task setting and proposed a classification of these relations for biomedical entities. We further found that – in terms of the taxonomy of Winston et al. (1987) – object-component and collection-member relations account for the the great majority of part-of relations relevant to the domain. For REL, we chose to omit collection-member relations in part to minimize overlap with the targets of the coreference task. Instead, we focused on two specific types of object-component relations, that holding between a gene or protein and its part (domain, regions, promoters, amino acids, etc.) and that between a protein and a complex that it is a subunit of. Following the biological motivation and the general practice in the shared task to term genes and gene products PROTEIN for simplicity, we named these two relations PROTEIN-COMPONENT and SUBUNIT-COMPLEX. Figure 1 shows an illustration of a simple relation with an associated event (not part of REL). Events with *Site* arguments such as that shown in the figure are targeted in the GE, EPI, and ID tasks (Kim et al., 2011b; Ohta et al., 2011; Pyysalo et al., 2011) that REL is intended to support.

3 Data

The task dataset consists of new annotations for the GENIA corpus (Kim et al., 2008), building on the existing biomedical term annotation (Ohta et al., 2002), the gene and gene product name annotation (Ohta et al., 2009) and the syntactic annotation (Tateisi et al., 2005) of the corpus. The general features of the annotation are presented by Pyysalo et al. (2009), describing a previous release of a subset of the data. The REL task annotation effort extended the coverage of the previously released annotation to all relations of the targeted types stated within sentence scope in the GENIA corpus.

For compatibility with the BioNLP ST'09 and its repeat as the GE task in 2011 (Kim et al., 2011b), the REL task training/development/test set division of the GENIA corpus abstracts matches that of the BioNLP ST'09 data. The statistics of the corpus are presented in Table 1. We note that both in terms of training examples and the data available in the given development set, the number of examples of the PROTEIN-COMPONENT relation is more than twice that for SUBUNIT-COMPLEX. Thus, at least for methods based on machine learning, we might generally expect to find higher extraction performance for the former relation.

Rank	Team	Org	NLP Word	NLP Parse	Extraction Entities	Extraction Relations	Other resources Corpora	Other resources Other
1	UTurku	1BI	Porter	McCCJ + SD	SVM	SVM	-	-
2	VIBGhent	1NLP, 1ML, 1BI	Porter	McCCJ + SD	SVM	SVM	GENIA, PubMed	word similarities
3	ConcordU	2NLP	-	McCCJ + SD	Dict	Rules	-	-
3	HCMUS	6L	OpenNLP	OpenNLP	Dict	Rules	-	-

Table 2: Participants and summary of system descriptions. Abbreviations: BI=Bioinformatician, NLP=Natural Language Processing researcher, ML=Machine Learning researcher, L=Linguist, Porter=Porter stemmer, McCCJ=McClosky-Charniak-Johnson parser, SD=Stanford Dependency conversion, Dict=Dictionary

	UTurku	VIBGhent	ConcordU	HCMUS
PROTEIN-COMPONENT	50.90 / 68.57 / **58.43**	47.31 / 36.53 / 41.23	23.35 / 52.05 / 32.24	20.96 / 21.63 / 21.29
SUBUNIT-COMPLEX	48.47 / 66.95 / **56.23**	47.85 / 38.12 / 42.43	26.38 / 39.81 / 31.73	4.91 / 66.67 / 9.14
Total	50.10 / 68.04 / **57.71**	47.48 / 37.04 / 41.62	24.35 / 46.85 / 32.04	15.69 / 23.26 / 18.74

Table 3: Primary evaluation results for the REL task. Results given as recall / precision / F-score.

4 Evaluation

The evaluation of the REL task is relation-based and uses the standard precision/recall/F_1-score metrics. Similarly to the BioNLP'09 ST and most of the 2011 main tasks, the REL task relaxes the equality criteria for matching text-bound annotations: for a submission entity to match an entity in the gold reference annotation, it is sufficient that the span of the submitted entity (i.e. its start and end positions in text) is entirely contained within the span of the gold annotation. This corresponds largely to the *approximate span matching* criterion of the 2009 task (Kim et al., 2009), although the REL criterion is slightly stricter in not involving testing against an extension of the gold entity span. Relation matching is exact: for a submitted relation to match a gold one, both its type and the related entities must match.

5 Results

5.1 Participation

Table 2 summarizes the participating groups and approaches. We find a remarkable number of similarities between the approaches of the systems, with all four utilizing full parsing and a dependency representation of the syntactic analysis, and the three highest-ranking further specifically the phrase structure parser of Charniak and Johnson (2005) with the biomedical domain model of McClosky (2009), converted into Stanford Dependency form using the Stanford tools (de Marneffe et al., 2006). These specific choices may perhaps be influenced by the success of systems building on them in the 2009 shared task (e.g. Björne et al. (2009)). While UTurku (Björne and Salakoski, 2011) and VIBGhent (Van Landeghem et al., 2011) further agree in the choice of Support Vector Machines for the recognition of entities and the extraction of relations, ConcordU (Kilicoglu and Bergler, 2011) and HCMUS (Le Minh et al., 2011) pursue approaches building on dictionary- and rule-based extraction. Only the VIBGhent system makes use of resources external to those provided for the task, extracting specific semantic entity types from the GENIA corpus as well as inducing word similarities from a large unannotated corpus of PubMed abstracts.

5.2 Evaluation results

Table 3 shows the results of the REL task. We find that the four systems diverge substantially in terms of overall performance, with all pairs of systems of neighboring ranks showing differences approaching or exceeding 10% points in F-score. While three of the systems notably favor precision over recall, VIBGhent shows a decided preference for recall, suggesting a different approach from UTurku in design details despite the substantial similarities in overall system architecture. The highest-performing system, UTurku, shows an F-score in the general range of state-of-the-art results in the main event extraction task, which could be taken as an indication that the reliability of REL task analyses created with presently available methods may not be high enough for direct use as a building block for the main tasks. However, the emphasis of the UTurku system on precision is encouraging for such applications: nearly 70% of the entity-relation pairs that the system predicts are correct. The two top-ranking systems show similar precision and recall results for the two relation types. The submission of HCMUS shows a decided advantage for PROTEIN-COMPONENT relation extraction as tentatively predicted from the relative numbers of training examples (Section 3 and Table 1), but their rule-based approach suggests training data size is likely not the decisive factor. While the limited amount of data available prevents strong conclusions from being drawn, overall the lack of correlation between training data size and extraction performance suggests that performance may not be primarily limited by the size of the available training data.

6 Discussion

The REL task was explicitly cast in a support role for the main event extraction tasks, and REL participants were encouraged to make their predictions of the task extraction targets for the various main task datasets available to main task participants. The UTurku team responded to this call for supporting analyses, running their top-ranking REL task system on all main task datasets and making its output available as a supporting resource (Stenetorp et al., 2011). In the main tasks, we are so far aware of one application of this data: the BMI@ASU team (Emadzadeh et al., 2011) applied the UTurku REL predictions as part of their GE task system for resolving the *Site* arguments in events such as BINDING and PHOSPHORYLATION (see Figure 1). While more extensive use of the data would have been desirable, we find this application of the REL analyses very appropriate to our general design for the role of the supporting and main tasks and hope to see other groups pursue similar possibilities in future work.

7 Conclusions

We have presented the preparation, resources, results and analysis of the Entity Relations (REL) task, a supporting task of the BioNLP Shared Task 2011 involving the recognition of two specific types of part-of relations between genes/proteins and associated entities. The task was run in a separate early stage in the overall shared task schedule to allow participants to make use of methods and analyses for the task as part of their main task submissions.

Of four teams submitting finals results, the highest-performing system, UTurku, achieved a precision of 68% at 50% recall (58% F-score), a promising level of performance given the relative novelty of the specific extraction targets and the short development period. Nevertheless, challenges remain for achieving a level of reliability that would allow event extraction systems to confidently build on REL analyses to address the main information extraction tasks. The REL task submissions, representing four independent perspectives into the task, are a valuable resource for further study of both the original task data as well as the relative strengths and weaknesses of the participating systems. In future work, we will analyse this data in detail to better understand the challenges of the task and effective approached for addressing them.

The UTurku team responded to a call for supporting analyses by providing predictions from their REL system for all BioNLP Shared Task main task datasets. These analyses were adopted by at least one main task participant as part of their system, and we expect that this resource will continue to serve to facilitate the study of the position of part-of relations in domain event extraction. The REL task will continue as an open shared challenge, with all task data, evaluation software, and analysis tools available to all interested parties from `http://sites.google.com/site/bionlpst/`.

Acknowledgments

We would like to thank the UTurku team for their generosity with their time and tools in providing REL task analyses for all the BioNLP Shared Task 2011 main task datasets. This work was supported by Grant-in-Aid for Specially Promoted Research (MEXT, Japan).

References

Jari Björne and Tapio Salakoski. 2011. Generalizing biomedical event extraction. In *Proceedings of the BioNLP 2011 Workshop Companion Volume for Shared Task*, Portland, Oregon, June. Association for Computational Linguistics.

Jari Björne, Juho Heimonen, Filip Ginter, Antti Airola, Tapio Pahikkala, and Tapio Salakoski. 2009. Extracting complex biological events with rich graph-based feature sets. In *Proceedings of the BioNLP 2009 Workshop Companion Volume for Shared Task*, pages 10–18, Boulder, Colorado, June. Association for Computational Linguistics.

Eugene Charniak and Mark Johnson. 2005. Coarse-to-Fine n-Best Parsing and MaxEnt Discriminative Reranking. In *Proceedings of the 43rd Annual Meeting of the Association for Computational Linguistics (ACL'05)*, pages 173–180.

Marie-Catherine de Marneffe, Bill MacCartney, and Christopher D. Manning. 2006. Generating Typed Dependency Parses from Phrase Structure Parses. In *Proceedings of the Fifth International Conference on Language Resources and Evaluation (LREC'06)*, pages 449–454.

Ehsan Emadzadeh, Azadeh Nikfarjam, and Graciela Gonzalez. 2011. Double layered learning for biological event extraction from text. In *Proceedings of the BioNLP 2011 Workshop Companion Volume for Shared Task*, Portland, Oregon, June. Association for Computational Linguistics.

J. Hakenberg, C. Plake, R. Leaman, M. Schroeder, and G. Gonzalez. 2008. Inter-species normalization of gene mentions with GNAT. *Bioinformatics*, 24(16):i126.

Halil Kilicoglu and Sabine Bergler. 2011. Adapting a general semantic interpretation approach to biological event extraction. In *Proceedings of the BioNLP 2011 Workshop Companion Volume for Shared Task*, Portland, Oregon, June. Association for Computational Linguistics.

Jin-Dong Kim, Tomoko Ohta, and Jun'ichi Tsujii. 2008. Corpus annotation for mining biomedical events from literature. *BMC Bioinformatics*, 9(10).

Jin-Dong Kim, Tomoko Ohta, Sampo Pyysalo, Yoshinobu Kano, and Jun'ichi Tsujii. 2009. Overview of BioNLP'09 Shared Task on Event Extraction. In *Proceedings of Natural Language Processing in Biomedicine (BioNLP) NAACL 2009 Workshop*, pages 1–9.

Jin-Dong Kim, Sampo Pyysalo, Tomoko Ohta, Robert Bossy, and Jun'ichi Tsujii. 2011a. Overview of BioNLP Shared Task 2011. In *Proceedings of the BioNLP 2011 Workshop Companion Volume for Shared Task*, Portland, Oregon, June. Association for Computational Linguistics.

Jin-Dong Kim, Yue Wang, Toshihisa Takagi, and Akinori Yonezawa. 2011b. Overview of the Genia Event task in BioNLP Shared Task 2011. In *Proceedings of the BioNLP 2011 Workshop Companion Volume for Shared Task*, Portland, Oregon, June. Association for Computational Linguistics.

M. Krallinger, A. Morgan, L. Smith, F. Leitner, L. Tanabe, J. Wilbur, L. Hirschman, and A. Valencia. 2008. Evaluation of text-mining systems for biology: overview of the Second BioCreative community challenge. *Genome biology*, 9(Suppl 2):S1.

Quang Le Minh, Son Nguyen Truong, and Quoc Ho Bao. 2011. A pattern approach for biomedical event annotation. In *Proceedings of the BioNLP 2011 Workshop Companion Volume for Shared Task*, Portland, Oregon, June. Association for Computational Linguistics.

R. Leaman and G. Gonzalez. 2008. Banner: an executable survey of advances in biomedical named entity recognition. *Pacific Symposium on Biocomputing*, pages 652–663.

David McClosky. 2009. *Any Domain Parsing: Automatic Domain Adaptation for Natural Language Parsing*. Ph.D. thesis, Department of Computer Science, Brown University.

A.A. Morgan, Z. Lu, X. Wang, A.M. Cohen, J. Fluck, P. Ruch, A. Divoli, K. Fundel, R. Leaman, J. Hakenberg, et al. 2008. Overview of BioCreative II gene normalization. *Genome biology*, 9(Suppl 2):S3.

Ngan Nguyen, Jin-Dong Kim, and Jun'ichi Tsujii. 2011. Overview of the Protein Coreference task in BioNLP Shared Task 2011. In *Proceedings of the BioNLP 2011 Workshop Companion Volume for Shared Task*, Portland, Oregon, June. Association for Computational Linguistics.

Tomoko Ohta, Yuka Tateisi, Hideki Mima, and Jun'ichi Tsujii. 2002. GENIA corpus: An annotated research abstract corpus in molecular biology domain. In *Proceedings of the Human Language Technology Conference (HLT'02)*, pages 73–77.

Tomoko Ohta, Jin-Dong Kim, Sampo Pyysalo, Yue Wang, and Jun'ichi Tsujii. 2009. Incorporating GENETAG-style annotation to GENIA corpus. In *Proceedings of BioNLP'09*, pages 106–107.

Tomoko Ohta, Sampo Pyysalo, Jin-Dong Kim, and Jun'ichi Tsujii. 2010. A re-evaluation of biomedical named entity-term relations. *Journal of Bioinformatics and Computational Biology (JBCB)*, 8(5):917–928.

Tomoko Ohta, Sampo Pyysalo, and Jun'ichi Tsujii. 2011. Overview of the Epigenetics and Post-translational Modifications (EPI) task of BioNLP Shared Task 2011. In *Proceedings of the BioNLP 2011 Workshop*

Companion Volume for Shared Task, Portland, Oregon, June. Association for Computational Linguistics.

Sampo Pyysalo, Tomoko Ohta, Jin-Dong Kim, and Jun'ichi Tsujii. 2009. Static Relations: a Piece in the Biomedical Information Extraction Puzzle. In *Proceedings of Natural Language Processing in Biomedicine (BioNLP) NAACL 2009 Workshop*, pages 1–9, Boulder, Colorado. Association for Computational Linguistics.

Sampo Pyysalo, Tomoko Ohta, Rafal Rak, Dan Sullivan, Chunhong Mao, Chunxia Wang, Bruno Sobral, Jun'ichi Tsujii, and Sophia Ananiadou. 2011. Overview of the Infectious Diseases (ID) task of BioNLP Shared Task 2011. In *Proceedings of the BioNLP 2011 Workshop Companion Volume for Shared Task*, Portland, Oregon, June. Association for Computational Linguistics.

Pontus Stenetorp, Goran Topić, Sampo Pyysalo, Tomoko Ohta, Jin-Dong Kim, and Jun'ichi Tsujii. 2011. BioNLP Shared Task 2011: Supporting Resources. In *Proceedings of the BioNLP 2011 Workshop Companion Volume for Shared Task*, Portland, Oregon, June. Association for Computational Linguistics.

Yuka Tateisi, Akane Yakushiji, Tomoko Ohta, and Jun'ichi Tsujii. 2005. Syntax annotation for the GENIA corpus. In *Proceedings of IJCNLP'05*, pages 222–227.

Sofie Van Landeghem, Sampo Pyysalo, Tomoko Ohta, and Yves Van de Peer. 2010. Integration of static relations to enhance event extraction from text. In *Proceedings of the 2010 Workshop on Biomedical Natural Language Processing*, pages 144–152.

Sofie Van Landeghem, Thomas Abeel, Bernard De Baets, and Yves Van de Peer. 2011. Detecting entity relations as a supporting task for bio-molecular event extraction. In *Proceedings of the BioNLP 2011 Workshop Companion Volume for Shared Task*, Portland, Oregon, June. Association for Computational Linguistics.

J. Wermter, K. Tomanek, and U. Hahn. 2009. High-performance gene name normalization with GeNo. *Bioinformatics*, 25(6):815.

Morton E. Winston, Roger Chaffin, and Douglas Herrmann. 1987. A taxonomy of part-whole relations. *Cognitive Science*, 11.

The Taming of Reconcile as a Biomedical Coreference Resolver

Youngjun Kim, Ellen Riloff, Nathan Gilbert
School of Computing
University of Utah
Salt Lake City, UT
{youngjun, riloff, ngilbert} @cs.utah.edu

Abstract

To participate in the Protein Coreference section of the BioNLP 2011 Shared Task, we use Reconcile, a coreference resolution engine, by replacing some pre-processing components and adding a new mention detector. We got some improvement from training two separate classifiers for detecting anaphora and antecedent mentions. Our system yielded the highest score in the task, F-score 34.05% in partial mention, protein links, and system recall mode. We witnessed that specialized mention detection is crucial for coreference resolution in the biomedical domain.

1 Introduction

Coreference resolution is a mechanism that groups entity mentions in a text into coreference chains based on whether they refer to the same real-world entity or concept. Like other NLP applications, which must meet the need for aggressive and sophisticated methods of detecting valuable information in emerging domains, numerous coreference resolvers have been developed, including JavaRap (Qiu et al., 2004), GuiTaR (Poesio and Kabadjov, 2004) and BART (Versley et al., 2008). Our research uses a recently released system, Reconcile (Stoyanov et al, 2009; 2010a; 2010b), which was designed as a general architecture for coreference resolution that can be used to easily create learning-based coreference resolvers. Reconcile is based on supervised learning approaches to coreference resolution and has showed relatively good performance compared with similar types of systems.

As a first step to adapting Reconcile for the biomedical domain, specifically the BioNLP Shared Task 2011 (Kim et al., 2011), we modified several subcomponents in Reconcile and revised the feature set for this task. Most importantly, we created a specialized mention detector trained for biomedical text. We trained separate classifiers for detecting anaphor and antecedent mentions, and experimented with several clustering techniques to discover the most suitable algorithm for producing coreference chains in this domain.

2 BioNLP 2011 Shared Task

Our system was developed to participate in a Protein Coreference (COREF) task (Nguyen et al., 2011), one of the supporting tasks in the BioNLP Shared Task 2011. The COREF task is to find all mentions participating in the coreference relation and to connect the anaphora-antecedent pairs. The corpus is based on the Genia-Medco coreference corpus. The Genia-Medco corpus was produced for the biomedical domain, and some comparative analysis with this corpus and other newswire domain data have been performed (Yang et al., 2004a; 2004b; Nguyen and Kim, 2008; Nguyen et al., 2008).

The COREF corpus consists of 800 text files for training, 150 for development, and 260 for testing, which all have gene/protein coreference annotations. The training set has 2,313 pairs of coreference links with 4,367 mentions. 2,117 mentions are antecedents, with an average of 4.21 tokens each (delimited by white space), and 2,301

Proceedings of BioNLP Shared Task 2011 Workshop, pages 89–93,

mentions are anaphora, with an average of 1.28 tokens each. The anaphora are much shorter because many of them are pronouns. The five most frequent anaphora are *that* (686 times), *which* (526), *its* (270), *their* (130), and *it* (100).

3 Our Coreference Resolver

Reconcile was designed to be a research testbed capable of implementing the most current approaches to coreference resolution. Reconcile is written in Java, to be portable across platforms, and was designed to be easily reconfigurable with respect to sub-components, feature sets, parameter settings, etc. A mention detector and an anaphora-antecedent pairs generator are added for the COREF task.

3.1 Preprocessing

For pre-processing, we used the Genia Tagger (Tsuruoka and Tsujii. 2005) for sentence splitting, tokenizing, and part-of-speech (POS) tagging. For parsing, we used the Enju parser (Miyao and Tsujii, 2008).

We replaced Reconcile's mention detection module with new classifiers because of poor performance on the biomedical domain with the provided classifiers. We reformatted the training data with IOB tags and trained a sequential classifier using CRF++ (Kudoh, 2007). For this sequence tagging, we borrowed the features generally used for named entity recognition in the biomedical literature (Finkel et al., 2005; Zhou et al., 2005; McDonald and Pereira, 2005), including word, POS, affix, orthographic features and combinations of these features. We extracted features from the target word, as well as two words to its left and two words to its right. Two versions of mention detectors were developed. The first (MD-I) trained one model without differentiating between anaphora and antecedents. For this method, we chose the longest mentions when multiple mentions overlapped. The other detector (MD-II) used two different models for the antecedent and anaphor, classifying them separately. MD-II's classification result was used when generating the anaphora-antecedent pairs. Table 1 shows the performance of exact matching by these detectors compared with the performance of the Genia Noun Phrase (NP) chunker. Our classifiers did much better, 81.31% precision and 64.78% recall (MD-II), than the Genia chunker, 6.58% precision and 72.67% recall. Only an average of six mentions occurred in each text, while the Genia chunker detected 66.27 noun phrases on average. The Genia annotation scheme was not limited to specific types of concepts, so the Genia NP chunker identifies every possible concept. In contrast, the COREF shared task only involves a subset of the concepts. Mention boundaries were also frequently mismatched. For example, "its" was annotated as a mention in the COREF task when it appears as a possessive inside a noun phrase (e.g., "its activity"), but the Genia NP Chunker tags the entire noun phrase as a mention.

	Prec	Rec	F
Genia NP Chunker	6.58	72.67	12.07
Mention Detector-I	80.85	63.33	71.03
Mention Detector-II	81.31	64.78	**72.11**
Antecedent	65.48	41.35	50.69
Anaphor	91.72	85.07	88.27

Table 1:Mention Detection Results on Dev. Set

3.2 Feature Generation

We used the following four types of features:

Lexical: String-based comparisons of the two mentions, such as exact string matching and head noun matching.

Proximity: Sentence measures of the distance between two mentions.

Grammatical: A wide variety of syntactic properties of the mentions, either individually or in pairs. These features are based on part-of-speech tags, or parse trees.

Semantic: Semantic information about one or both mentions, such as tests for gender and animacy.

Due to the unavailability of paragraph information in our training data, we excluded Reconcile's paragraph features. Also, named entity and dependency parsing features were not used for training. Table 2 shows the complete feature set used for this task. In total, we excluded nine existing Reconcile features, mostly semantic features: WordNetClass, WordNetDist, WordNetSense, Subclass, ParNum, SameParagraph, IAntes, Prednom, WordOverlap. Full descriptions of these features can be found in Stoyanov (2010a).

Lexical	HeadMatch, PNStr, PNSubstr, ProStr, SoonStr, WordsStr, WordsSubstr
Proximity	ConsecutiveSentences, SentNum, SameSentence
Syntactic	Binding, BothEmbedded, BothInQuotes, BothPronouns, BothProperNouns, BothSubjects, ContainsPN, Contraindices, Definite1, Definite2, Demonstrative2, Embedded1, Embedded2, Indefinite, Indefinite1, InQuote1, InQuote2, MaximalNP, Modifier, PairType, Pronoun, Pronoun1, Pronoun2, ProperNoun, ProResolve, RuleResolve, Span, Subject1, Subject2, Syntax
Semantic	Agreement, Alias, AlwaysCompatible, Animacy, Appositive, ClosestComp, Constraints, Gender, instClass, Number, ProComp, ProperName, Quantity, WNSynonyms

Table 2: Feature Set for Coreference Resolution

3.3 Clustering

After Reconcile makes pairwise decisions linking each anaphor and antecedent, it produces a clustering of the mentions in a document to create coreference chains. Because the format of the COREF task submission was not chains but anaphora-antecedent pairs, it would have been possible to submit the direct results of Reconcile's pairwise decisions. However, it was easier to use Reconcile as a black-box and post-process the chains to reverse-engineer coreferent pairs from them. Reconcile supports three clustering algorithms:

Single-link Clustering (SL) (Transitive Closure) groups together all mentions that are connected by a path of coreferent links.

Best-first (BF) clustering uses the classifier's confidence value to cluster each noun phrase with its most confident antecedent.

Most Recent First (MRF) pairs each noun phrase with the single most recent antecedent that is labeled as coreferent.

Table 3 shows the MUC scores of each clustering method with gold standard mentions and with the mentions automatically detected by each of our two mention detectors. Not surprisingly, using gold mentions produced the highest score of 87.32%. Automatically detected mentions yielded much lower performance. MD-I performed best, in this evaluation, achieving 49.65%. The *most recent first* clustering algorithm produced the best results for both gold mentions and MD-I. The *single link* clustering algorithm, which is the default method used by Reconcile, produced the lowest results for both gold mentions and MD-I.

	SL	BF	MRF
Gold Mention	85.34	86.87	**87.32**
Mention Detector-I	48.64	48.82	**49.65**
Mention Detector-II	48.31	**48.62**	48.07

Table 3: MUC Scores of Dev. Set by Three Different Clustering Methods (SL: *Single-link*, BF: *Best-first*, MRF: *Most recent first*)

3.4 Pair Generation from Chains

Reconcile generates coreference chains, but the output for the shared task required anaphora-antecedent pairs. Therefore, we needed to extract individual pairs from the chains. We used the chains produced by the *most recent first* clustering algorithm for pair generation. When using MD-I output, we took the earliest mention (i.e., the one occurring first in the source document) in the chain and paired it with each of the subsequent mentions in the same chain. Thus, each chain of size N produced N-1 pairs. When using the MD-II predictions, the classifiers gave us two separate lists of antecedent and anaphora mentions. In this case, we paired each anaphor in the chain with every antecedent in the same chain that preceded it in the source document.

3.5 Evaluation and Analysis

The mention linking can be evaluated using three different scores: *atom* coreference links, *protein* coreference links, and *surface* coreference links. In the *atom* link option, only links containing given gene/protein annotations are considered while in the *surface* link option, every link is a target for the evaluation. *Protein* links are similar to *atom* links but loosen the boundary of gene/protein annotations. There were 202 protein links out of 469 surface links in development set.

For mention detection, *exact* match and *partial* match are supported in the task evaluation. Recall is measured in two modes. In *system* mode, every link is calculated for the linking evaluation. In *algorithm* mode, only links with correctly detected mentions are considered for evaluation. For detailed information refer to Nguyen et al. (2011) or the task web site.[1] Table 4 shows the mention linking results (F-score) for the COREF task evaluation using *partial* match and *system* recall. The *surface* link score on gold mentions reached 90.06%. For automatic mention detection, MD-I achieved a score of 45.38% score, but MD-II produced a substantially better score of 50.41%. MD-II, which was trained separately for antecedent and anaphora detection, performed about 5% higher than MD-I in every link mode.

	Atom	Protein	Surface
Gold Mention	84.09	**84.09**	90.06
Mention Detector-I	28.67	**34.41**	45.38
Mention Detector-II	33.45	**39.27**	50.41

Table 4: Dev. Set Results by Three Different Evaluation Options

Table 5 shows the recall and precision breakdown for the *protein* evaluation results. Looking behind the composite F-score reveals that our system produced higher precision than recall. Looking back at Table 1, we saw that our anaphor detector performed much better than our antecedent detector. Since every coreference link requires one of each, the relatively poor performance of antecedent detection (especially in terms of recall) is a substantial bottleneck.

	Prec	Rec	F
Gold Mention	98.67	73.27	**84.09**
Mention Detector-I	62.34	23.76	**34.41**
Mention Detector-II	73.97	26.73	**39.27**

Table 5: Precision and Recall Breakdown for *Protein* Evaluation Coreference Links

3.6 Results: Submission for COREF Task

We merged the training and development sets to use as training data for Reconcile. We used MD-II for mention detection and the *most recent first* algorithm for clustering to submit the final output on the test data. Table 6 shows the results of our final submission along with the five other participating teams for the *protein* evaluation coreference links (Nguyen et al., 2011). Our system produced a 34.05% F-score (73.26% precision and 22.18% recall) in *protein* coreference links and 25.41% F-score in *atom* links.

Team	Prec	Rec	F
University of Utah	73.26	22.18	**34.05**
University of Zurich	55.45	21.48	**30.96**
Concordia University	63.22	19.37	**29.65**
University of Turku	67.21	14.44	**23.77**
University of Szeged	3.47	3.17	**3.31**
University College Dublin	0.25	0.70	**0.37**

Table 6: Evaluation Results of Final Submissions (*Protein* Coreference Links)

4 Conclusions

The effort to tame Reconcile as a coreference engine for the biomedical domain was successful and our team's submission obtained satisfactory results. However, there is ample room for improvement in coreference resolution. We observed that mention detection is crucial - the MUC score reached 87.32% with gold mentions on the development set but only 49.65% with automatically detected mentions (Table 3). One possible avenue for future work is to develop domain-specific features to better identify mentions in biomedical domains.

Acknowledgments

We thank the BioNLP Shared Task 2011 organizers for their efforts, and gratefully acknowledge the support of the National Science Foundation under grants IIS-1018314 and DBI-0849977 and the Defense Advanced Research Projects Agency (DARPA) Machine Reading Program under Air Force Research Laboratory (AFRL) prime contract no. FA8750-09-C-0172. Any opinions, findings, and conclusion or recommendations expressed in this material are those of the authors and do not necessarily express the view of the DARPA, AFRL, NSF, or the U.S. government.

References

Jenny Finkel, Shipra Dingare, Christopher D Manning, Malvina Nissim, Beatrice Alex, and Claire Grover. 2005. Exploring the Boundaries: Gene and Protein Identification in Biomedical Text. BMC Bioinformatics. 6:S5.

[1] http://sites.google.com/site/bionlpst/home/protein-gene-coreference-task

Jin-Dong Kim, Sampo Pyysalo, Tomoko Ohta, Robert Bossy, and Jun'ichi Tsujii. 2011. Overview of BioNLP Shared Task 2011. Proceedings of the BioNLP 2011 Workshop Companion Volume for Shared Task, Portland, Oregon, June. ACL 2011.

Taku Kudoh. 2007. CRF++. http://crfpp.sourceforge.net/.

Ryan McDonald and Fernando Pereira. 2005. Identifying Gene and Protein Mentions in Text Using Conditional Random Fields. BMC Bioinformatics. 6:S6.

Yusuke Miyao and Jun'ichi Tsujii. 2008. Feature Forest Models for Probabilistic HPSG Parsing. Computational Linguistics, 34(1):35–80.

Ngan L. T. Nguyen and Jin-Dong Kim. 2008. Exploring Domain Differences for the Design of Pronoun Resolution Systems for Biomedical Text. Proceedings of COLING 2008:625-632

Ngan L. T. Nguyen, Jin-Dong Kim, and Jun'ichi Tsujii. 2008. Challenges in Pronoun Resolution System for Biomedical Text. Proceedings of LREC 2008.

Ngan L. T. Nguyen, Jin-Dong Kim, and Jun'ichi Tsujii. 2011. Overview of the Protein Coreference task in BioNLP Shared Task 2011. Proceedings of the BioNLP 2011 Workshop Companion Volume for Shared Task, Portland, Oregon, June. ACL 2011.

Massimo Poesio and Mijail A. Kabadjov. 2004. A general-purpose, off-the-shelf anaphora resolution module: implementation and preliminary evaluation. Proceedings of LREC 2004.

Long Qiu, Min-Yen Kan, and Tat-Seng Chua. 2004. A Public Reference Implementation of the Rap Anaphora Resolution Algorithm. Proceedings of LREC 2004.

Veselin Stoyanov, Claire Cardie, Nathan Gilbert, Ellen Riloff, David Buttler, and David Hysom. 2010a. Reconcile: A Coreference Resolution Platform. Tech Report. Cornell University.

Veselin Stoyanov, Claire Cardie, Nathan Gilbert, Ellen Riloff, David Buttler, and David Hysom. 2010b. Coreference Resolution with Reconcile. Proceedings of ACL 2010.

Veselin Stoyanov, Nathan Gilbert, Claire Cardie, and Ellen Riloff. 2009. Conundrums in Noun Phrase Coreference Resolution: Making Sense of the State-of-the-Art. Proceedings of ACL-IJCNLP 2009.

Yoshimasa Tsuruoka and Jun'ichi Tsujii. 2005. Bidirectional Inference with the Easiest-First Strategy for Tagging Sequence Data. Proceedings of HLT/EMNLP 2005:467-474.

Yannick Versley, Simone P. Ponzetto, Massimo Poesio, Vladimir Eidelman, Alan Jern, Jason Smith, Xiaofeng Yang, and Alessandro Moschitti. 2008. BART: A modular toolkit for coreference resolution. Proceedings of LREC 2008.

Xiaofeng Yang, Jian Su, Guodong Zhou, and Chew Lim Tan. 2004a. A NP-Cluster Based Approach to Coreference Resolution. Proceedings of COLING 2004:226-232.

XiaoFeng Yang, GuoDong Zhou, Jian Su, and Chew Lim Tan. 2004b. Improving Noun Phrase Coreference Resolution by Matching Strings. Proceedings of IJCNLP 2004:226-333.

GuoDong Zhou, Dan Shen, Jie Zhang, Jian Su, and SoonHeng Tan. 2005. Recognition of Protein/Gene Names from Text Using an Ensemble of Classifiers. BMC Bioinformatics. 6:S7.

Extracting Bacteria Biotopes with Semi-supervised Named Entity Recognition and Coreference Resolution

Nhung T. H. Nguyen and **Yoshimasa Tsuruoka**
School of Information Science
Japan Advanced Institute of Science and Technology
1-1 Asahidai, Nomi, Ishikawa 923-1292 Japan
{nthnhung,tsuruoka}@jaist.ac.jp

Abstract

This paper describes our event extraction system that participated in the bacteria biotopes task in BioNLP Shared Task 2011. The system performs semi-supervised named entity recognition by leveraging additional information derived from external resources including a large amount of raw text. We also perform coreference resolution to deal with events having a large textual scope, which may span over several sentences (or even paragraphs). To create the training data for coreference resolution, we have manually annotated the corpus with coreference links. The overall F-score of event extraction was 33.2 at the official evaluation of the shared task, but it has been improved to 33.8 thanks to the refinement made after the submission deadline.

1 Introduction

In this paper, we present a machine learning-based approach for bacteria biotopes extraction of the BioNLP Shared Task 2011 (Bossy et al. , 2011). The task consists of extracting bacteria localization events, namely, mentions of given species and the place where it lives. Places related to bacteria localization events range from plant or animal hosts for pathogenic or symbiotic bacteria to natural environments like soil or water[1]. This task also targets specific environments of interest such as medical environments (hospitals, surgery devices, etc.), processed food (dairy) and geographical localizations.

The task of extracting bacteria biotopes involves two steps: Named Entity Recognition (NER) and event detection. The current dominant approach to NER problems is to use supervised machine learning models such as Maximum Entropy Markov Models (MEMMs), Support Vector Machines (SVMs) and Conditional Random Fields (CRFs). These models have been shown to work reasonably well when a large amount of training data is available (Nadeau and Sekine, 2007). However, because the annotated corpus delivered for this particular subtask in the shared task is very small (78 documents with 1754 sentences), we have decided to use a semi-supervised learning method in our system. Our NER module uses a CRF model with enhanced features created from external resources. More specifically, we use additional features created from the output of HMM clustering performed on a large amount of raw text, and word senses from WordNet for tagging.

The target events in this shared task are divided into two types. The first is Localization events which relates a bacterium to the place where it lives. The second is PartOf events which denotes an organ that belongs to an organism. As in Bossy et al. (2010), the largest possible scope of the mention of a relation is the whole document, and thus it may span over several sentences (or even paragraphs). This observation motivated us to perform coreference resolution as a pre-processing step, so that each event can be recognized within a narrower textual scope. There are two common approaches to coreference resolution: one mainly relies on heuristics, and the other employs machine learning. Some

[1]https://sites.google.com/site/bionlpst/home/bacteria-biotopes

Proceedings of BioNLP Shared Task 2011 Workshop, pages 94–101,
Portland, Oregon, USA, 24 June, 2011.

instances of the heuristics-based approach are described in (Harabagiu et al., 2001; Markert and Nissim, 2005; Yang and Su, 2007), where they use lexical and encyclopedic knowledge. Machine learning-based methods (Soon and Ng, 2001; Ng and Cardie, 2002; Yang et al. , 2003; Luo et al. , 2004; Daume and Marcu, 2005) train a classifier or search model using a corpus annotated with anaphoric pairs. In our system, we employ the simple supervised method presented in Soon and Ng (2001). To create the training data, we have manually annotated the corpus with coreference information about bacteria.

Our approach, consequently, has three processes: NER, coreference resolution of bacterium entities, and event extraction. The latter two processes can be formulated as classification problems. Coreference resolution is to determine the relation between candidate noun phrases and bacterium entities, and the event extraction is to detect the relation between two entities. It should be noted that our official submission in the shared task was carried out without using a coreference resolution module, and the system has been improved after the submission deadline.

Our contribution in this paper is two-fold. In the methodology aspect, we use an unsupervised learning method to create additional features for the CRF model and perform coreference resolution to narrow the scope of events. In the resource aspect, the manual annotations for training our coreference resolution module will be made available to the research community.

The remainder of this paper is organized as followed. Section 2, 3 and 4 describe details about the implementation of our system. Section 5 presents the experimental results with some error analysis. Finally, we conclude our approach and discuss future work in section 6.

2 Semi-supervised NER

According to the task description, the NER task consists of detecting the phrases that denote bacterial taxon names and localizations which are broken into eight types: Host, HostPart, Geographical, Food, Water, Soil, Medical and Environment. In this work, we use a CRF model to perform NER. CFRs (Lafferty et. al., 2001) are a sequence modeling framework that not only has all the advantages of MEMMs but also solves the label bias problem in a principled way. This model is suitable for labeling sequence data, especially for NER. Based on this model, our CRF tagger is trained with a stochastic gradient descent-based method described in Tsuruoka et al. (2009), which can produce a compact and accurate model.

Due to the small size of the training corpus and the complexity of their category, the entities cannot be easily recognized by standard supervised learning. Therefore, we enhance our learning model by incorporating related information from other external resources. On top of the lexical and syntactic features, we use two additional types of information, which are expected to alleviate the data sparseness problem. In summary, we use four types of features including lexical and syntactic features, word cluster and word sense features as the input for the CRF model.

2.1 Word cluster features

The idea of enhancing a supervised learning model with word cluster information is not new. Kamaza et. al. (2001) use a hidden Markov model (HMM) to produce word cluster features for their maximum entropy model for part-of-speech tagging. Koo et al. (2008) implement the Brown clustering algorithm to produce additional features for their dependency parser. For our NER task, we use an HMM to produce word cluster features for our CRF model.

We employed an open source library[2] for learning HMMs with the online Expectation Maximization (EM) algorithm proposed by Liang and Klein (2009). The online EM algorithm is much more efficient than the standard batch EM algorithm and allows us to use a large amount of data. For each hidden state, words that are produced by this state with the highest probability are written. We use this result of word clustering as a feature for NER. The optimal number of hidden states is selected by evaluating its effectiveness on NER using the development set.

To prepare the raw text for HMM clustering, we downloaded 686 documents (consisting of both full documents and abstracts) about bacteria biotopes

[2] http://www-tsujii.is.s.u-tokyo.ac.jp/~hillbig/ohmm.htm

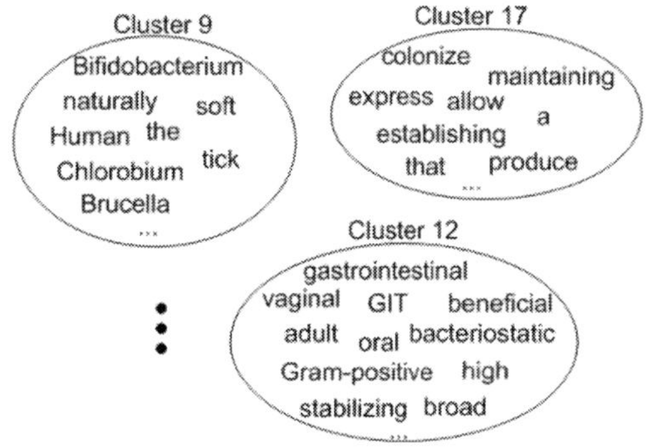

Figure 1: Sample of HMM clustering result.

Word	POS	Sense
chromosome	NN	body
colonize	VBP	social
detected	VBN	perception
fly	NN	animal
gastrointestinal	JJ	pert
infant	NN	person
longum	FW	bacterium
maintaining	VBG	stative
milk	NN	food
onion	NN	plant
proterins	NNS	substance
USA	NNP	location

Table 1: Sample of word sense features given by WordNet and NCBI dictionary.

from MicrobeWiki, JGI Genome Portal, Genoscope, 2Can bacteria pages at EBI and NCBI Genome Project (the training corpus is also downloaded from these five webpages). In addition, we use the 100,000 latest MEDLINE abstracts containing the string "bacteri" in our clustering. In total, the raw text consists of more than 100,000 documents with more than 2 million sentences.

A part of the result of HMM clustering is shown in Figure 1. According to this result, the word "*Bifidobacterium*" belongs to cluster number 9, and its feature value is "*Cluster-9*". The word cluster features of the other words are extracted in the same way.

2.2 Word sense features

We used WordNet to produce additional features on *word senses*. Although WordNet[3] is a large lexical database, it only comprises words in the general genre, to which only the localization entities belong. Since it does not contain the bacterial taxon names, the most important entities in this task, we used another dictionary for bacteria names. The dictionary was extracted from the genomic BLAST page of NCBI [4]. To connect these two resources, we simply place all entries from the NCBI dictionary under the 'bacterium' sense of WordNet. Table 1 illustrates some word sense features employed in our model.

2.3 Pre-processing for bacteria names

In biomedical documents, the bacteria taxon names are written in many forms. For example, they are presented in a full name like "Bacillius cereus", or in a short form such as "B. cereus", or even in an abbreviation as "GSB" (green sulfur bacteria). Moreover, the bacteria names are often modified with some common strings such as "strain", "spp.", "sp.", etc. "Borrelia hermsii ***strain*** DAH", "Bradyrhizobium ***sp.*** BTAi1", and "Spirochaeta ***spp.***" are examples of this kind. In order to tackle this problem, we apply a pre-processing step before NER. Although there are many previous studies solving this kind of problem, in our system, we apply a simple method for this step.

- *Retrieving the full form of bacteria names.* We assume that (a) both short form and full form must occur in the same document; (b) a token is considered as an abbreviation if it is written in upper case and its length is shorter than 4 characters. When a token satisfies condition (b) (which means it is an abbreviation), the processing retrieves its full form by identifying all sequences containing tokens initialized by its abbreviated character. In case of short form like "B. cereus", the selected sequence must include the right token (which is "cereus" in "B. cereus").

- *Making some common strings transparent.* As our observation on the training data, there are 8 common strings in bacteria names, including "strain", "str", "str.", "subsp", "spp.", "spp", "sp.", "sp". All of these strings will be removed before NER and recovered after that.

[3] http://wordnet.princeton.edu/

[4] http://www.ncbi.nlm.nih.gov/sutils/genom_table.cgi

3 Coreference Resolution as Binary Classification

Coreference resolution is the process of determining whether different nominal phrases are used to refer to the same real world entity or concept. Our approach basically follows the learning method described in Soon and Ng (2001). In this approach, we build a binary classifier using the coreferring entities in the training corpus. The classifier takes a pair of candidates and returns *true* if they refer to the same real world entity and *false* otherwise. In this paper, we limit our module to detecting the bacteria's coreference, and hence the candidates consist of noun phrases (NPs) (starting by a determiner), pronouns, possessive adjective and name of bacteria.

In addition to producing the candidates, the preprocessing step creates a set of features for each anaphoric pair. These features are used by the classifier to determine if two candidates have a coreference relation or not.

The following features are extracted from each candidate pair.

- *Pronoun*: 1 if one of the candidates is a pronoun; 0 otherwise.
- *Exact or Partial Match*: 1 if the two strings of the candidates are identical, 2 if they are partial matching; 0 otherwise.
- *Definite Noun Phrase*: 1 if one of the candidates is a definite noun phrases; 0 otherwise.
- *Demonstrative Noun Phrase*: 1 if one of the candidates is a demonstrative noun phrase; 0 otherwise.
- *Number Agreement*: 1 if both candidates are singular or plural; 0 otherwise.
- *Proper Name*: 1 if both candidates are bacterium entities or proper names; 0 otherwise.
- *Character Distance*: count the number of the characters between two candidates.
- *Possessive Adjective*: 1 if one of the candidates is possessive adjective; 0 otherwise.

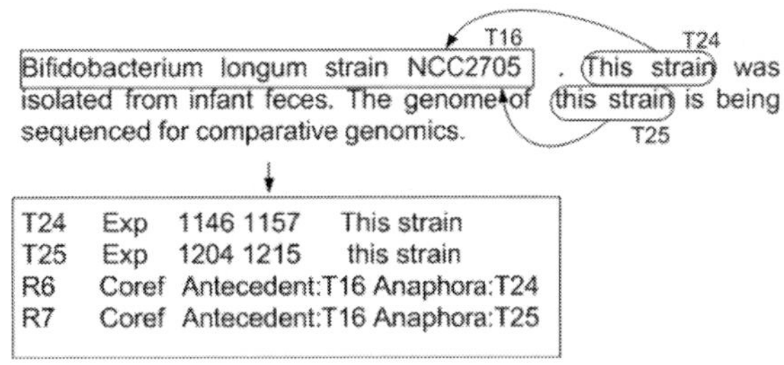

Figure 2: Example of annotating coreference resolution. T16 is a bacterium which is delivered in *.a2 file, T24 and T25 are anaphoric expressions. There are two coreference relations of T16 and T24, T16 and T25.

- *Exist in Coreference Dictionary*: 1 if the candidate exists in the dictionary extracted from the training data; 0 otherwise. This feature aims to remove noun phrases which are unlikely to be related to the bacterium entities.

The first five features are exactly the same as those in Soon and Ng (2001), while the others are refined or added to make it suitable for our specific task.

In the testing phase, we used the ***best-first clustering*** as in Ng and Cardie (2002). Rather than performing a right-to-left search from each anaphoric NP for the first coreferent NP, a right-to-left search for a *highly likely antecedent* was performed. Hence, the classifier was modified to select the antecedent of NP with the coreference likelihood score above a threshold. This threshold was tuned by evaluating it on the development set.

3.1 Corpus annotation

To create the training data for coreference resolution, we have manually annotated the corpus based on the gold-standard named entity annotations delivered by the organizer. Due to our decision to focus on bacteria names, only the coreference of these entities are labeled. We use a format similar to those of the organizer, i.e. the standoff presentation and text-bound annotations. The coreference annotation file consists of two parts, one part for anaphoric expressions and the other for coreference relation. Figure 2 shows an example of a coreference annotation with the original text.

4 Event Extraction

The bacteria biotopes, as mentioned earlier, are divided into two types. The first type of events, namely localization events, relates a bacterium to the place where it lives, and has two mandatory arguments: a Bacterium type and a localization type. The second type of events, i.e. PartOf events, denote an organ that belongs to an organism, and has two mandatory arguments of type HostPart and Host respectively. We view this step as determining the relationship between two specific entities. Because of no ambiguity between the two types of event, the event extraction can be solved as the binary classification of pairs of entities. The classifier is trained on the training data with four types of feature extracted from the context between two entities: distance in sentences, the number of entities, the nearest left and right verbs.

Generating Training Examples. Given the coreference information on bacterium entities, the system considers all the entities belonging to the coreference chains as real bacteria and generates event instances. Since about 96% of all annotated events occur in the same paragraph, we restrict our method to detecting events within one paragraph.

- ***Localization Event.*** The system creates a relationship between a bacterium and a localization entity with ***minimum distance*** between them by the following priorities:

  (1) The bacterium *precedes* the localization entity *in the same sentence.*

  (2) The bacterium *precedes* the localization entity *in the same paragraph.*

- ***PartOf Event.*** All possible relationships between Host and HostPart entities are generated if they are in the same paragraph.

5 Experiments and Discussion

The training and evaluation data used in these experiments are provided by the shared task organizers. The token and syntactic information are extracted from the supporting resources (Stenetorp et. al. , 2011). More detail, the tokenized text was done by GENIA tools, and the syntactic analyses was created by the McClosky-Charinak parser (McClosky and Charniak, 2008), trained on the GENIA Treebank corpus (Tateisi et al., 2005), which is one of the most accurate parsers for biomedical documents.

Experiment	Acc.	Pre.	Re.	F-score
Baseline	94.28	76.32	35.51	48.47
Word cluster	94.46	78.23	39.59	52.57
Word sense	94.63	74.15	44.49	55.61
All Features	**94.70**	**77.62**	**45.31**	**57.22**

Table 2: Performance of Named Entity Recognition in terms of Accuracy, Precision, Recall and F-score with different features on the development set.

For both classification of anaphoric pairs in coreference resolution and determining relationship of two entites, we used the SVM^{light} library [5], a state-of-the-art classifier, with the linear kernel.

In order to find the best parameters and features for our final system, we conducted a series of experiments at each step of the approach.

5.1 Named Entity Recognition

We evaluated the impact of additional featues on NER by running four experiments. The Baseline experiment was conducted by using the original CRF tagger, which did not use any additional features derived from external resources. The other three experiments were conducted by incrementally adding more features to the CRF tagger. Table 2 shows the results on the development set[6].

Through these experiments we have realized that using the external resources is very effective. The *word cluster* and *word sense* features are used like a dictionary. The first one can be considered as the dictionary of specific classes of entity in the same domain with this task, which mainly supports the precision, whereas the latter is a general dictionary boosting the recall. With regard to F-score, the *word sense* features outperform the *word cluster* features. When we combine all of them, the F-score is improved significantly by nearly 9 points.

The detailed results of individual classes in Table 3 show that the Environment entities are the hardest to recognize. Because of their general characteristic, these entities are often confused with Host

[5] `http://svmlight.joachims.org/`

[6] These scores were generated by using the CoNLL 2000 evaluation script.

Class	Gold	Pre.	Re.	F-score
Bacterium	86	70.00	40.23	51.09
Host	78	78.57	56.41	65.67
HostPart	44	91.67	50.00	64.71
Geographical	8	71.43	62.50	66.67
Environment	8	0.00	0.00	0.00
Food	0	N/A	N/A	N/A
Medical	2	100.00	50.00	66.67
Water	17	100.00	17.65	30.00
Soil	1	100.00	100.00	100.00
All	244	**77.62**	**45.31**	**57.22**

Table 3: Results of NER using all features on the development set. The "Gold" column shows the number of entities of that class in the gold-standard corpus. The score of Food entities is not available because there is no positive instance in the development set.

	Detection	Linking
Precision	24.18	20.48
Recall	91.36	33.71
F-score	38.24	25.48

Table 4: Result of coreference resolution on the development set achieved with gold-standard named entity annotations.

or Water. In contrast, the Geographical category is easier than the others if we have gazetteers and administrative name lists.

5.2 Coreference Resolution

We next evaluated the accuracy of coreference resolution for bacterium entities. The evaluation[7] is carried out in two steps: evaluation of mention detection, and evaluation of mention linking to produce coreference links. The exact matching criterion was used when evaluating the accuracy of the two steps. Table 4 shows the performance of the coreference resolution module when taking annotated entites as input. As mentioned in section 3, the first step of this module considers all NPs beginning with a determiner and bacterium entities as candidates. Therefore, the number of the candidate NPs is vastly larger than that of the positive ones. This is the reason why the precision of mention detection is low, while the recall is high. This high recall leads to a large number of generated linkings and raises the complexity of linking detection. In order to obtain more accurate results, we had to remove weak linkings whose classification score is under 0.7 (this is the best threshold on the development set). However, as shown in Table 4, the performance of mention linking was not satisfactory.

[7] http://sites.google.com/site/bionlpst/home/protein-gene-coreference-task

Experiment	Pre.	Re.	F-score
No Coref.	42.11	27.34	33.15
With Coref.	43.40	27.64	33.77

Table 5: Comparative results of event extraction with and without coreference information on the test set.

Type of event	Num. of addition		Num. of ruled out	
	True	False	True	False
Localization	17	1	6	20
PartOf	6	5	1	0
Total	29		27	

Table 6: Contribution of coreference resolution to event extraction.

5.3 Event Extraction

Finally, we carried out two experiments on the test set to investigate the effect of coreference resolution on event extraction. The results shown in Table 5 indicate that the contribution of coreference resolution in this particular experiment is not significant. The coreference information helps the module to add 29 more events (23 true and 6 false events) and rule out 27 events (20 false and 7 true events) compared with the experiment with no coreference resolution. Detail about this contribution is presented in Table 6.

We further analyzed the result of event extraction and found that there exist two kinds of Localization events, which we call *direct* and *indirect* events. The *direct* events are the ones that are easily recognizable on the surface level of textual expressions. The three Localization events in Figure 3 belong to this type. Our module is able to detect most of the direct events, especially when we have the coreference information on bacteria – it is straight-forward because the two arguments of the event occur in the same sentence. In constrast, the *indirect* events are more complicated. They appear implicitly in the document and we need to infer them through an intermediate agent. For example, a bacterium causes a disease, and this disease infects the humans or an-

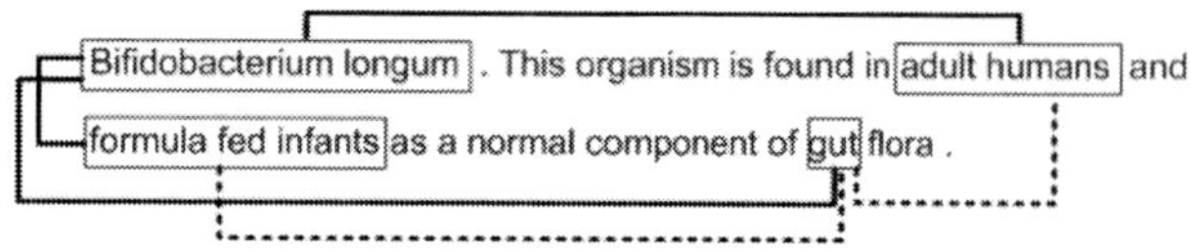

Figure 3: Example of direct events. The solid line is the Localization event, the dash line is the PartOf event.

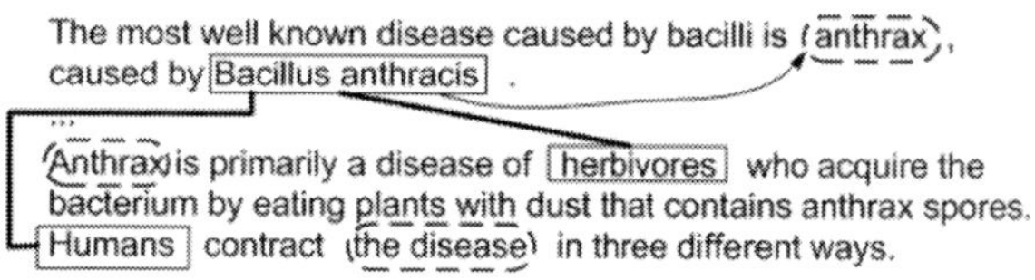

Figure 4: Example of indirect events. The solid line is the Localization event, the arrow shows the causative relation.

imals. Therefore, it can be considered that the bacterium locates in the humans or animals. Figure 4 illustrates this case. In this example, the Bacillus anthracis causes Anthrax, Humans contract the disease (which refers to Anthrax), and the Bacillus anthracis locates in Humans. These events are very difficult to recognize since, in this context, we do not have any information about the disease. Events of this type provide an interesting challenge for bacteria biotopes extraction.

6 Conclusion and Future Work

We have presented our machine learning-based approach for extracting bacteria biotopes. The system is implemented with modules for three tasks: NER, coreference resolution and event extraction.

For NER, we used a CRF tagger with four types of features: lexical and syntactic features, the word cluster and word sense extracted from the external resources. Although we achieved a significant improvement by employing WordNet and the HMM clustering on raw text, there is still much room for improvement. For example, because all extracted knowledge used in this NER module belongs to the general knowlegde, its performance is not as good as our expectation. We envisage that the performance of the module will be improved if we can find useful biological features.

We have attempted to use the information obtained from the coreference resolution of bacteria to narrow the event's scope. On the test set, although it does not improve the system significantly, the coreference information has shown to be useful in event extraction. [8]

In this work, we simply used binary classifiers with standard features for both coreference resolution and event detection. More advanced machine learning approaches for structured prediction may lead to better performance, but we leave it for future work.

References

Robert Bossy, Claire Nedellec, and Julien Jourde. 2010. Guidelines for Annotation of Bacteria Biotopes.

Robert Bossy, Julien Jourde, Philippe Bessières, Marteen van de Guchte, and Claire Nédellec. 2011. BioNLP Shared Task 2011 - Bacteria Biotope, In *Proceedings of the BioNLP 2011 Workshop Companion Volume for Shared Task*. Portland, Oregon, Association for Computational Linguistics.

Hal Daumé III and Daniel Marcu. 2005. A Large-scale Exploration of Effective Global Features for a Joint Entity Detection and Tracking Model. In *Proceedings of HLT-EMNLP 2005*, pp. 97-104.

Sanda M. Harabagiu, Razvan C. Bunescu and Steven J. Maiorano. 2001. Text and Knowlegde Mining for Coreference Resolution. In *Proceedings of NAACL 2001*, pp. 1-8.

Jun'ichi Kazama, Yusuke Miyao, and Jun'ichi Tsujii. 2001. A Maximum Entropy Tagger with Unsupervised Hidden Markov Models. In *Proceedings of NLPRS 2001*, pp. 333-340.

Terry Koo, Xavier Carreras, and Michael Collins. 2008. Simple Semi-supervised Dependency Parsing. In *Proceedings of ACL-08: HLT*, pp. 595-603.

John Lafferty, Andrew McCallum and Fernando Pereira. 2001. Conditional Random Fields: Probabilistic Models for Segmenting and Labeling Sequence Data. In *Proceedings of ICML'01*, pp. 282-289.

Percy Liang and Dan Klein. 2009. Online EM for Unsupervised Models. In *Proceedings of NAACL 2009*, pp. 611-619.

Xiaoqiang Luo, Abe Ittycheriah, Hongyan Jing, Nanda Kambhatla and Salim Roukos. 2004. A Mention-Synchronous Co-reference Resolution Algorithm based on the Bell Tree. In *Proceedings of ACL 2004*, pp. 135-142.

Katja Markert and Malvina Nissim. 2005. Comparing Knowledge Sources for Nominal Anaphora Resolution. In *Computational Linguistics, Volume 31 Issue 3*, pp. 367-402.

[8] If you are interesting in the annotated corpus used for our coreference resolution model, please request us by email.

David McClosky and Eugene Charniak. 2008. Self-Training for Biomedical Parsing. *Proceedings of the Association for Computational Linguistics (ACL 2008, short papers)*, Columbus, Ohio, pp. 101-104.

David Nadeau and Satoshi Sekine. 2007. A survey of named entity recognition and classification. *Linguisticae Investigationes, Volume 30(1)*, pp. 326.

Vincent Ng and Claire Cardie. 2002. Improving Machine Learning Approach to Co-reference Resolution. In *Proceedings of ACL 2002*, pp. 104-111.

Wee Meng Soon and Hwee Tou Ng. 2001. A Machine Learning Approach to Co-reference Resolution of Noun Phrases. *Computational Linguistics 2001, Volume 27 Issue 4*, pp. 521-544.

Pontus Stenetorp, Goran Topić, Sampo Pyysalo, Tomoko Ohta, Jin-Dong Kim, and Jun'ichi Tsujii. 2011. BioNLP Shared Task 2011: Supporting Resources. In*Proceedings of the BioNLP 2011 Workshop Companion Volume for Shared Task*, Portland, Oregon, Association for Computational Linguistics.

Yuka Tateisi, Akane Yakushiji, Tomoko Ohta and Junichi Tsujii. 2005. Syntax Annotation for the GENIA corpus. In *Proceedings of IJCNLP 2005 (Companion volume)*, pp. 222-227.

Yoshimasa Tsuruoka, Jun'ichi Tsujii, and Sophia Ananiadou. 2009. Stochastic Gradient Descent Training for L1-regularized Log-linear Models with Cumulative Penalty. In *Proceedings of ACL-IJCNLP*, pp. 477-485.

Xiaofeng Yang, Guodong Zhou, Jian Su and Chew Lim Tan. 2003. Co-reference Resolution using Competition Learning Approach. In *Proceedings of ACL 2003*, pp. 176-183.

Xiaofeng Yang and Jian Su. 2007. Coreference Resolution Using Semantic Relatedness Information from Automatically Discovered Patterns. In *Proceedings of ACL 2007*, pp. 528-535.

BioNLP 2011 Task Bacteria Biotope – The Alvis system

Zorana Ratkovic[1,2] Wiktoria Golik[1] Pierre Warnier[1] Philippe Veber[1] Claire Nédellec[1]

1 MIG INRA UR1077, Domaine de Vilvert
F-850 Jouy-en-Josas, France
forename.name@jouy.inra.fr

2 LaTTiCe UMR 8094 CNRS Univ. Paris 3
1 rue Maurice Arnoux
F-92120 MONTROUGE

Abstract

This paper describes the system of the INRA Bibliome research group applied to the Bacteria Biotope (BB) task of the BioNLP 2011 shared tasks. Bacteria, geographical locations and host entities were processed by a pattern-based approach and domain lexical resources. For the extraction of environment locations, we propose a framework based on semantic analysis supported by an ontology of the biotope domain. Domain-specific rules were developed for dealing with Bacteria anaphora. Official results show that our Alvis system achieves the best performance of participating systems.

1 Introduction

Given a set of Web pages, the information extraction goal of the Bacteria Biotope (BB) task is to precisely identify bacteria and their locations and to relate them. The type of the predicted locations has to be selected among eight types. Among them the host and host-part locations have to be related by the part-of relation. Three teams participated in the challenge.

BB task example

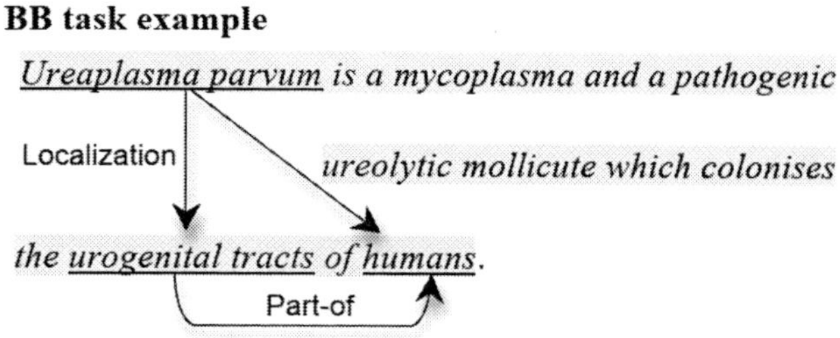

One of the specificities of the BB task is that the bacteria location vocabulary is very large and various as opposed to protein subcellular locations in biology challenges (Kim et al., 2010) and geographical locations (Zhou et al., 2005). Locations include natural environments and hosts as well as food and medical locations. In order to deal with this heterogeneity, we propose a framework based on a term analysis of the test corpus and a shallow mapping of these terms to a bacteria biotope (BB) termino-ontology. This mapping derives the type of location terms and filters out non-location terms. Large external dictionaries of host names (*i.e.* NCBI taxonomy) and geographical names (*i.e.* Agrovoc thesaurus) complete the lexical resources.

The high frequency of bacteria anaphora and ambiguous antecedent candidates in the corpus was also a difficulty. Our Alvis system implements an anaphora resolution algorithm that takes into consideration the anaphoric distance and the position of the antecedent in the sentence. Alvis predicts the bacteria names and their relation to the locations with the help of hand-made patterns based on linguistic analysis and lexical resources.

The methods for predicting and typing locations (section 2) and bacteria (section 3) are first described. Section 4 details the method for relating them. Section 5 comments the experimental results.

2 Location

Our system handles separately the recognition of host and geographical names by dictionary mappings, while the recognition of locations of the environment and host part types is based on linguistic analysis and ontology inference.

Host names and geographical names appeared to be easier to predict by using a named-entity recognition strategy than the other types of location. They are less subject to variation than environmental locations, which can include any physical feature. For host name extraction, we used the NCBI taxonomy as the major source. Only the eukaryote subtree was considered for host detection.

Proceedings of BioNLP Shared Task 2011 Workshop, pages 102–111,
Portland, Oregon, USA, 24 June, 2011. ©2011 Association for Computational Linguistics

Our system filters out the ambiguous names such as Indicator (honeyguides) or Dialysis (xylophage insect) by comparing them to a list of common words in English. The host name list was enriched with additional common names including non-taxonomic host groups (*e.g.* herbivores), progeny names (*e.g.* calf) and human categories (*e.g.* patient). The resulting host name list contains more than 1,800,000 scientific names and 60,000 common names. The geographical name recognition component uses a small dictionary of all geographic terms from the Agrovoc thesaurus sub-vocabularies. At first, we considered using the very rich resource GeoNames. However, it contains too many ambiguous names to be directly usable by short-term development.

2.1 Location of Environment type

The identification of environment locations is done in two steps. First, the automatic extraction of all candidate terms from the test corpus, then the assignment of a location type to these terms with the help of the Bacteria Biotope (BB) termino-ontology. The type assigned to a given term is the type of the closest concept label in the ontology. Since the BB termino-ontology was originally not structured according to the eight types, in order to be usable it first had to be enriched by the new concepts and then mapped to this topology.

Corpus term extraction. The corpus terms were automatically extracted by the AlvisNLP/ML pipeline (Nedellec et al., 2008) with BioYatea (Nedellec et al., 2010). BioYatea is the version of Yatea (Hamon & Aubin, 2006) adapted to the biology domain. We modified BioYatea setting according to the training dataset study. We observed that most of the location terms in the training dataset are noun phrases with adjective modifiers (*e.g. rodent nests*) while prepositional phrases are rather rare (*e.g. breaks in the skin*). We set the term boundaries of BioYatea to include all prepositions except the *of* preposition. Considering other prepositions such as *with* may yield syntactic attachment errors, thus we prefer the risk of incomplete terms to incorrect prepositional attachments.

Bacteria Biotope ontology. We used the Bacteria Biotope (BB) termino-ontology for typing the extracted terms. It is under development for the study of bacteria phenotypes and habitats. The high level of the habitat part is structured in a manner similar to that proposed by the one level classification by Floyd (Floyd et al., 2005). It has a fine-grained structure with the same goal as the generalist EnvO habitat ontology (Field et al., 2008), but it focuses on bacteria phenotype and biotope modeling. It includes a terminological level that records lexical forms of the concepts including terms, synonyms and variations.

For the purpose of the challenge, the initial ontology was manually completed using location concepts. The training corpus, as well as the habitat and isolation site fields of the GOLD database on sequenced prokaryotes (Liolios et al., 2009) are the main sources of location terms and synonyms. The analysis of the training corpus mainly led to the addition of adjectival forms of host parts (*e.g. lymphatic, intracellular*) and human references (*e.g. patient, infant, progeny*).

The GOLD database isolation site field is a very rich source of bacteria location terms. It is filled by natural language descriptions of matters, natural habitats, hosts and geographical locations. For instance, the isolation site of *Anoxybacillus flavithermus* bacterium is *waste water drain at the Wairakei geothermal power station in New Zealand*. The term analysis of GOLD isolation site entries yielded 3,415 location terms including 1,050 geographical names. Hundreds of these terms were manually added to the BB termino-ontology. The lack of time as well as the full sentence structure of the GOLD resource prevented us from correctly handling them in a fully automatic way. We are currently developing a method for the automatic alignment of the terms extracted from GOLD to the BB termino-ontology. Additionally, the GOLD habitat field provided around a hundred different terms that have been directly integrated into the BB termino-ontology.

The current version of the habitat subpart of the BB termino-ontology contains 1,247 concepts and 266 synonyms.

Location types in Bacteria Biotope ontology. The BB termino-ontology has been developed previous to the BB task and the structure of its habitat subpart does not reflect the eight location types of the task. In order to reuse the ontology for the BB task, we assigned types to each location concept. We manually associated the high level nodes of the location hierarchies to the eight BB task types. The types of the lower level concepts were then automatically inferred. For instance, the concept *aquatic environment* is tagged Water in the ontology and all of its descendants *lake*, *sea*, *ocean* are of type Water as well. Local type exceptions were manually tagged. For instance, the *waste* tree includes water-carried wastes of type Water and solid industrial residues of type Environment. This way all concepts in the resulting typed ontology were assigned a unique type. The concept types are then propagated to their associated term classes at the terminological level. For instance, *underground water* and its synonym *subterranean water* are both typed as Water. The resulting typed BB termino-ontology is then usable for deriving the types of the terms extracted from the test corpus.

Derivation of location type. The BB termino-ontology scope is too limited for the correct prediction of all candidate term types by Boolean and exact comparison. From the 2,290 candidate terms of the test corpus, only 152 belong as such to the BB termino-ontology. We propose a method based on the head comparison of the candidate and BB terms for the derivation of the candidate term type.

The quality of the ontology-based annotation depends to a large extent on an accurate match between the resource and the terms extracted from the corpus. Our method targets the syntactic structure of terms (candidate and BB terms) in order to gather the most of semantically similar terms. This approach differs from the ontology alignment and population methods that also use the information from the ontology structure in order to infer semantic relationships (*e.g.* hyponyms, meronyms) (Euzenat, 2007). It also differs from semantic annotation supported by context analysis such as distributional semantics (Grefenstette, 1994) or Hearst patterns (Hearst, 1992). It belongs to the class of methods that focus on the morphology of the corpus terms, which use string-based (Levensthein, 1966, Jaro, 1989) or linguistic-based methods (Jacquemin & Tzoukermann, 1999).

Even though the context-based approach should produce very good results, we chose a less time-consuming method that is easier and faster to set up, which is based on morphosyntactic analysis. In our case, string similarity measures turn out to be irrelevant (*laboratory rat* does not mean *rat laboratory*). We observed that in candidate and BB terms, the head is very often the most informative element. Thus, the linguistic-based analysis of terms, in particular the head-similarity analysis (Hamon & Nazarenko, 2001), represents a promising alternative. Our method is inspired by MetaMap (Aronson, 2001). MetaMap tags biomedical corpora with the UMLS Metathesaurus by syntactic analysis that takes into account lexical heads of terms. The similarity scores computed by linguistically-based metrics are higher for terms whose heads have previously been analyzed.

The MetaMap method includes a variant computation that maps acronyms, abbreviations, synonyms as well as derivational, inflectional and spelling variants. Our term typing method is less sophisticated and uses a few lexical variants due to the lack of a complete resource. Some ontology enrichment applications also use head-supported term matching, as in Desmontils (Desmontils et all, 2003). In Desmontils, new concepts belonging to WordNet (Fellbaum, 1998) are automatically added to the ontology in order to improve the indexing process. However, the analysis of the results shows that a great number of concepts found in the texts are not considered because they do not exist in WordNet. Our typing task uses a similar head-based method, but only for type derivation.

Our system derives the location type of candidate terms in several steps. First, if there is a term in the BB termino-ontology that is strictly equal to the candidate term, it is assigned the same type. Then, the other candidate terms are assigned types according to the comparison of their heads to the BB term heads. We assume that in most of the cases the term head conveys the information about the type and is non-ambiguous. A given head *H* is non-ambiguous if all BB terms with head *H* are of the same type. The location term head set is the set of all habitat term heads found in the BB termino-ontology. The current version contains 693 different heads. Let T_e denote the extracted term to be typed. If the head of T_e does not belong to the BB term head set, then the type of T_e is simply *not* Location (e.g. *high metabolic diversity*). If T_e head *does* belong to the BB term head set and the head is non-ambiguous, then T_e is assigned the associated type. For instance, the head of the extracted term *stratified lake* is *lake*. The type of all the BB terms with *lake* head is *Water* (e.g. *meromictic lake*). *Stratified lake* is therefore typed as *Water*.

Specific processing is applied to terms with ambiguous heads. The associative set of BB term heads and types exhibits some cases of ambiguous heads with multiple types that we analyzed in detail. There are two kinds of ambiguities that were processed in different ways. In the first, multiple types reflect different roles of the same object. In the second, the head is non-informative with respect to the type. In the latter case the type is conveyed by the subterm (term after head removal). We qualify non-informative BB term heads as *neutral*. They mainly denote habitats (*habitat, environment, medium, zone*) and extracts (*sample, surface, isolate, material, content*). In this case, the type is derived from the subterm. For instance, the head *isolate* of the extracted term *marine isolate* is neutral. After head removal, it is assigned the type Water since *marine* is of type Water. *Freshwater* has the same type as *freshwater medium* or *freshwater environment* since *medium* and *environment* are neutral heads.

Some heads have more than one type although they denote specific locations. Their multiple types reflect different uses or states. For instance, the head *bottle* has two types: Food and Medical. The type Food is derived from the BB concept *water bottle* and the type Medical is derived from *bedside water bottles* in a hospital environment. The correct type for the extracted terms is then selected by a set of patterns based on the context of the term in the document. For instance, many vegetables and meats could be either of type Host or Food. The type is Host by default. One pattern states that if a term includes or is preceded by a food processing-related word (*e.g. cooked, grilled, fermented*), then the term is reassigned the type Food. Another pattern states that if a host is preceded by a death-related adjective (*dead, decaying*), then its type should be revised as Environment.

Our system currently includes nine disambiguation/retyping patterns. The first version of the type derivation method was automatically applied to the 1,263 GOLD terms after head analysis. Manual examination of the results yielded an extension of the two lists of neutral heads and heads with ambiguous types. There are 20 neutral heads and 21 ambiguous heads in the current version of the BB termino-ontology. The head-matching algorithm appears to be quite productive for the biotope terms. The procedure applied to the test corpus yielded the following figures: BioYatea extracted 2,290 terms. 416 terms matching the post-processing filters were discarded. This includes terms which are too general (i.e. *approach, diversity*), terms containing irrelevant or non desirable adjectives (i.e. *numerous deficiencies, known species*) and terms containing forbidden words according to the annotation location rules (i.e. *bacteria, pathogen, contaminated, parasite*). Finally, 1,873 candidate terms were kept.

Among these figures:

- 152 terms belong to the BB termino-ontology
- 90 terms were typed using the ontology heads
- 6 terms with several types were handled by disambiguation patterns.

We plan to extend the list of neutral heads and discriminate adjectives for type disambiguation by machine learning classification applied to the BB termino-ontology modifiers.

Location entity boundary. The analysis of term extraction result from the training corpus shows that the predicted boundaries of locations were not fully consistent with the task annotation guidelines. Post-processing adjusts incorrect boundaries by filtering irrelevant words, packing and merging terms. Irrelevant words (*e.g. contaminated, infected, host species, disease, inflammation*) were removed from the location candidate terms independently of their types (e.g. *contaminated Bachman Road site* vs. *Bachman Road* ; *host plant* vs. *plant*). Note that BioYatea extracts not only the maximum terms (e.g. *contaminated Bachman Road site*), but also their constituents (*Bachman Road site, Bachman Road* and *site*). Boundary adjustment often consists in selecting the relevant alternative among the subterms.

Other boundary issues are handled by several patterns, which are applied after the typing stage. These patterns are type-dependent: each pattern only applies to one type or a subset of location types. When necessary, they shift the boundaries in order to include relevant modifiers. They also split location terms or join adjacent location terms. BioYatea may have missed relevant modifiers because of POS-tagging errors. For instance, if a nationality name precedes a location, then it is included (*e.g. German oil field*). Also, it frequently happens that hosts are modifiers of host parts (*e.g. insect gut*). BioYatea extracts the whole term and its constituents. The term is correctly typed as *Host-part* and the host modifier as *Host*. In order to avoid embedded locations, a specific pattern is devoted to the splitting of these terms. In this way *insect gut* (Host-part) becomes *insect* (Host) and *gut* (Host-part).

Most of these patterns involve several specific lexicons, including cardinal directions, relevant and irrelevant modifiers for each type of location, as well as types, which can be merged and split. The current resources were manually built by examining the location terms of the training set and GOLD isolation fields. The acquisition of relevant and irrelevant modifiers could be automated by machine learning. Some linguistic phenomena could be better handled by the customization of BioYatea. For instance BioYatea considers the preposition *with* as a term boundary so it cannot extract terms containing *with*, like *areas with high sulfur and salt concentrations*.

3 Extraction of Bacteria names

We observed in the training corpus that not only were bacteria names tagged, but also higher level taxa (families) and lower level taxa (strains). We used the NCBI taxonomy as the main bacteria taxon resource since it includes all organism levels and is kept up-to-date. This bacteria dictionary was enriched by taxa from the training corpus, in particular by non standard abbreviations (*e.g. Chl. = Chlorobium, ssp. = subsp*) and plurals, (*Vibrios* as the plural for *Vibrio*) that were hopefully rather rare.

Determining the boundaries of the bacteria names was one of the main issues because corpus strain names do not always follow conventional nomenclature rules. Also, the recognition of bacteria name is evaluated using a strict exact match. Patterns were developed to account for such cases. They handle inversion (*LB400 of Burkholderia xenovorans* instead of *Burkholderia xenovorans LB400*) and parenthesis (*Tropheryma whipplei (the Twist strain)* instead of *Tropheryma whipplei strain* Twist). The corpus also mentions names of bacteria that contain modifiers not found in the NCBI dictionary, such as *antimicrobial-resistant C. coli* or *L. pneumophila serogroup 1*. Such cases, as well as abbreviations (e.g. *GSB* for *green sulfur bacteria*) and partial strain names (*e.g. strain DSMZ 245 T* for *Chlorobium limicola strain DSMZ 245 T*) were also specifically handled.

The main source of error in bacteria name prediction is due to the mixture of family names and strain name abbreviations in the same text. It frequently happens that the strain name is abbreviated into the first word of the name. For instance *Bartonella henselae* is abbreviated as *Bartonella*. Unfortunately, *Bartonella* is a genus mentioned in the same text, thus yielding ambiguities between the anaphora and the family name, which are identical.

3.1 Bacteria anaphora resolution

Anaphors are frequent in the text, especially for bacteria reference and to a smaller extent for host reference. Our effort focused on bacteria anaphora resolution ignoring host anaphora. The extraction method of location relations (section 4) assumes that the relation arguments, location and bacterium (or anaphora of the bacterium) occur in the same sentence. From a total of 2,296 sentences in the training corpus, only 363 sentences contain both the location and the explicit bacterium, while 574 mention only the location. Two thirds of the locations do not co-occur with bacteria. This demonstrates the importance of recovering the bacteria for these cases, which is potentially referred to by an explicit anaphora.

The manual examination of the training corpus showed that the most frequent anaphora of bacteria are not pronouns but higher level taxa, often preceded by a demonstrative determinant, (i.e. *This bacteria, This Clostridium*) and sortal anaphora (i.e. *genus*, *organism*, *species* and *strain*), both of which are commonly found in biological texts (Torri & Vijay-Shanker, 2007). The style of some of the documents is rather relaxed and the antecedent may be ambiguous even for a human reader. We observed three types of anaphora in the corpus. First, the standard anaphora which includes both pronouns and sortal anaphora, which requires a unique bacterial antecedent. Second, *bi-anaphora* or an anaphora that requires two bacteria antecedents. This happens when the properties of two strains are compared in the document. Finally, the case of a higher taxon being used to refer to a lower taxon, which we named *name taxon anaphora*.

Anaphora with a unique antecedent

C. coli is pathogenic in animals and humans. People usually get infected by eating poultry that contained the bacteria, eating raw food, drinking raw milk, and drinking bottle water [...].

Anaphora with two antecedents

C. coli is usually found hand in hand with its bacteria relative, C. jejuni. These two organisms are recognized as the two most leading causes of acute inflammation of intestine in the United States and other nations.

Name taxon anaphora

Ticks become infected with Borrelia duttonii while feeding on an infected rodent. Borrelia then multiplies rapidly, causing a generalized infection throughout the tick.

For anaphora detection and resolution a pattern-based approach was preferred to machine learning because the constraints for relating anaphora to antecedent candidates of the same taxonomy level were mainly semantic and domain-dependent and the annotation of anaphora was not provided in the training corpus.

Anaphora detection consists of identifying potential anaphora in the corpus, given a list of pronouns, sortal anaphora and taxa and then filtering out irrelevant cases (Segura-Bedmar *et al.*, 2010, Lin & Lian, 2004) before anaphora resolution. Not all the pronouns, sortal anaphora terms and higher taxon bacteria are anaphoric. For example, if a higher taxon is preceded or followed by the word *genus*, this signals that it is not anaphoric but that the text is actually about the higher taxon.

Non-anaphoric higher taxon

Burkholderia cenocepacia HI2424[...]
The genus Burkholderia consists of some 35 bacterial species, most of which are soil saprophytes and phytopathogens that occupy a wide range of environmental niches.

The anaphora resolution algorithm takes into account two features: the distance to the antecedent candidate and its position in the sentence. The antecedent is usually found in proximity to the anaphora, in order to maintain the coherence of the text. Therefore, our method ranks the antecedent candidates according to the anaphoric distance counted in sentences.

If more than one bacterium is found in a given sentence, their position is discriminate. Centering theory states that in a sentence the most prominent entities and therefore the most probable antecedent candidates are in the order: subject > object > other position (Grosz *et al.*, 1995). In English, due to the SVO order of the language the subject is most often found at the beginning of the sentence, followed by the object and the others. Therefore, the method retains the leftmost bacterium in the sentence when searching for the best antecedent candidate.

More precisely, the method selects the first antecedent that it finds according to the following precedence list:

- First bacterium in the current sentence (s)
- First bacterium in the previous sentence (s-1)
- First bacterium in sentence s-2
- First bacterium in sentence s-3
- First bacterium in the current paragraph
- Last bacterium in the previous paragraph
- First bacterium in the first sentence of the document
- The first bacterium ever mentioned.

The method only relates anaphora to antecedents that are found before. It does not handle cataphors since they are rarely found in the corpus. For anaphors that require two antecedents we use the same criteria but search for two bacteria in each sentence or paragraph, instead of one. For taxon anaphora we look for the presence of a lower taxon in the document found before the anaphora that is compatible according to the species taxonomy.
The counts of anaphora detected by the patterns are given in Table 1.

Corpus	Single ante	Bi ante	Taxon ante
Train	933	4	129
Dev	204	3	22
Test	240	0	18
Total	1,377	7	169

Table 1. The count of the types of anaphora per corpus.

The anaphora resolution algorithm allowed us to retrieve more sentences that contain both a bacterium and a location. Out of the 574 sentences that contain only a location, 436 were found to contain an anaphora related to at least one bacterium. The remaining 138 sentences are cases where there is no bacterial anaphora or the bacterium name is implicit. It frequently happens that the bacterium is referred to through its action. For example in the sentence below, the bacterium name could be derived from the name of the disease that it causes.

In the 1600s anthrax was known as the "Black bane" and killed over 60,000 cows.

One of the questions we had about the resolution of anaphora is whether anaphora that are found in the same sentence together with a bacterium (therefore potentially its antecedent) should be considered or not. We tested this on the development set. We found that removing such anaphora from consideration improved the overall score. It yielded an F-score of 53.22% (precision: 46.17%, recall: 62.81%), compared to the original F-score of 50.15% (precision: 41.06%, recall: 64.44%). This improvement in F-score is solely due to an increase in precision, which shows that while resolving anaphora is important and required, the incorrect recognition of terms as anaphora and incorrect anaphora resolution can introduce noise.

4 Relation extraction

In this work we concentrated most of our effort on the prediction of entities. For the prediction of events we used a strategy based on the co-occurrence of arguments and trigger words within a sentence:

- If a bacteria name, a location and a trigger word are present in a sentence, then the system predicts a Localization event between the bacterium and the location.
- If a bacteria anaphora, a location and a trigger word are present in a sentence, then the system predicts a Localization event between each anaphora antecedent and the location.
- If a host, a host part, a bacterium and at least one trigger word are present in a sentence, then the system predicts a *PartOf* event between the host and the host part.

The list of trigger words contains 20 verbs (*e.g. inhabit*, *colonize*, but also *discover*, *isolate*), 16 disease markers (*e.g. chronic*, *pathogen*) and 19 other relevant words (*e.g. ingest*, *environment*, *niche*). This list was designed by ranking words in the sentences of the training corpus containing both a bacteria name and a location. The ranking criterion used was the information gain with respect to whether the sentence contained an event or not. The ranked list was adjusted by removing spurious words and adding domain knowledge words.

By removing the constraint of the occurrence of a trigger word in the sentence, we can determine that the maximum recall the method can achieve with this strategy is 47% (precision: 41%, F-score: 44%). The selected trigger word list yielded a recall close to the maximum, thus it seems that the trigger words do not affect the recall and are suitable for the task.

5 Results

Table 2 summarizes the official scores that the Bibliome Alvis system achieved for the Bacteria Biotope Task. It ranked first among three participants. The first column gives the recall of entity prediction. The prediction of hosts and bacteria named-entities achieved a good recall of 84 and 82, respectively.

	Entity recall	**Event recall**	**Event Precis.**	**F-score**
Bacteria	84	-	-	-
Host	82	61	48	53
Host part	72	53	42	47
Env.	53	29	24	26
Geo.	29	13	38	19
Food	-	-	29	41
Medical	100	50	33	40
Water	83	60	55	57
Soil	86	69	59	63
Total		**45**	**45**	**45**

Table 2. Bibliome system scores at Bacteria Biotope Task in BioNLP shared tasks 2011.

However, geographical locations based on a similar strategy were poorly predicted (29%). Our system predicted only 15 countries. A more appropriate resource of geographical names than the Agrovoc thesaurus would certainly increase the recall of geographical locations.

The host parts, medical, water and soil locations predicted with the same ontology-based method were surprisingly good with a recall of 72, 100, 83 and 86, respectively. The small size of the ontology and the small number of different term heads (*i.e.* 51 different heads) initially appeared as a limitation factor for reuse on new corpora. The good recall shows that the location vocabulary of the test set has similarities with the training set compared to potential space of location names. The potential space is reflected by the richness of the GOLD isolation site field. This demonstrates the robustness of the type derivation approach based on term heads. The correctness of the derivation type cannot be calculated without a corpus where all the locations and not only bacteria ones are annotated. The recall of the environment location prediction is a little bit lower, 53%. The environment type includes many different types that cannot all be anticipated. Therefore the coverage of the BB termino-ontology environment part is limited except for water and soil, which are more focused topics.

The localization event recall (column 2) is on average 20% lower for all types than the location entity recall. The regularity of the difference may suggest that once the argument is identified, the localization relation is equally harder to find by our method independently of the type. The localization event precision (column 3) is more difficult to analyze because many sources of error may be involved, such as an incorrect arguments, incorrect anaphora resolution, relation to the wrong bacterium among several or the absence of a relation.

The prediction precision of localization events involving soil, water and host is better than environment and food. The manual analysis of the test corpus shows that in some cases environmental locations were mentioned as potential sources of industrial applications without actually being bacteria isolation places. For instance, in *Other fields of application for thermostable enzymes are starch-processing, organic synthesis, diagnostics, waste treatment, pulp and paper manufacture, and animal feed and human food,* the Alvis system erroneously predicted *waste treatment, paper manufacture, animal feed* and *human food.* This is due to the fact that the system does not handle modalities. Such hypotheses are specific to the BB task text genre, *i.e.* Bacteria sequencing projects. Such projects contain details for potential industrial applications, which are absent from academic literature.

Ambiguous types are also a source of error. Despite the host dictionary cleaning, some ambiguities remained. For example, the head *canal* in *tooth root canal* is erroneously typed as water and should be disambiguated with its *tooth* host-part modifier.

After test publication we measured the gain of anaphora resolution by using the on-line service. The anaphora resolution algorithm was found to have a strong impact on the final result. Running the test set using all of the modules *except for* the anaphora resolution algorithm yielded a decrease in the F-score by almost 13% (F-score: 32.5%, precision: 48.5%, 24.4%). This shows that the addition of an anaphora resolution algorithm significantly increases the precision and that a resolution algorithm adapted to the Bacteria domain is necessary for the Biotope corpus.

The part-of event prediction relies on the strict co-occurrence of a bacterium, trigger word, host and host part within a sentence. An additional run with the more relaxed constraint where the bacterium can be denoted by an anaphora as well yielded a gain of 6 recall points, a loss of 5 precision points with a net benefit of 1 F-measure point.

6 Discussion

The use of trigger words for the selection of sentences for relation extraction does not take into account the structure or syntax of the sentence for the prediction of relation arguments. The system predicts all combinations of bacteria and locations as localization events and all combination of host and host parts as part-of event. This has a negative effect on the precision measure since some pairs are irrelevant as in the sentence below.

> *Baumannia cicadellinicola. This newly discovered organism is an obligate endosymbiont of the leafhopper insect Homalodisca coagulata (Say), also known as the Glassy-Winged Sharpshooter, which feeds on the xylem of plants.*

It has been shown that the use of syntactic dependencies to extract biological events (such as protein-protein interactions) improves the results of such systems (Erkan *et al.*, 2007, Manine *et al.*, 2008, Airola *et al.* 2008). The use of syntactic dependencies could offer a more in depth examination of the syntax and the semantics and therefore allow for a more refined extraction of bacteria-localization and host-host part relations.

Term extraction appears to be a good method for predicting locations including unseen terms, but it is limited by the typing strategy that filters out all terms with unknown heads (with respect to the BB termino-ontology). In the future, we will study the effect of linguistic markers such as enumeration and exemplification structures for recovering additional location terms. For instance, in *heated organic materials such as compost heaps, rotting hay, manure piles or mushroom growth medium*, our system has correctly typed *heated organic materials* as environment but not the other examples because of their unknown heads.

The promising performance of the Alvis system on the BB task shows that a combination of semantic analysis and domain-adapted resources is a good strategy for information extraction in the biology domain.

References

Agrovoc: http://aims.fao.org/website/AGROVOC-Thesaurus

Antti Airola, Sampo Pyysalo, Jari Björne, Tapio Pahnikkala, Filip Ginter, and Tapio Salakoski. 2008. A Graph Kernel for Protein-Protein Interaction Extraction. BioNLP2008: Current Trends in Biomedical Natural Language Processing, pages 1-9.

Alan R. Aronson. 2001. Effective mapping of biomedical text to the UMLS Metathesaurus: The MetaMap program. Proceedings of AMIA Symposium 2001, pages 17-21.

Emmanuel Desmontils, Christine Jacquin and Laurent Simon. 2003. Ontology enrichment and indexing process. Research report RR-IRIN-03.05, Institut de Recherche en Informatique de Nantes, Nantes, France.

Jérôme Euzenat and Pavel Shvaiko. 2007. Ontology matching, Springer Verlag, Heidelberg (DE),page 333.

Dawn Field et al. 2008. Towards a richer description of our complete collection of genomes and metagenomes: the Minimum Information about a Genome Sequence (MIGS) specification. Nature Biotechnology 26, pages 541-547.

GeoNames: http://www.geonames.org/

Gregory Grefenstette. 1994. Explorations in Automatic Thesaurus Discovery. Natural Language Processing and Machine Translation. London: Kluwer Academic Publishers.

Barbara J. Grosz, Araving K. Joshi and Scott Weinstein. 1995. Centering: A Framework for Modelling the Local Coherence of Discourse. University of Pennsylvania Institute for Research in Cognitive Science Technical Reports Series.

Güneş Erkan, Arzucan Özgür and Dragomir R. Radev. 2007. Semi-Supervised Classification for Extracting Protein Interaction Sentences using Dependency Parsing. Proceedings of the 2007 Joint Conference on Empirical Methods in Natural Language Processing and Computational Natural Language Learning, pages 228-237.

Thierry Hamon and Sophie Aubin. 2006. Improving term extraction with terminological resources. In Salakoski, T. et al., editors, Advances in Natural Language Processing 5th International Conference on NLP (Fin- TAL'06), pages 380–387. Springer.

Thierry Hamon and Adeline Nazarenko. 2001. Detection of synonymy links between terms: experiment and results, Recent Advances in Computational Terminology. Pages 185-208. John Benjamins.

Marti A. Hearst. 1992. Automatic acquisition of hyponyms from large text corpora. In Zampolli, A.(ed.), Proceedings of the 14 th COLING, pages 539–545, Nantes, France.

Christian Jacquemin and Evelyne Tzoukermann. 1999. NLP for term variant extraction: A synergy of morphology, lexicon, and syntax. In Strzalkowski, T. (ed.), Natural language information retrieval, volume 7 of Text, speech and language technology, chapter 2, pages 25–74. Dordrecht & Boston: Kluwer Academic Publishers.

Matthew A. Jaro. 1989. Advances in record linkage methodology as applied to matching the 1985 census of Tampa, Florida. Journal of the American Statistical Association 84(406), pages 414-20.

Jin-Dong Kim, Tomoko Ohta, Sampo Pyysalo, Yoshinobu Kano and Jun'ichi Tsujii. (to appear). Extracting bio-molecular events from literature - the BioNLP'09 shared task. Special issue of the International Journal of Computational Intelligence.

Vladimir I. Levenshtein. 1966. Binary codes capable of correcting deletions, insertions and reversals. Doklady akademii nauk SSSR, 163(4):845-848, 1965. In Russian. English translation in Soviet Physics Doklady, 10(8), pages 707-710.

Yu-Hsiang Lin and Tyne Liang. 2004. Pronomial and Sortal Anaphora Resolution for Biomedical Literature. In Proceedings of ROCLING XVI: Conference on Computational Linguistics and Speech Processing.

Konstantinos Liolios, I-Min A. Chen., Konstantinos Mavromatis, Nektarios Tavernarakis, Philip Hugenholtz, Victor M. Markowitz and Nikos C. Kyrpides. 2009. The Genomes On Line Database (GOLD) in 2009: status of genomic and metagenomic projects and their associated metadata. NAR Epub.

Alain-Pierre Manine, Erick Alphonse and Philippe Bessières. 2008. Information extraction as an ontology population task and its application to genic interactions, 20th IEEE Intl. Conf. Tools with Artificial Intelligence, ICTAI'08., vol. II, pp. 74-81.

NCBI taxonomy: http://www.ncbi.nlm.nih.gov/Taxonomy/

Claire Nédellec, Wiktoria Golik, Sophie Aubin and Robert Bossy. 2010. Building Large Lexicalized Ontologies from Text: a Use Case in Indexing Biotechnology Patents, International Conference on Knowledge Engineering and Knowledge Management (EKAW 2010), Lisbon, Portugal.

Isabel Segura-Bedmar, Mario Crespo, César de De Pablo-Sánchez and Paloma Martínez. 2010. Resolving anaphoras for the extraction of drug-drug interactions in pharmacological documents. BMC Bioinformatics 11(Supl 2):S1.

Manabu Torii and K. Vijay-Shanker. 2007. Sortal Anaphora Resolution in Medline Abstracts. Computational Intelligence 23, pages 15-27.

Zhou GuoDong, Su Jian, Zhang Jie and Zhang Min. 2005. Exploring Various Knowledge in Relation Extraction. In Proceedings of the 43rd Annual Meeting of the ACL, pages 427-434, Ann Arbor. Association for Computational Linguistics.

BioNLP Shared Task 2011: Supporting Resources

Pontus Stenetorp[*†] **Goran Topić**[*] **Sampo Pyysalo**[*]
Tomoko Ohta[*] **Jin-Dong Kim**[‡] **and Jun'ichi Tsujii**[$]
[*]Tsujii Laboratory, Department of Computer Science, University of Tokyo, Tokyo, Japan
[†]Aizawa Laboratory, Department of Computer Science, University of Tokyo, Tokyo, Japan
[‡] Database Center for Life Science,
Research Organization of Information and Systems, Tokyo, Japan
[$]Microsoft Research Asia, Beijing, People's Republic of China
{pontus,goran,smp,okap}@is.s.u-tokyo.ac.jp
jdkim@dbcls.rois.ac.jp
jtsujii@microsoft.com

Abstract

This paper describes the supporting resources provided for the BioNLP Shared Task 2011. These resources were constructed with the goal to alleviate some of the burden of system development from the participants and allow them to focus on the novel aspects of constructing their event extraction systems. With the availability of these resources we also seek to enable the evaluation of the applicability of specific tools and representations towards improving the performance of event extraction systems. Additionally we supplied evaluation software and services and constructed a visualisation tool, *stav*, which visualises event extraction results and annotations. These resources helped the participants make sure that their final submissions and research efforts were on track during the development stages and evaluate their progress throughout the duration of the shared task. The visualisation software was also employed to show the differences between the gold annotations and those of the submitted results, allowing the participants to better understand the performance of their system. The resources, evaluation tools and visualisation tool are provided freely for research purposes and can be found at http://sites.google.com/site/bionlpst/

1 Introduction

For the BioNLP'09 Shared Task (Kim et al., 2009), the first in the ongoing series, the organisers provided the participants with automatically generated syntactic analyses for the sentences from the annotated data. For evaluation purposes, tools were made publicly available as both distributed software and online services. These resources were well received. A majority of the participants made use of one or more of the syntactic analyses, which have remained available after the shared task ended and have been employed in at least two independent efforts studying the contribution of different tools and forms of syntactic representation to the domain of information extraction (Miwa et al., 2010; Buyko and Hahn, 2010). The evaluation software for the BioNLP'09 Shared Task has also been widely adopted in subsequent studies (Miwa et al., 2010; Poon and Vanderwende, 2010; Björne et al., 2010).

The reception and research contribution from providing these resources encouraged us to continue providing similar resources for the BioNLP Shared Task 2011 (Kim et al., 2011a). Along with the parses we also encouraged the participants and external groups to process the data with any NLP (Natural Language Processing) tools of their choice and make the results available to the participants.

We provided continuous verification and evaluation of the participating systems using a suite of in-house evaluation tools. Lastly, we provided a tool for visualising the annotated data to enable the participants to better grasp the results of their experiments and to help gain a deeper understanding of the underlying concepts and the annotated data. This paper presents these supporting resources.

2 Data

This section introduces the data resources provided by the organisers, participants and external groups for the shared task.

Proceedings of BioNLP Shared Task 2011 Workshop, pages 112–120,
Portland, Oregon, USA, 24 June, 2011. ©2011 Association for Computational Linguistics

Task	Provider	Tool
CO	University of Utah	Reconcile
CO	University of Zürich	UZCRS
CO	University of Turku	TEES
REL	University of Turku	TEES

Table 1: Supporting task analyses provided, TEES is the Turku Event Extraction System and UZCRS is the University of Zürich Coreference Resolution System

2.1 Supporting task analyses

The shared task included three Supporting Tasks: Coreference (CO) (Nguyen et al., 2011), Entity relations (REL) (Pyysalo et al., 2011b) and Gene renaming (REN) (Jourde et al., 2011). In the shared task schedule, the supporting tasks were carried out before the main tasks (Kim et al., 2011b; Pyysalo et al., 2011a; Ohta et al., 2011; Bossy et al., 2011) in order to allow participants to make use of analyses from the systems participating in the Supporting Tasks for their main task event extraction systems.

Error analysis of BioNLP'09 shared task submissions indicated that coreference was the most frequent feature of events that could not be correctly extracted by any participating system. Further, events involving statements of non-trivial relations between participating entities were a frequent cause of extraction errors. Thus, the CO and REL tasks were explicitly designed to support parts of the main event extraction tasks where it had been suggested that they could improve the system performance.

Table 1 shows the supporting task analyses provided to the participants. For the main tasks, we are currently aware of one group (Emadzadeh et al., 2011) that made use of the REL task analyses in their system. However, while a number of systems involved coreference resolution in some form, we are not aware of any teams using the CO task analyses specifically, perhaps due in part to the tight schedule and the somewhat limited results of the CO task. These data will remain available to allow future research into the benefits of these resources for event extraction.

2.2 Syntactic analyses

For syntactic analyses we provided parses for all the task data in various formats from a wide range of parsers (see Table 2). With the exception of the Pro3Gres[1] parser (Schneider et al., 2007), the parsers were set up and run by the task organisers. The emphasis was put on availability for research purposes and variety of parsing models and frameworks to allow evaluation of their applicability for different tasks.

In part following up on the results of Miwa et al. (2010) and Buyko and Hahn (2010) regarding the impact on performance of event extraction systems depending on the dependency parse representation, we aimed to provide several dependency parse formats. Stanford Dependencies (SD) and Collapsed Stanford Dependencies (SDC), as described by de Marneffe et al. (2006), were generated by converting Penn Treebank (PTB)-style (Marcus et al., 1993) output using the Stanford CoreNLP Tools[2] into the two dependency formats. We also provided Conference on Computational Natural Language Learning style dependency parses (CoNLL-X) (Buchholz and Marsi, 2006) which were also converted from PTB-style output, but for this we used the conversion tool[3] from Johansson and Nugues (2007). While this conversion tool was not designed with converting the output from statistical parsers in mind (but rather to convert between treebanks), it has previously been applied successfully for this task (Miyao et al., 2008; Miwa et al., 2010).

The text from all documents provided were split into sentences using the Genia Sentence Splitter[4] (Sætre et al., 2007) and then postprocessed using a set of heuristics to correct frequently occurring errors. The sentences were then tokenised using a tokenisation script created by the organisers intended to replicate the tokenisation of the Genia Tree Bank (GTB) (Tateisi et al., 2005). This tokenised and sentence-split data was then used as input for all parsers.

We used two deep parsers that provide phrase structure analysis enriched with deep sentence struc-

[1] https://files.ifi.uzh.ch/cl/gschneid/parser/

[2] http://nlp.stanford.edu/software/corenlp.shtml

[3] http://nlp.cs.lth.se/software/treebank_converter/

[4] http://www-tsujii.is.s.u-tokyo.ac.jp/~y-matsu/geniass/

Name	Format(s)	Model	Availability	BioNLP'09
Berkeley	PTB, SD, SDC, CoNLL-X	News	Binary, Source	No
C&C	CCG, SD	Biomedical	Binary, Source	Yes
Enju	HPSG, PTB, SD, SDC, CoNLL-X	Biomedical	Binary	No
GDep	CoNLL-X	Biomedical	Binary, Source	Yes
McCCJ	PTB, SD, SDC, CoNLL-X	Biomedical	Source	Yes
Pro3Gres	Pro3Gres	Combination	–	No
Stanford	PTB, SD, SDC, CoNLL-X	Combination	Binary, Source	Yes

Table 2: Parsers, the formats for which their output was provided and which type of model that was used. The availability column signifies public availability (without making an explicit request) for research purposes

tures, for example predicate-argument structure for Head-Driven Phrase Structure Grammar (HPSG). First we used the C&C Combinatory Categorial Grammar (CCG) parser[5] (C&C) by Clark and Curran (2004) using the biomedical model described in Rimell and Clark (2009) which was trained on GTB. Unlike all other parsers for which we supplied SD and SDC dependency parses, the C&C output was converted from its native format using a separate conversion script provided by the C&C authors. Regrettably we were unable to provide CoNLL-X format output for this parser due to the lack of PTB-style output. The other deep parser used was the HPSG parser Enju[6] by Miyao and Tsujii (2008), also trained on GTB.

We also applied the frequently adopted Stanford Parser[7] (Klein and Manning, 2003) using a mixed model which includes data from the biomedical domain, and the Charniak Johnson re-ranking parser[8] (Charniak and Johnson, 2005) using the self-trained biomedical model from McClosky (2009) (McCCJ).

For the BioNLP'09 shared task it was observed that the Bikel parser[9] (Bikel, 2004), which used a non-biomedical model and can be argued that it uses the somewhat dated Collins' parsing model (Collins, 1996), did not contribute towards event extraction performance as strongly as other parses supplied for the same data. We therefore wanted to supply a parser that can compete with the ones above in a domain which is different from the biomedical domain to see whether conclusions could be drawn as to the importance of using a biomedical model. For this we used the Berkeley parser[10] (Petrov et al., 2006). Lastly we used a native dependency parser, the GENIA Dependency parser (GDep) by Sagae and Tsujii (2007).

At least one team (Choudhury et al., 2011) performed experiments on some of the provided lexical analyses and among the 14 submissions for the EPI and ID tasks, 13 submissions utilised tools for which resources were provided by the organisers of the shared task. We intend to follow up on whether or not the majority of the teams ran the tools themselves or used the provided analyses.

2.3 Other analyses

The call for analyses was open to all interested parties and all forms of analysis. In addition to the Supporting Task analyses (CO and REL) and syntactic analyses provided by various groups, the University of Antwerp CLiPS center (Morante et al., 2010) responded to the call providing negation/speculation analyses in the BioScope corpus format (Szarvas et al., 2008).

Although this resource was not utilised by the participants for the main task, possibly due to a lack of time, it is our hope that by keeping the data available it can lead to further development of the participating systems and analysis of BioScope and BioNLP ST-style hedging annotations.

3 Tools

This section presents the tools produced by the organisers for the purpose of the shared task.

[5]http://svn.ask.it.usyd.edu.au/trac/candc/

[6]http://www-tsujii.is.s.u-tokyo.ac.jp/enju/

[7]http://nlp.stanford.edu/software/lex-parser.shtml

[8]ftp://ftp.cs.brown.edu/pub/nlparser/

[9]http://www.cis.upenn.edu/~dbikel/software.html

[10]http://code.google.com/p/berkeleyparser/

```
1  10411007-E1      Regulation     <Exp>regulate[26-34] <Theme>TNF-alpha[79-88]    ↵
     ↳<Excerpt>[regulate] an enhancer activity in the third intron of [TNF-alpha]
2  10411007-E2      Gene_expression       <Exp>activity[282-290] <Theme>TNF-alpha[252-261]    ↵
     ↳<Excerpt>[TNF-alpha] gene displayed weak [activity]
3  10411007-E3      +Regulation    <Exp>when[291-295] <Theme>E2     <Excerpt>[when]
```

Figure 1: Text output from the BioNLP'09 Shared Event Viewer with line numbering and newline markings

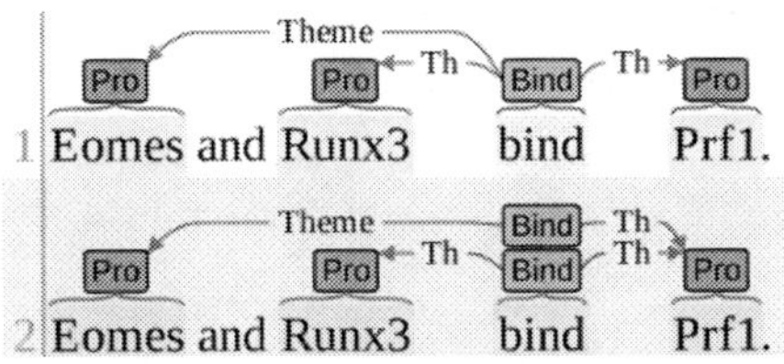

Figure 2: An illustration of collective (sentence 1) and distributive reading (sentence 2). "Theme" is abbreviated as "Th" and "Protein" as "Pro" when there is a lack of space

3.1 Visualisation

The annotation data in the format specified by the shared task is not intended to be human-readable – yet researchers need to be able to visualise the data in order to understand the results of their experiments. However, there is a scarcity of tools that can be used for this purpose. There are three available for event annotations in the BioNLP ST format that we are aware of.

One is the BioNLP'09 Shared Task Event Viewer[11], a simple text-based annotation viewer: it aggregates data from the annotations, and outputs it in a format (Figure 1) that is meant to be further processed by a utility such as `grep`.

Another is What's Wrong with My NLP[12], which visualises relation annotations (see Figure 3a) – but is unable to display some of the information contained in the Shared Task data. Notably, the distributive and collective readings of an event are not distinguished (Figure 2). It also displays all annotations on a single line, which makes reading and analysing longer sentences, let alone whole documents, somewhat difficult.

The last one is U-Compare[13] (Kano et al., 2009), which is a comprehensive suite of tools designed for managing NLP workflows, integrating many available services. However, the annotation visualisation component, illustrated in Figure 3b, is not optimised for displaying complex event structures. Each annotation is marked by underlining its text segment using a different colour per annotation type, and a role in an event is represented by a similarly coloured arc between the related underlined text segments. The implementation leaves some things to be desired: there is no detailed information added in the display unless the user explicitly requests it, and then it is displayed in a separate panel, away from the text it annotates. The text spacing makes no allowance for the annotations, with opaque lines crossing over it, with the effect of making both the annotations and the text hard to read if the annotations are above a certain degree of complexity.

As a result of the difficulties of these existing tools, in order to extract a piece of annotated text and rework it into a graph that could be embedded into a publication, users usually read off the annotations, then create a graph from scratch using vector drawing or image editing software.

To address these issues, we created a visualisation tool named *stav* (*stav* Text Annotation Visualizer), that can read the data formatted according to the Shared Task specification and aims to present it to the user in a form that can be grasped at a glance. Events and entities are annotated immediately above the text, and the roles within an event by labelled arcs between them (Figure 3c). In a very complex graph, users can highlight the object or association of interest to follow it even more easily. Special features of annotations, such as negation or speculation, are shown by unique visual cues, and more in-depth, technical information that is usually not required can be requested by floating the mouse cursor over the annotation (as seen in Figure 5).

We took care to minimise arc crossovers, and to

[11] http://www-tsujii.is.s.u-tokyo.ac.jp/GENIA/SharedTask/downloads.shtml

[12] http://code.google.com/p/whatswrong/

[13] http://u-compare.org/bionlp2009.html

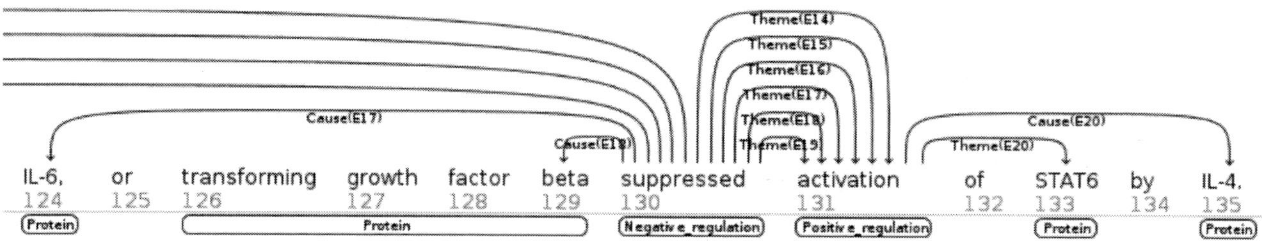

(a) Visualisation using What's Wrong with My NLP

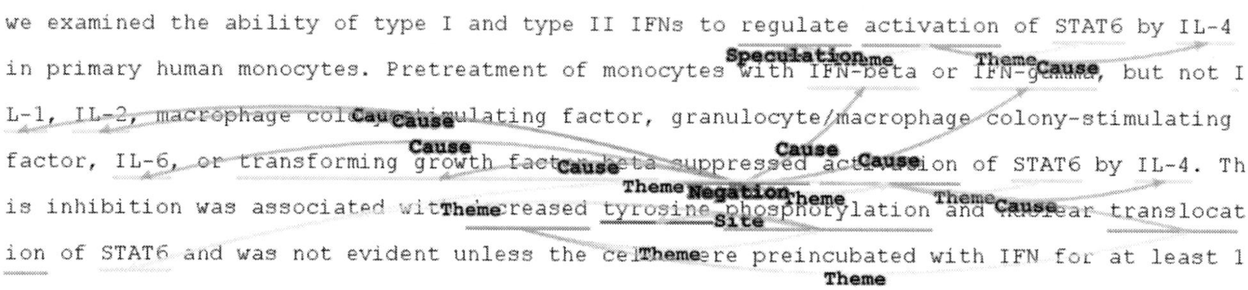

(b) Visualisation using U-Compare

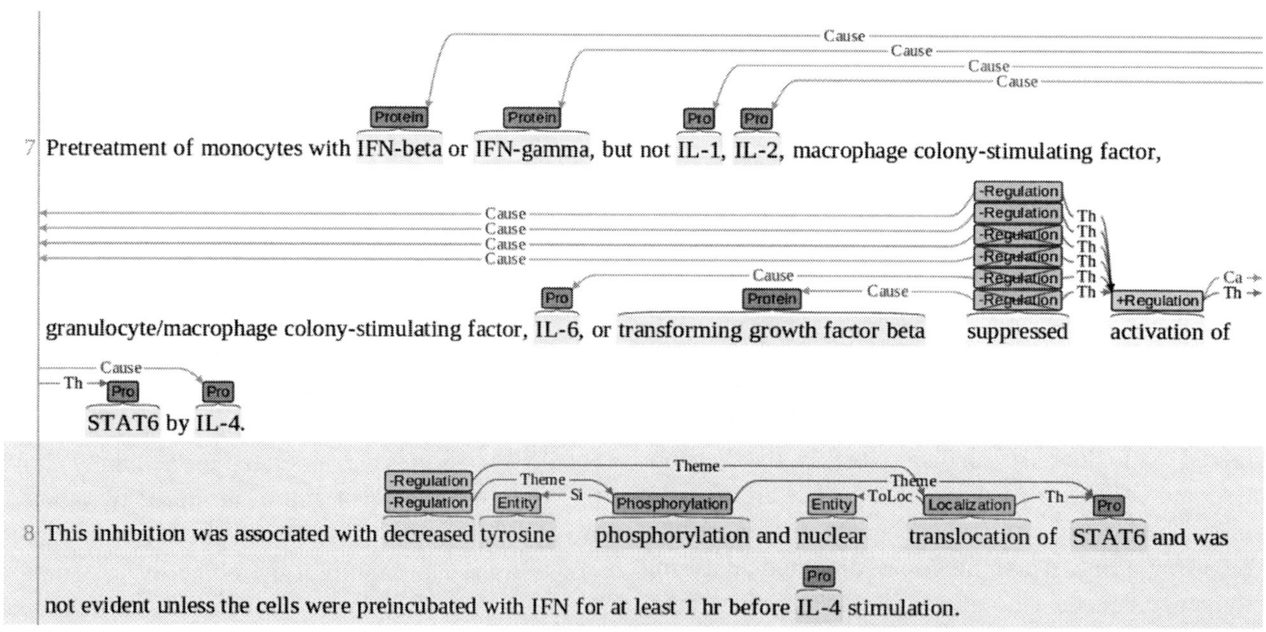

(c) Visualisation using *stav*

Figure 3: Different visualisations of complex textual annotations of Dickensheets et al. (1999)

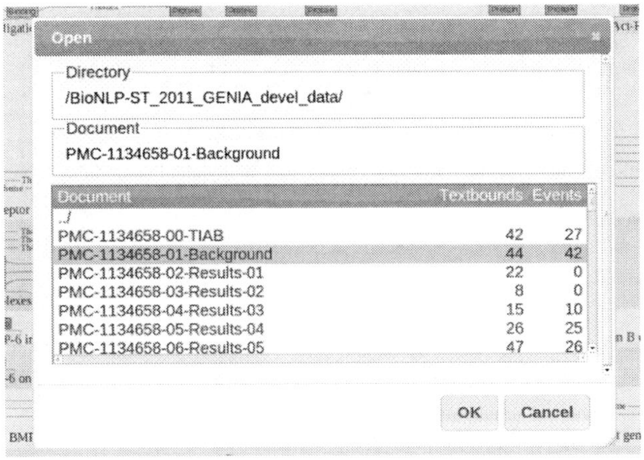

Figure 4: A screenshot of the *stav* file-browser

keep them away from the text itself, in order to maintain text readability. The text is spaced to accommodate the annotations between the rows. While this does end up using more screen real-estate, it keeps the text legible, and annotations adjacent to the text. The text is broken up into lines, and each sentence is also forced into a new line, and given a numerical identifier. The effect of this is that the text is laid out vertically, like an article would be, but with large spacing to accomodate the annotations. The arcs are similarly continued on successive lines, and can easily be traced – even in case of them spanning multiple lines, by the use of mouseover highlighting. To preserve the distributionality information of the annotation, any event annotations are duplicated for each event, as demonstrated in the example in Figure 2.

stav is not limited to the Shared Task datasets with appropriate configuration settings, it could also visualise other kinds of relational annotations such as: frame structures (Fillmore, 1976) and dependency parses (de Marneffe et al., 2006).

To achieve our objectives above, we use the Dynamic Scalable Vector Graphics (SVG) functionality (i.e. SVG manipulated by JavaScript) provided by most modern browsers to render the WYSIWYG (What You See Is What You Get) representation of the annotated document. An added benefit from this technique is that the installation process, if any, is very simple: although not all browsers are currently supported, the two that we specifically tested against are Safari[14] and Google Chrome[15]; the former comes preinstalled with the Mac OS X operating system, while the latter can be installed even by relatively non-technical users. The design is kept modular using a dispatcher pattern, in order to allow the inclusion of the visualiser tool into other JavaScript-based projects. The client-server architecture also allows centralisation of data, so that every user can inspect an uploaded dataset without the hassle of downloading and importing into a desktop application, simply by opening an URL which can uniquely identify a document, or even a single annotation. A screenshot of the *stav* file browser can be seen in Figure 4.

3.2 Evaluation Tools

The tasks of BioNLP-ST 2011 exhibit very high complexity, including multiple non-trivial subproblems that are partially, but not entirely, independent of each other. With such tasks, the evaluation of participating systems itself becomes a major challenge. Clearly defined evaluation criteria and their precise implementation is critical not only for the comparison of submissions, but also to help participants follow the status of their development and to identify the specific strengths and weaknesses of their approach.

A further challenge arising from the complexity of the tasks is the need to process the relatively intricate format in which annotations are represented, which in turn carries a risk of errors in submissions. To reduce the risk of submissions being rejected or the evaluation showing poor results due to formatting errors, tools for checking the validity of the file format and annotation semantics are indispensable.

For these reasons, we placed emphasis in the organisation of the BioNLP-ST'11 on making tools for format checking, validation and evaluation available to the participants already during the early stages of system development. The tools were made available in two ways: as downloads, and as online services. With downloaded tools, participants can perform format checking and evaluation at any time without online access, allowing more efficient optimisation processes. Each task in BioNLP-ST also

[14] http://www.apple.com/safari

[15] http://www.google.com/chrome

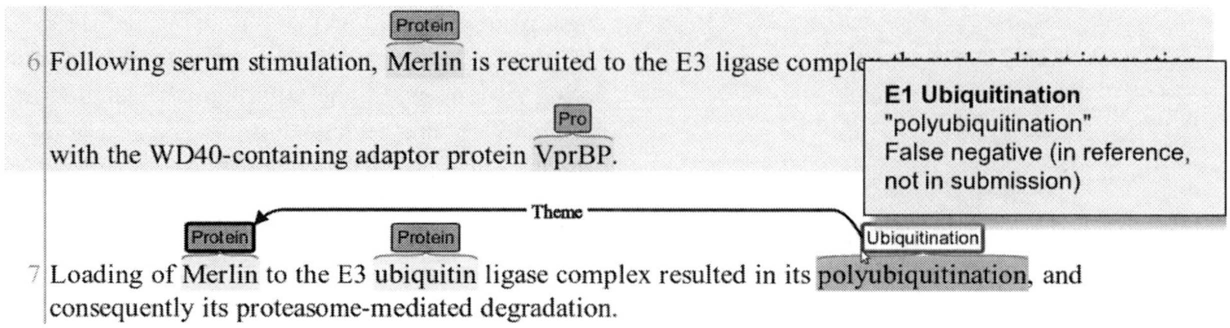

Figure 5: An example of a false negative illustrated by the evaluation tools in co-ordination with *stav*

maintained an online evaluation tool for the development set during the development period. The online evaluation is intended to provide an identical interface and criteria for submitted data as the final online submission system, allowing participants to be better prepared for the final submission. With online evaluation, the organisers could also monitor submissions to ensure that there were no problems in, for example, the evaluation software implementations.

The system logs of online evaluation systems show that the majority of the participants submitted at least one package with formatting errors, confirming the importance of tools for format checking. Further, most of the participants made use of the online development set evaluation at least once before their final submission.

To enhance the evaluation tools we drew upon the *stav* visualiser to provide a view of the submitted results. This was done by comparing the submitted results and the gold data to produce a visualisation where errors are highlighted, as illustrated in Figure 5. This experimental feature was available for the EPI and ID tasks and we believe that by doing so it enables participants to better understand the performance of their system and work on remedies for current shortcomings.

4 Discussion and Conclusions

Among the teams participating in the EPI and ID tasks, a great majority utilised tools for which resources were made available by the organisers. We hope that the continued availability of the parses will encourage further investigation into the applicability of these and similar tools and representations.

As for the analysis of the supporting analyses provided by external groups and the participants, we are so far aware of only limited use of these resources among the participants, but the resources will remain available and we are looking forward to see future work using them.

To enable reproducibility of our resources, we provide a publicly accessible repository containing the automated procedure and our processing scripts used to produce the released data. This repository also contains detailed instructions on the options and versions used for each parser and, if the software license permits it, includes the source code or binary that was used to produce the processed data. For the cases where the license restricts redistribution, instructions and links are provided on how to obtain the same version that was used. We propose that using a multitude of parses and formats can benefit not just the task of event extraction but other NLP tasks as well.

We have also made our evaluation tools and visualisation tool *stav* available along with instructions on how to run it and use it in coordination with the shared task resources. The responses from the participants in relation to the visualisation tool were very positive, and we see this as encouragement to advance the application of visualisation as a way to better reach a wider understanding and unification of the concept of events for biomedical event extraction.

All of the resources described in this paper are available at http://sites.google.com/site/bionlpst/.

Acknowledgements

We would like to thank Jari Björne of the University of Turku BioNLP group; Gerold Schneider, Fabio Rinaldi, Simon Clematide and Don Tuggener of the Univerity of Zurich Computational Linguistics group; Roser Morante of University of Antwerp CLiPS center; and Youngjun Kim of the University of Utah Natural Language Processing Research Group for their generosity with their time and expertise in providing us with supporting analyses.

This work was supported by Grant-in-Aid for Specially Promoted Research (MEXT, Japan) and the Royal Swedish Academy of Sciences.

References

Daniel M. Bikel. 2004. Intricacies of Collins' Parsing Model. *Computational Linguistics*, 30(4):479–511.

J. Björne, F. Ginter, S. Pyysalo, J. Tsujii, and T. Salakoski. 2010. Complex event extraction at PubMed scale. *Bioinformatics*, 26(12):i382.

Robert Bossy, Julien Jourde, Philippe Bessières, Marteen van de Guchte, and Claire Nédellec. 2011. BioNLP Shared Task 2011 - Bacteria Biotope. In *Proceedings of the BioNLP 2011 Workshop Companion Volume for Shared Task*, Portland, Oregon, June. Association for Computational Linguistics.

S. Buchholz and E. Marsi. 2006. CoNLL-X shared task on multilingual dependency parsing. In *Proceedings of the Tenth Conference on Computational Natural Language Learning*, pages 149–164. Association for Computational Linguistics.

E. Buyko and U. Hahn. 2010. Evaluating the impact of alternative dependency graph encodings on solving event extraction tasks. In *Proceedings of the 2010 Conference on Empirical Methods in Natural Language Processing*, pages 982–992. Association for Computational Linguistics.

Eugene Charniak and Mark Johnson. 2005. Coarse-to-Fine n-Best Parsing and MaxEnt Discriminative Reranking. In *Proceedings of the 43rd Annual Meeting of the Association for Computational Linguistics (ACL'05)*, pages 173–180.

Pallavi Choudhury, Michael Gamon, Chris Quirk, and Lucy Vanderwende. 2011. MSR-NLP entry in BioNLP Shared Task 2011. In *Proceedings of the BioNLP 2011 Workshop Companion Volume for Shared Task*, Portland, Oregon, June. Association for Computational Linguistics.

S. Clark and J.R. Curran. 2004. Parsing the WSJ using CCG and log-linear models. In *Proceedings of the 42nd Annual Meeting on Association for Computational Linguistics*, page 103. Association for Computational Linguistics.

Michael John Collins. 1996. A new statistical parser based on bigram lexical dependencies. In *Proceedings of the 34th Annual Meeting of the Association for Computational Linguistics*, pages 184–191, Santa Cruz, California, USA, June. Association for Computational Linguistics.

Marie-Catherine de Marneffe, Bill MacCartney, and Christopher D. Manning. 2006. Generating Typed Dependency Parses from Phrase Structure Parses. In *Proceedings of the Fifth International Conference on Language Resources and Evaluation (LREC'06)*, pages 449–454.

H.L. Dickensheets, C. Venkataraman, U. Schindler, and R.P. Donnelly. 1999. Interferons inhibit activation of STAT6 by interleukin 4 in human monocytes by inducing SOCS-1 gene expression. *Proceedings of the National Academy of Sciences of the United States of America*, 96(19):10800.

Ehsan Emadzadeh, Azadeh Nikfarjam, and Graciela Gonzalez. 2011. A generalizable and efficient machine learning approach for biological event extraction from text. In *Proceedings of the BioNLP 2011 Workshop Companion Volume for Shared Task*, Portland, Oregon, June. Association for Computational Linguistics.

Charles J. Fillmore. 1976. Frame semantics and the nature of language. *Annals of the New York Academy of Sciences*, 280(1):20–32.

R. Johansson and P. Nugues. 2007. Extended constituent-to-dependency conversion for English. In *Proc. of the 16th Nordic Conference on Computational Linguistics (NODALIDA)*.

Julien Jourde, Alain-Pierre Manine, Philippe Veber, Karën Fort, Robert Bossy, Erick Alphonse, and Philippe Bessières. 2011. BioNLP Shared Task 2011 - Bacteria Gene Interactions and Renaming. In *Proceedings of the BioNLP 2011 Workshop Companion Volume for Shared Task*, Portland, Oregon, June. Association for Computational Linguistics.

Yoshinobu Kano, William Baumgartner, Luke McCrohon, Sophia Ananiadou, Kevin Cohen, Larry Hunter, and Jun'ichi Tsujii. 2009. U-Compare: share and compare text mining tools with UIMA. *Bioinformatics*, 25(15):1997–1998, May.

Jin-Dong Kim, Tomoko Ohta, Sampo Pyysalo, Yoshinobu Kano, and Jun'ichi Tsujii. 2009. Overview of BioNLP'09 Shared Task on Event Extraction. In *Proceedings of Natural Language Processing in Biomedicine (BioNLP) NAACL 2009 Workshop*, pages 1–9.

Jin-Dong Kim, Sampo Pyysalo, Tomoko Ohta, Robert Bossy, and Jun'ichi Tsujii. 2011a. Overview of BioNLP Shared Task 2011. In *Proceedings of the BioNLP 2011 Workshop Companion Volume for Shared Task*, Portland, Oregon, June. Association for Computational Linguistics.

Jin-Dong Kim, Yue Wang, Toshihisa Takagi, and Akinori Yonezawa. 2011b. Overview of the Genia Event task in BioNLP Shared Task 2011. In *Proceedings of the BioNLP 2011 Workshop Companion Volume for Shared Task*, Portland, Oregon, June. Association for Computational Linguistics.

D. Klein and C.D. Manning. 2003. Fast exact inference with a factored model for natural language parsing. *Advances in neural information processing systems*, pages 3–10.

M.P Marcus, B. Santorini, and M.A Marcinkiewicz. 1993. Building a large annotated corpus of English: The Penn Tree Bank. *Computational Linguistics*, pages 313–318.

D. McClosky. 2009. *Any Domain Parsing: Automatic Domain Adaptation for Natural Language Parsing*. Ph.D. thesis, Ph. D. thesis, Department of Computer Science, Brown University.

M. Miwa, S. Pyysalo, T. Hara, and J. Tsujii. 2010. Evaluating Dependency Representation for Event Extraction. In *In the 23rd International Conference on Computational Linguistics (COLING 2010)*, pages 779–787.

Y. Miyao and J. Tsujii. 2008. Feature forest models for probabilistic HPSG parsing. *Computational Linguistics*, 34(1):35–80.

Yusuke Miyao, Rune Sætre, Kenji Sagae, Takuya Matsuzaki, and Jun'ichi Tsujii. 2008. Task-oriented evaluation of syntactic parsers and their representations. In *Proceedings of ACL-08: HLT*, pages 46–54, Columbus, Ohio, June. Association for Computational Linguistics.

R. Morante, V. Van Asch, and W. Daelemans. 2010. Memory-based resolution of in-sentence scopes of hedge cues. *CoNLL-2010: Shared Task*, page 40.

Ngan Nguyen, Jin-Dong Kim, and Jun'ichi Tsujii. 2011. Overview of the Protein Coreference task in BioNLP Shared Task 2011. In *Proceedings of the BioNLP 2011 Workshop Companion Volume for Shared Task*, Portland, Oregon, June. Association for Computational Linguistics.

Tomoko Ohta, Sampo Pyysalo, and Jun'ichi Tsujii. 2011. Overview of the Epigenetics and Post-translational Modifications (EPI) task of BioNLP Shared Task 2011. In *Proceedings of the BioNLP 2011 Workshop Companion Volume for Shared Task*, Portland, Oregon, June. Association for Computational Linguistics.

S. Petrov, L. Barrett, R. Thibaux, and D. Klein. 2006. Learning accurate, compact, and interpretable tree annotation. In *Proceedings of the 21st International Conference on Computational Linguistics and the 44th annual meeting of the Association for Computational Linguistics*, pages 433–440. Association for Computational Linguistics.

H. Poon and L. Vanderwende. 2010. Joint inference for knowledge extraction from biomedical literature. In *Human Language Technologies: The 2010 Annual Conference of the North American Chapter of the Association for Computational Linguistics*, pages 813–821. Association for Computational Linguistics.

Sampo Pyysalo, Tomoko Ohta, Rafal Rak, Dan Sullivan, Chunhong Mao, Chunxia Wang, Bruno Sobral, Jun'ichi Tsujii, and Sophia Ananiadou. 2011a. Overview of the Infectious Diseases (ID) task of BioNLP Shared Task 2011. In *Proceedings of the BioNLP 2011 Workshop Companion Volume for Shared Task*, Portland, Oregon, June. Association for Computational Linguistics.

Sampo Pyysalo, Tomoko Ohta, and Jun'ichi Tsujii. 2011b. Overview of the Entity Relations (REL) supporting task of BioNLP Shared Task 2011. In *Proceedings of the BioNLP 2011 Workshop Companion Volume for Shared Task*, Portland, Oregon, June. Association for Computational Linguistics.

Laura Rimell and Stephen Clark. 2009. Porting a lexicalized-grammar parser to the biomedical domain. *Journal of Biomedical Informatics*, 42(5):852 – 865. Biomedical Natural Language Processing.

R. Sætre, K. Yoshida, A. Yakushiji, Y. Miyao, Y. Matsubyashi, and T. Ohta. 2007. AKANE system: protein-protein interaction pairs in BioCreAtIvE2 challenge, PPI-IPS subtask. In *Proceedings of the Second BioCreative Challenge Workshop*, pages 209–212.

Kenji Sagae and Jun'ichi Tsujii. 2007. Dependency parsing and domain adaptation with LR models and parser ensembles. In *Proceedings of the CoNLL 2007 Shared Task*.

G. Schneider, M. Hess, and P. Merlo. 2007. Hybrid long-distance functional dependency parsing. *Unpublished PhD thesis, Institute of Computational Linguistics, University of Zurich*.

G. Szarvas, V. Vincze, R. Farkas, and J. Csirik. 2008. The BioScope corpus: annotation for negation, uncertainty and their scope in biomedical texts. In *Proceedings of the Workshop on Current Trends in Biomedical Natural Language Processing*, pages 38–45. Association for Computational Linguistics.

Y. Tateisi, A. Yakushiji, T. Ohta, and J. Tsujii. 2005. Syntax Annotation for the GENIA corpus. In *Proceedings of the IJCNLP*, pages 222–227.

Sentence Filtering for BioNLP: Searching for Renaming Acts

Pierre Warnier[1,2] **Claire Nédellec**[1]
[1]MIG INRA UR 1077, F78352 Jouy-en-Josas, France
[2]LIG Université de Grenoble, France
forename.lastname@jouy.inra.fr

Abstract

The Bacteria Gene Renaming (RENAME) task is a supporting task in the BioNLP Shared Task 2011 (BioNLP-ST'11). The task consists in extracting gene renaming acts and gene synonymy reminders in scientific texts about bacteria. In this paper, we present in details our method in three main steps: 1) the document segmentation into sentences, 2) the removal of the sentences exempt of renaming act (false positives) using both a gene nomenclature and supervised machine learning (feature selection and SVM), 3) the linking of gene names by the target renaming relation in each sentence. Our system ranked third at the official test with 64.4% of F-measure. We also present here an effective post-competition improvement: the representation as SVM features of regular expressions that detect combinations of trigger words. This increases the F-measure to 73.1%.

1 Introduction

The Bacteria Gene Renaming (Rename) supporting task consists in extracting gene renaming acts and gene synonymy reminders in scientific texts about bacteria. The history of bacterial gene naming has led to drastic amounts of homonyms and synonyms that are often missing in gene databases or even worse, erroneous (Nelson et al., 2000). The automatic extraction of gene renaming proposals from scientific papers is an efficient way to maintain gene databases up-to-date and accurate. The present work focuses on the recognition of renaming acts in the literature between gene synonyms that are recorded in the *Bacillus subtilis* gene databases. We assume that renaming acts do not involve unknown gene names. Instead, our system verifies the accuracy of synonymy relations as reported in gene databases by insuring that the literature attests these synonymy relations.

1.1 Example

This positive example of the training corpus is representative of the IE task:
*"Thus, a separate **spoVJ** gene as defined by the 517 mutation does not exist and is instead identical with **spoVK**."*

There are 2 genes in this sentence:

ID	Start	End	Name
T1	17	22	spoVJ
T2	104	109	spoVK

Table 1: Example of provided data.

There is also a renaming act: **R1** Renaming Former:T1 New:T2

Given all gene positions and identifications (Tn), the Rename task consists in predicting all renaming acts (Rn) between *Bacillus subtilis* genes in multi-sentence documents. The gene names involved are all acronyms or short names. Gene and protein names often have both a short and a long form. Linking short to long names is a relatively well-known task but linking short names together remains little explored (Yu et al., 2002). Moreover, specifying some of these synonymy relations as renaming appears quite rare (Weissenbacher, 2004). This task relates to the more general search of relations of

Proceedings of BioNLP Shared Task 2011 Workshop, pages 121–129,
Portland, Oregon, USA, 24 June, 2011.

synonymous nicknames, aliases or pseudonyms of proper nouns from definitory contexts in encyclopedia or dictionaries. For instance, in *Alexander III of Macedonia commonly known as Alexander the Great* the synonymy relation is supported by *commonly known as* between the proper noun *Alexander III of Macedonia* and the nickname *Alexander the Great*. Renaming act extraction differs from the search of coreferences or acronyms by the linguistic markers involved.

1.2 Datasets

The renaming corpus is a set of 1,648 PubMed references of bacterial genetics and genome studies. The references include the title and the abstract. The annotations provided are: the position and name of genes (see Table 1) for all sets and the renaming acts in the training and the development sets only.

	Train	Dev.	Test
Documents	1146	246	252
Genes	14372	3331	3375
Unique Genes	3415	1017	1126
New genes	0	480	73
Relations	308	65	88
Words / Doc	209	212	213
Genes / Doc	12.5	12.7	13.4
Unique Genes / Doc	3.0	4.1	4.5
Relations / Doc	0.27	0.26	0.35

Table 2: Datasets of the Rename task corpus.

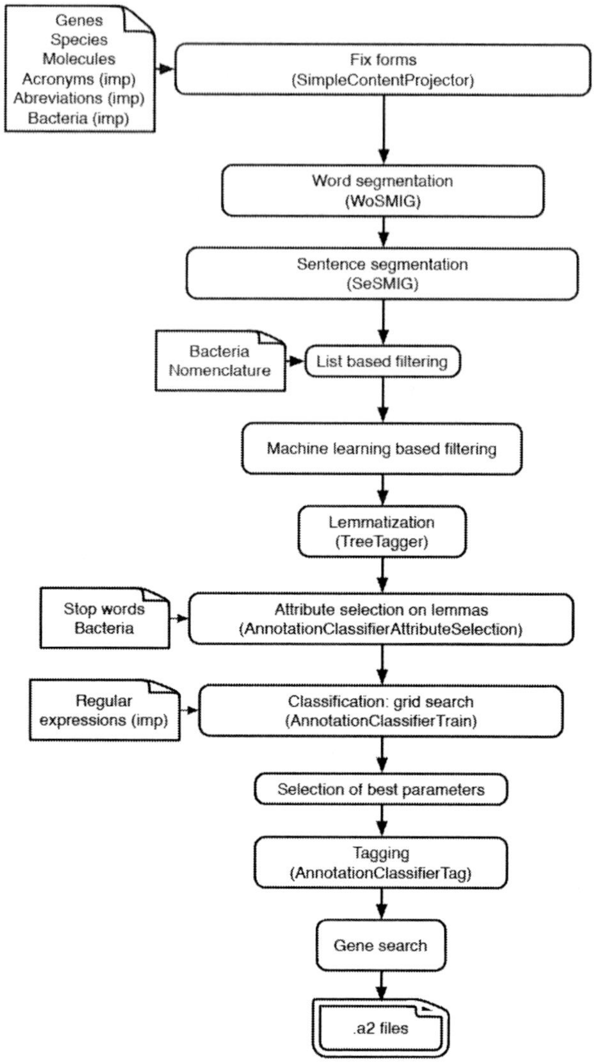

Figure 1: Flowchart: Notes represent the resources used and (imp) represent later improvements not used for the official submission.

2 Methods

An early finding is that renaming acts very seldom span several sentences (i.e. *former* and *new* are in the same sentence). For the training set, 95.4% of the relations verify this claim and in the development set, 96.1%. Thus, it is decided to first segment the documents into sentences and then to look for renaming acts inside independent sentences. Thus the maximum expected recall is then 96.1% on the development set. This is done by automatically filtering all the sentences out that do not contain evidence of a renaming act and then to relate the gene names occurring in the renaming sentences. The AlvisNLP pipeline (Nédellec et al., 2009) is used throughout this process (see Fig. 1).

2.1 Word and sentence segmentation

Word and sentence segmentation is achieved by the Alvis NLP pipeline. Named entity recognition supplements general segmentation rules.

2.1.1 Derivation of boundaries from named entities

Named entities often contains periods that should not be confused with sentence ends. Species abbreviations with periods are specially frequent in the task corpus. First, dictionaries of relevant named entities from the molecular biology domain (e.g.

genes, species and molecules) are projected onto the documents before sentence segmentation, so that periods that are part of named entities are disambiguated and not interpreted as sentence ends. Moreover, named enties are frequently multi-word. Named entity recognition prior to segmentation prevents irrelevant word segmentation. For example, the projection of named entity dictionaries on the excerpt below reveals the framed multi-word entities: *"Antraformin, a new inhibitor of* *Bacillus subtilis* *transformation.* [...] *During this screening program,* *Streptomyces sp.* *7725-CC1 was found to produce a specific inhibitor of* *B. subtilis* *transformation."*

2.1.2 Word segmenter

The word segmenter (WosMIG in Fig. 1) has the following properties: 1) primary separator: space, 2) punctuation isolation: customized list, 3) custom rules for balanced punctuation, 4) fixed words: not splittable segments The following list of terms is obtained from the example:
['Antraformin' , ',', 'a', 'new', 'inhibitor', 'of', 'Bacillus subtilis', 'transformation', '.', [...], 'During', 'this', 'screening', 'program', ',', 'Streptomyces sp.', '7725-CC1', 'was', 'found', 'to', 'produce', 'a', 'specific', 'inhibitor', 'of', 'B. subtilis', 'transformation', '.']

2.1.3 Sentence segmenter

The sentence segmenter (SeSMIG in Fig. 1) has the following properties: 1) strong punctuation: customized list; 2) tokens forcing the end of a sentence (e.g. *etc...*); 3) an upper case letter must follow the end of a sentence. The system works very well but could be improved with supervised machine learning to improve the detection of multi-word named entities. Finally, the list of words is split into sentences:
[['Antraformin' , ',', 'a', 'new', 'inhibitor', 'of', 'Bacillus subtilis', 'transformation', '.'],
[...],
['During', 'this', 'screening', 'program', ',', 'Streptomyces sp.', '7725-CC1', 'was', 'found', 'to', 'produce', 'a', 'specific', 'inhibitor', 'of', 'B. subtilis', 'transformation', '.']]

2.2 Sentence filtering

Once the corpus is segmented into sentences, the system filters out the numerous sentences that most likely do not contain any renaming act. This way, the further relation identification step focuses on relevant sentences and increases the precision of the results (Nedellec et al., 2001). Before the filtering, the recall is maximum (not 100% due to few renaming acts spanning two sentences), but the precision is very low. The sentence filters aim at keeping the recall as high as possible while gradually increasing the precision. It is composed of two filters. The first filter makes use of an a priori knowledge in the form of a nomenclature of known synonyms while the second filter uses machine learning to filter the remaining sentences. In the following, the term Bacillus subtilis gene nomenclature is used in the sense of an exhaustive inventory of names for Bacillus subtilis genes.

2.2.1 Filtering with a gene nomenclature

We developed a tool for automatically building a nomenclature of *Bacillus subtilis* gene and protein names. It aggregates the data from various gene databases with the aim of producing the most exhaustive nomenclature. The result is then used to search for pairs of synonyms in the documents. Among various information on biological sequences or functions, the entries of gene databases record the identifiers of the genes and proteins as asserted by the biologist community of the species. *Bacillus subtilis* community as opposed to other species has no nomenclature committee. Each database curator records unilateral naming decisions that may not reported elsewhere. The design of an exhaustive nomenclature require the aggregation of multiple sources.

Databases Our sources for the *Bacillus subtilis* nomenclature are six publicly available databases plus an in-house database. The public databases are generalist (1 to 3) or devoted to *Bacillus subtilis* genome (4 to 6) (see Table 3):

GenBank The genetic sequence database managed by the National Center for Biotechnology Information (NCBI) (Benson et al., 2008). It contains the three official versions of the annotated genome of *B. subtilis* with all gene canonical names;

UniProt the protein sequence database managed by the Swiss Institute of Bioinformatics (SIB), the European Bioinformatics Institute (EBI) and the Protein Information Resource (PIR) (Bairoch et al., 2005). It contains manual annotated protein sequences (Swiss-Prot) and automatically annotated protein sequences (TrEMBL (Bairoch and Apweiler, 1996)). Its policy is to conserve a history of all information relative to these sequences and in particular all names of the genes that code for these sequences.

Genome Reviews The genome database managed by the European Bioinformatics Institute (EBI) (Sterk et al., 2006). It contains the re-annotated versions of the two first official versions of the annotated genome of *B. subtilis*;

BSORF The Japanese *Bacillus subtilis* genome database (Ogiwara et al., 1996);

Genetic map the original genetic map of *Bacillus subtilis*;

GenoList A multi-genome database managed by the Institut Pasteur (Lechat et al., 2008). It contains an updated version of the last official version of the annotated genome of *B. subtilis*;

SubtiWiki A wiki managed by the Institute for Microbiology and Genetics in Göttingen (Flórez et al., 2009) for *Bacillus subtilis* reannotation. It is a free collaborative resource for the *Bacillus* community;

EA_List a local lexicon manually designed from papers curation by Anne Goelzer and Élodie Marchadier (MIG/INRA) for Systems Biology modeling (Goelzer et al., 2008).

Nomenclature merging We developed a tool for periodically dumping the content of the seven source databases through Web access. With respect to gene naming the entries of all the databases contain the same type of data per gene:

- a unique identifier (required);
- a canonical name, which is the currently recommended name (required);
- a list of synonyms considered as deprecated names (optional).

The seven databases are handled one after the other. The merging process follows the rules:

- the dump of the first database (SubtiWiki, see Table 3 for order) in the list is considered the most up-to-date and is used as the reference for the integration of the dumps of the other databases;
- for all next dumps, if the unique gene identifier is new, the whole entry is considered as new and the naming data of the entry is added to the current merge;
- else, if the unique identifier is already present into the merge, the associated gene names are compared to the names of the merge. If the name does not exist in the merge, it is added to the merge as a new name for this identifier and synonym of the current names. The synonym class is not ordered.

Order	Databases	AE	AN
1	SubtiWiki	4 261	5920
2	GenoList	0	264
3	EA_List	33	378
4	BSORF	0	42
5	UniProt	0	74
6	Genome Reviews	0	0
7	GenBank	0	7
8	Genetic Map	0	978
	Total	4 294	7 663

Table 3: Database figures. AE: number of added entries, AN: number of added names.

Synonym pair dictionary: The aggregated nomenclature is used to produce a dictionary of all combinations of pairs in the synonym classes.

Sentence filtering by gene cooccurrence: For each sentence in the corpus, if a pair of gene synonyms according to the lexicon is found inside then the sentence is kept for the next stage. Otherwise, it is definitively discarded. The comparison is a case-insensitive exact match preserving non alphanumeric symbols. The recall at this step is respectively 90.9% and 90.2% on the train and development sets. The recall loss is due to typographic errors in gene names in the nomenclature. The precision at this stage is respectively 38.9% and 38.1% on the train and development sets. There are still many false positives due to gene homologies or renaming acts concerning other species than *Bacillus subtilis* for instance.

2.2.2 Sentence filtering by SVM

Feature selection The second filtering step aims at improving the precision by machine learning classification of the remaining sentences after the first filtering step. Feature selection is applied to enhance the performances of the SVM as it is shown to suffer from high dimensionality (Weston et al., 2001). Feature selection is applied to a bag-of-word representation using the Information Gain metrics of the Weka library (Hall et al., 2009). Words are lemmatized by TreeTagger (Schmid, 1994). A manual inspection of the resulting sorting highly ranks words such as *formerly* or *rename* and parentheses while ranking other words such as *cold* or *encode* surprisingly certainly due to over-fitting. Although the feature selection is indeed not particularly efficient compared to the manual selection of relevant features but does help filtering out unhelpful words and then drastically reducing the space dimension from 1919 to 141 for the best run.

Sentence classification and grid search: A SVM algorithm (LibSVM) with a RBF kernel is applied to the sentences encoded as bag of words. The two classes are: "contains a renaming act" (True) or not (False). There are 4 parameters to tune: 1) the number of features to use ($N \in 1, 5, 10, ..., 150$) meaning the N first words according to the feature selection, 2) the weight of the classes: True is fixed to 1 and False is tuned ($W \in 0.2, 0.4, ..., 5.0$), 3) the errors weight ($C \in 2^{-5,-7,...,9}$), 4) the variance of the Gaussian kernel ($G \in 2^{-11,-9,...,1}$). Thus, to find the best combination of parameters for this problem, $\#N * \#W * \#C * \#G = 31 * 25 * 8 * 7 = 43,400$ models are trained using 10-fold cross-validation on the training and development sets together (given the relatively small size of the training set) and ranked by F-measure. This step is mandatory because the tuning of C and G alone yield variations of F-measure from 0 to the maximum. The grid search is run on a cluster of 165 processors and takes around 30 minutes. The best model is the model with the highest F-measure found by the grid search.

Test sentence filtering: Finally the test set is submitted to word and sentence segmentation, feature filtering and tagged by the best SVM model (AnnotationClassifierTag in Fig. 1). The sentences that are assumed to contain a renaming act are kept and the others are discarded (see Fig. 2).

2.3 Gene position searching

At this step, all remaining sentences are assumed to be true positives. They all contain at least one pair of genes that are synonymous according to our gene nomenclature. The other gene names are not considered. The method for relating gene candidates by a renaming relation, relies on the assumption that all gene names are involved in at least one relation. Most of the time, sentences contain only two genes. We assume in this case that they are related by a renaming act. When there are more than two genes in a sentence, the following algorithm is applied: 1) compute all combinations of couples of genes; 2) look-up the lexicon for those couples and discard those that are not present; 3) if a given gene in a couple has multiple occurrences, take the nearest instance from the other gene involved in the renaming act.

3 Discussion

The system ranks $3^{rd}/3$ among three participants in the Rename task official evaluation with a F-measure of 64.4% (see Fig. 4), five points behind the second. The general approach we used for this task is pragmatic: 1) simplify the problem by focusing on sentences instead of whole documents for a minimal loss, 2) then use a series of filters to improve the precision of the sentence classification while keeping the recall to its maximum, 3) and finally relate gene names known to be synonymous inside sentences for a minimal loss (around 2% of measure). As opposed to what is observed in Gene Normalization tasks (Hirschman et al., 2005), the Rename task is characterised by the lack of morphological resemblance of gene synonyms. The gene synonyms are not typographic variants and the recognition of renaming act requires gene context analysis. The clear bottleneck of our system is the sentence filtering part and in particular the feature selection that brings a lot of noise by ranking statistically spurious terms. On the plus side, the whole system is fully automated to the exception of the resources used for the word segmentation that were designed manually for other tasks. Moreover, our strategy does not assume that the gene pairs from the nomenclature may be mentioned for other reasons than renaming, it then tends to overgeneralize. However, many occurrences of the predicted gene pairs are not involved in renaming acts because the reasons for mentioning synonyms may be different than renaming. In particular, equivalent genes of other species (orthologues) with high sequence similarities may have the same name as in *Bacillus subtilis*. An obvious improvement of our method would consists in first relating the genes to their actual species before relating the only *Bacillus subtilis* gene synonyms by the renaming relation.

Team	Pre.	Rec.	F-M.
U. of Turku	95.9	79.6	87.0
Concordia U.	74.4	65.9	69.9
INRA	**57.0**	**73.9**	**64.4**

Table 4: Official scores in percentage on the test set.

3.1 Method improvement by IE patterns

After the official submission and given the result of our system compared to competitors, a simple modification of the feature selection was tested with significant benefits: the addition of regular expressions as additional features. Intuitively there are words or patterns that strongly appeal to the reader as important markers of renaming acts. For example, variations of *rename* or adverbs such *originally* or *formerly* would certainly be reasonable candidates. Fifteen such shallow patterns were designed (see Table 5) supplemented by six more complex ones, originally designed to single out gene names. In appendix A, one of them is presented, the precision of which is 95.3% and recall 27.5%. That is, more than a quarter of renaming acts in the training and development sets together. Interestingly, in table 5 the word *formerly* (3rd in feature selection ranking) alone recalls 10.7% of the renaming acts with a precision of 96.9%. In contrast, the words *originally* and *reannotated* although having 100% precision are respectively ranked 33rd and 777th. In total, 21 patterns are represented as boolean features of the classification step in addition to the ones selected by feature selection. Unsurprisingly, the best classifiers, according to the cross-validation F-measure after the grid search, only used the regular expressions as features neglecting the terms chosen by feature selection. A significant improvement is achieved: +8.7% of F-measure on the test set (see Fig. 2).

Pattern	Pre.	Rec.	F-M.
(reannotated)	100.0	0.4	0.7
(also called)	100.0	0.4	0.7
(formerly)	96.9	10.7	19.2
(originally)	100.0	1.4	2.8
((also)? known as)	100.0	1.8	3.4
(were termed)	100.0	0.4	0.7
(identity of)	100.0	0.7	1.4
(be referred (to|as)?)	100.0	0.4	0.7
(new designation)	100.0	0.4	0.7
(allel\w+)	80.0	2.8	5.4
(split into)	100.0	0.4	0.7
(rename)	83.4	1.8	3.4
(renamed)	88.5	8.0	14.6
(renaming)	100.0	0.4	0.7
(E(\.|scherichia) coli)	11.3	4.5	6.4

Table 5: Handwritten patterns. Scores are in percentage on the training and development sets together **after** the gene nomenclature filtering step. A very low precision means the pattern could be used to filter out rather than in.

3.2 Error analysis

The false positive errors of the sentence filtering step, using hand-written patterns can be classified as follows: 1) omission: *Characterization of* ***abn2 (yxiA)****, encoding a Bacillus subtilis GH43 arabinanase, Abn2, and its role in arabino-*

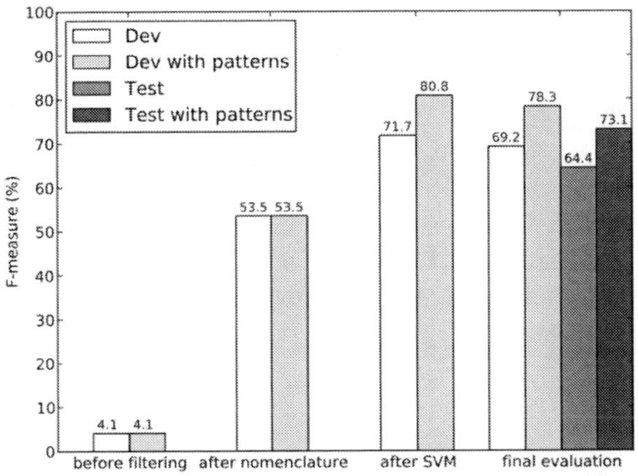

Figure 2: Evolution of F-measure at different measure points for the Rename task. Dev: training on train set and testing on dev set. Test: training on train + dev sets and testing on test set (no intermediary measure). 64.4% is the official submitted score. 73.1% is the best score achieved by the system on the test set.

polysaccharide degradation. (PMID 18408032). In this case the sentence has been filtered out by the SVM and then the couple **abn2/yxia** was not annotated as a renaming act, 2) incorrect information in the nomenclature: *These results substantiate the view that* ***sigE*** *is the distal member of a 2-gene operon and demonstrate that the upstream gene (****spoIIGA****) is necessary for sigma E formation.* (PMID 2448286). Here, the integration of the Genetic Map to the nomenclature has introduced a wrong synonymy relation between **spoIIGA** and **sigE**, 3) homology with another species: *We report the cloning of the wild-type allele of* ***divIVB1*** *and show that the mutation lies within a stretch of DNA containing two open reading frames whose predicted products are in part homologous to the products of the Escherichia coli minicell genes* ***minC*** *and minD.* (PMID 1400224). The name pair actually exists in the nomenclature but here, **divIVB1** is a gene of *B. subtilis* and **minC** is a gene of *E. Coli*, 4) another problem linked to the lexicon is the fact the synonym classes are not disjoint. Some deprecated names of given genes are reused as canonical names of other genes. For example, **purF** and **purB** referred to two different genes of *B. subtilis* but **purB** was also formerly known as **purF**: *The following gene order has been established: pbuG-****purB****-****purF****-purM-purH-purD-tre* (PMID 3125411). Hence, **purF** and **purB** are uncorrectly recognized as synonyms while they refer to two different genes in this context. Possible solutions for improving the system could be: 1) the inclusion of species names as SVM features, 2) the removal of some couples from the nomenclature (**PurF/purB** for instance), 3) evaluate the benefits of each resource part of the nomenclature.

4 Conclusion

Our system detects renaming acts of *Bacillus subtilis* genes with a final F-measure of 64.4%. After sentence segmentation, the emphasis is on sentence filtering using an exhaustive nomenclature and a SVM. An amelioration of this method using patterns as features of the machine learning algorithm was shown to improve significantly (+8.7%) the final performance. It was also shown that the bag of words representation is sub-optimal for text classification experiments (Fagan, 1987; Caropreso and Matwin, 2006) With the use of such patterns, the filtering step is now very efficient. The examination of the remaining errors showed the limits of the current shallow system. A deeper linguistic approach using syntactic parsing seems indicated to improve the filtering step further.

Acknowledgments

The authors would like to thank Julien Jourde for granting them the permission to use the *Bacteria subtilis* synonym nomenclature that he is currently building and Philippe Veber for his insightful advices on text classification. This research is partly founded by the French Oseo QUAERO project.

References

A. Bairoch and R. Apweiler. 1996. The SWISS-PROT protein sequence data bank and its new supplement TREMBL. *Nucleic Acids Research*, 24(1):21.

A. Bairoch, R. Apweiler, C.H. Wu, W.C. Barker, B. Boeckmann, S. Ferro, E. Gasteiger, H. Huang, R. Lopez, M. Magrane, and Others. 2005. The universal protein resource (UniProt). *Nucleic Acids Research*, 33(suppl 1):D154.

D.A. Benson, I. Karsch-Mizrachi, D.J. Lipman, J. Ostell, and D.L. Wheeler. 2008. GenBank. *Nucleic acids research*, 36(suppl 1):D25.

M. Caropreso and S. Matwin. 2006. Beyond the Bag of Words: A Text Representation for Sentence Selection. *Advances in Artificial Intelligence*, pages 324–335.

J.L. Fagan. 1987. Experiments in automatic phrase indexing for document retrieval: a comparison of syntactic and non-syntactic methods.

LA Flórez, SF Roppel, A.G. Schmeisky, C.R. Lammers, and J. Stülke. 2009. A community-curated consensual annotation that is continuously updated: the Bacillus subtilis centred wiki SubtiWiki. *Database: The Journal of Biological Databases and Curation*, 2009.

A Goelzer, B Brikci, I Martin-Verstraete, P Noirot, P Bessières, S Aymerich, and V Fromion. 2008. Reconstruction and analysis of the genetic and metabolic regulatory networks of the central metabolism of Bacillus subtilis. *BMC systems biology*, 2(1):20.

M Hall, E Frank, G Holmes, B Pfahringer, P Reutemann, and I H Witten. 2009. The WEKA data mining software: an update. *ACM SIGKDD Explorations Newsletter*, 11(1):10–18.

Lynette Hirschman, Alexander Yeh, Christian Blaschke, and Alfonso Valencia. 2005. Overview of BioCreAtIvE: critical assessment of information extraction for biology. *BMC bioinformatics*, 6 Suppl 1:S1, January.

P. Lechat, L. Hummel, S. Rousseau, and I. Moszer. 2008. GenoList: an integrated environment for comparative analysis of microbial genomes. *Nucleic Acids Research*, 36(suppl 1):D469.

C. Nedellec, M. Abdel Vetah, and Philippe Bessières. 2001. Sentence filtering for information extraction in genomics, a classification problem. *Principles of Data Mining and Knowledge Discovery*, pages 326–337.

C Nédellec, A Nazarenko, and R Bossy. 2009. Information Extraction. *Handbook on Ontologies*, pages 663–685.

K E Nelson, I T Paulsen, J F Heidelberg, and C M Fraser. 2000. Status of genome projects for nonpathogenic bacteria and archaea. *Nature biotechnology*, 18(10):1049–54, October.

A. Ogiwara, N. Ogasawara, M. Watanabe, and T. Takagi. 1996. Construction of the Bacillus subtilis ORF database (BSORF DB). *Genome Informatics*, pages 228–229.

H Schmid. 1994. Probabilistic part-of-speech tagging using decision trees.

P. Sterk, P.J. Kersey, and R. Apweiler. 2006. Genome reviews: standardizing content and representation of information about complete genomes. *Omics: a journal of integrative biology*, 10(2):114–118.

Davy Weissenbacher. 2004. La relation de synonymie en Génomique. *RECITAL*.

J. Weston, S. Mukherjee, O Chapelle, M. Pontil, T. Poggio, and V. Vapnik. 2001. Feature selection for SVMs. *Advances in neural information processing systems*, pages 668–674.

Hong Yu, Vasileios Hatzivassiloglou, Carol Friedman, Andrey Rzhetsky, and W.J. Wilbur. 2002. Automatic extraction of gene and protein synonyms from MEDLINE and journal articles. In *Proceedings of the AMIA Symposium*, page 919. American Medical Informatics Association.

A Gene or operon couple matching pattern

Pattern that uses bacteria gene naming rules (3 lower case + 1 upper case letters), short genes (3 lower case letters), long gene names, factorized operons (3 lower case + several upper case letters), gene names including special and/or numerical characters in presence or not of signal words such as *named*, *renamed*, *formerly*, *formally*, *here*, *herein*, *hereafter*, *now*, *previously*, *as*, *designated*, *termed* and/or *called*, only if the pattern does not begin with *and* or *orf*. Although this pattern could be used to directly filter in sentences containing a renaming act, its recall is too low thus it is used as a feature of the classifier instead.

```
and|orf\
GENE|OPERON-fact\
[|((now|as|previously|formerly|formally|here(in|after))\
((re)named|called|designated|termed) (now|as|previously|formerly|formally|here(in|after))\
GENE|OPERON-fact)|]
```

Table 6: Long pattern used for gene pair matching.

Terms matched	Pattern	PMID
short-GENE (short-GENE)	***cotA*** *(formerly **pig**)*	8759849
long-GENE (long-GENE)	***cotSA*** *(**ytxN**)*	10234840
fact-OPERON (fact-OPERON)	***ntdABC*** *(formally **yhjLKJ**)*	14612444
spe-GENE (spe-GENE)	***lpa-8*** *(**sfp**)*	10471562
GENE (GENE)	***cwlB*** [***lytC***]	8759849
GENE (now designated GENE)	***yfiA*** *(now designated **glvR**)*	11489864
GENE (previously GENE)	***nhaC*** *(previously **yheL**)*	11274110
GENE (formerly called GENE)	***bkdR*** *(formerly called **yqiR**)*	10094682
GENE (now termed GENE)	***yqgR*** *(now termed **glcK**)*	9620975
GENE (GENE) other forms	***fosB****(**yndN**)*	11244082
GENE (hereafter renamed GENE)	***yhdQ*** *(hereafter renamed **cueR**)*	14663075
GENE (herein renamed GENE)	***yqhN*** *(herein renamed **mntR**)*	10760146
GENE (formally GENE)	***ntdR*** *(formally **yhjM**)*	14612444
GENE (formerly GENE)	***mtnK*** *(formerly **ykrT**)*	11545674
GENE (renamed GENE)	***yfjS*** *(renamed **pdaA**)*	12374835
GENE (named GENE)	***yvcE*** *(named **cwlO**)*	16233686
GENE (GENE)	***pdaA*** *(**yfjS**)*	14679227

Table 7: Examples matched with the long pattern.

Complex Biological Event Extraction from Full Text using Signatures of Linguistic and Semantic Features

Liam R. McGrath and **Kelly Domico** and **Courtney D. Corley** and **Bobbie-Jo Webb-Robertson**
Pacific Northwest National Laboratory
902 Battelle BLVD, PO BOX 999
Richland, WA 99352
{liam | kelly.domico | court | bj}@pnl.gov

Abstract

Building on technical advances from the BioNLP 2009 Shared Task Challenge, the 2011 challenge sets forth to generalize techniques to other complex biological event extraction tasks. In this paper, we present the implementation and evaluation of a signature-based machine-learning technique to predict events from full texts of infectious disease documents. Specifically, our approach uses novel signatures composed of traditional linguistic features and semantic knowledge to predict event triggers and their candidate arguments. Using a leave-one out analysis, we report the contribution of linguistic and shallow semantic features in the trigger prediction and candidate argument extraction. Lastly, we examine evaluations and posit causes for errors in our complex biological event extraction.

1 Introduction

The BioNLP 2009 Shared Task (Kim et al., 2009) was the first shared task to address fine-grained information extraction for the bio-molecular domain, by defining a task involving extraction of event types from the GENIA ontology. The BioNLP 2011 Shared Task ((Kim et al., 2011)) series generalized this defining a series of tasks involving more text types, domains and target event types. Among the tasks for the new series is the Infection Disease task, proposed and investigated by (Pyysalo et al., 2011; Pyysalo et al., 2010; Bjorne et al., 2010).

Like the other tasks for the BioNLP Shared Task series, the goal is to extract mentions of relevant events from biomedical publications. To extract an event, the event trigger and all arguments must be identified in the text by exact offset and typed according to a given set of event and argument classes (Miwa et al., 2010). Entity annotations are given for a set of entity types that fill many of the arguments.

Here we describe Pacific Northwest National Laboratory's (PNNL) submission to the BioNLP 2011 Infectious Disease shared task. We describe the approach and then discuss results, including an analysis of errors and contribution of various features.

2 Approach

Our system uses a signature-based machine-learning approach. The system is domain-independent, using a primary task description vocabulary and training data to learn the task, but domain resources can be incorporated as additional features when available, as described here. The approach can be broken down into 4 components: an automated annotation pipeline to provide the basis for features, classification-based trigger identification and argument identification components, and a post-processing component to apply semantic constraints. The UIMA framework[1] is used to integrate the components into a pipeline architecture.

2.1 Primary Tasks

A definition of the events to be extracted is used to define candidates for classification and post-process the results of the classification. First a list of domain-specific entity classes is given. Entities of

[1]http://uima.apache.org/

Proceedings of BioNLP Shared Task 2011 Workshop, pages 130–137,

Event Class	Arguments
Gene_expression	Theme(Protein|Regulon-operon)
Transcription	Theme(Protein|Regulon-operon)
Protein_catabolism	Theme(Protein)
Phosphorylation	Theme(Protein), Site(entity)?
Localization	Theme(core_entity), AtLoc(entity)?, ToLoc(entity)?
Binding	Theme(core_entity)+, Site(entity)*
Regulation	Theme(core_entity|event), Cause(core_entity|event)?, Site(entity)?, CSite(entity)?
Positive_regulation	Theme(core_entity|event), Cause(core_entity|event)?, Site(entity)?, CSite(entity)?
Negative_regulation	Theme(core_entity|event), Cause(core_entity|event)?, Site(entity)?, CSite(entity)?
Process	Participant(core_entity)?

Table 1: Summary of the target events. Type restrictions on fillers of each argument type are shown in parenthesis. Multiplicity of each argument type is also marked (+ = one-to-many, ? = zero-to-one, * = zero-to-many, otherwise = one).

these classes are assumed to be annotated in the data, as is the case for the Infectious Disease task. Then, each event class is given, with a list of argument types for each. Each argument is marked with its multiplicity, indicating how many of this argument type is valid for each event, either: one – exactly one is required, one-to-many – one or more is required, zero-to-one – one is optional, and zero-to-many – one or many are optional. Also, restrictions on the classes of entities that can fill each argument are given, by listing: one or more class names – indicating the valid domain-specific entity classes from the definition, core_entity – indicating that any domain-specific entity in the definition is valid, event – indicating that any event in the definition is valid, or entity – indicating that any span from the text is valid. Table 1 shows the summary of the event extraction tasks for the Infectious Disease track.

2.2 Annotation

Linguistic and domain annotations are automatically applied to the document to be used for trigger and argument identification in framing the tasks for classification and generating features for each instance. Linguistic annotations include sentence splits, tokens, parts of speech, tree parses, typed dependencies (deMarneffe et al., 2006; MacKinlay et al., 2009), and stems. For the Infectious Disease task, the parses from the Stanford Parser (Klein and Manning, 2003) provided by the Supporting Analysis (Stenetorp et al., 2011) was used to obtain all of these linguistic annotations, except for the stems, which were obtained from the Porter Stemmer (van Rijsbergen et al., 1980).

For the Infectious Disease task, two sets of domain specific annotations are included: known trigger words for each event class and semantic tags from the Unified Medical Language System (UMLS) (Bodenreider, 2004). Annotations for known trigger words are created using a dictionary of word stem-event class pairs created from annotated training data. An entry is created in the dictionary every time a new stem is seen as a trigger for an event class. When a word with one of these stems is seen during processing, it is annotated as a typical trigger word for that event class.

Semantic tags are calculating using MetaMap 2010 (Aronson and Lang, 2010). MetaMap provides semantic tags for terms in a document with up to three levels of specificity, from most to least specific: concept, type and group (Torii et al., 2011). Word sense disambiguation is used to identify the best tags for each term. For example, consider the tags identified by MetaMap for the phrase *Human peripheral B cells*:

Human
  concept: Homo sapiens
  type: Human
  group: Living Beings
Peripheral
  type: Spatial Concept
  group: Concepts & Ideas
B-Cells
  concept: B-Lymphocytes
  type: Cell

group: Anatomy

In this example, semantic mappings were found for three terms: *Human*, *Peripheral* and *B-Cells*. *Human* and *B-Cells* were mapped to specific concepts, but *Peripheral* was mapped to a more general group.

Entities are also annotated at this point. For the Infectious Disease task, annotations for five entity types are given: Protein, Two-component system, Chemical, Organism, or Regulon/Operon.

2.3 Trigger Identification

Triggers are identified using an SVM classifier (Vapnik, 1995; Joachims, 1999). Candidate triggers are chosen from the words in the text by part-of-speech. Based on known triggers seen in the training data, all nouns, verbs, adjectives, prepositions and adverbs are selected as candidates. A binary model is trained for each event type, and candidate triggers are tested against each classifier.

The following features are used to classify candidate event triggers:

- **term** – the candidate trigger
- **stem** – the stem of the term
- **part of speech** – the part of speech of the term
- **capitalization** – capitalization of the term
- **punctuation** – individual features for the presence of different punctuation types
- **numerics** – the presence of a number in the term
- **ngrams** – 4-grams of characters from the term
- **known trigger types** – tags from list of known trigger terms for each event type
- **lexical context** – terms in the same sentence
- **syntactic dependencies** – the type and role (governor or dependent) of typed dependencies involving the trigger
- **semantic type** – type mapping from MetaMap
- **semantic group** – group mapping from MetaMap

For training data, both the Infectious Disease training set and the GENIA training set were used. Although the GENIA training set represents a different genre and is annotated with a slightly different vocabulary than the Infectious Disease task data, it is similar enough to provide some beneficial supervision. The Infectious Disease training data is relatively small at 154 documents so including the larger GENIA training set at 910 documents results in a much more larger training set. Testing on the Infectious Disease development data, a 1 point improvement in fscore in overall results is seen with the additional training data.

2.4 Argument Identification

Arguments are also identified using an SVM classifier. For each predicted trigger, candidate arguments are selected based on the argument types. For arguments that are restricted to being filled by some set of specific entity and event types, each annotated entity and predicted event is selected as a candidate. For arguments that can be filled by any span of text, each span corresponding to a constituent of the tree parse is selected as a candidate. Each pair of an event trigger and a candidate argument serves as an instance for the classification. A binary model is trained for each event type, and each pair is tested against each classifier.

Many of the features used are inspired by those used in semantic role labeling systems (Gildea and Jurafsky, 2002). Given an event trigger and a candidate argument, the following features are used to classify event arguments:

- **trigger type** – the predicted event type of the trigger
- **argument terms** – the text of the argument
- **argument type** – entity or event type annotation on the argument
- **argument super-type** – core entity or core argument
- **trigger and argument stems** – the stems of each
- **trigger and argument parts of speech** – the part of speech of each
- **parse tree path** – from the trigger to argument via least common ancestor in tree parse, as a list of phrase types
- **voice of sentence** – active or passive
- **trigger and argument partial paths** – from the trigger or argument to the least common ancestor in tree parse, as a list of phrase types

- **relative position of argument to trigger** – before or after
- **trigger sub-categorization** – representation of the phrase structure rule that describes the relationship between the trigger, its parent and its siblings.

The training data used is the same as for trigger identification: the Infectious Disease training set plus the Genia training set.

2.5 Post-processing

A post-processing component is used to turn output from the various classifiers into semantically valid output according to the target task. For each predicted trigger, the positive predictions for each argument model are collected, and the set is compared to the argument restrictions in the target task description.

For example, the types on argument predictions are compared to the argument restrictions in the target task, and non-conforming ones are dropped. Then the multiplicity of the arguments for each predicted event is checked against the task vocabulary. Where there were not sufficient positive argument predictions to make a full event, the best negative predictions from the model are tried. When a compliant set of arguments can not be created for a predicted event, it is dropped.

3 Results and Discussion

Results for the system on both the development data and the official test data for the task are shown in Table 2 and Table 5, respectively. For the development data, a system using gold-standard event triggers is included, to isolate the performance of argument identification. In all cases, the total fscore for non-regulation events were much higher than regulation events. On the official test data, the system performed the best in predicting Phosphorylation (fscore = 71.43), Gene Expression (fscore = 53.33) and Process events (fscore = 51.04), but was unable to find any Transcription and Regulation events. This is also evident in the results on the development data using predicted triggers; additionally, no matches were found for localization and binding events. The total fscore on the development data using gold triggers was 55.33, more than 13 points higher than when using predicted triggers. In the discussion that follows, we detail the importance of individual features and their contribution to evaluation fscores.

3.1 Feature Importance

The effect of each argument and trigger feature type on the Infectious Disease development data was determined using a leave-one-out approach. The argument and trigger feature effect results are shown in Table 3 and Table 4, respectively. In a series of experiments, each feature type is left out of the full feature set one-by-one. The difference in fscore between each of these systems and the full feature set system is the effect of the feature type; a high negative effect indicates a significant contribution to the system since the removal of the feature resulted in a lower fscore.

Features	**fscore**	**effect**
all features	41.66	
w/o argument terms	36.16	-5.50
w/o argument type	39.50	-2.16
w/o trigger partial path	40.65	-1.01
w/o argument part of speech	40.98	-0.68
w/o argument partial path	41.16	-0.50
w/o trigger sub-categorization	41.45	-0.21
w/o argument stem	41.48	-0.18
w/o argument super-type	41.63	-0.03
w/o trigger type	41.63	-0.03
w/o trigger part of speech	41.81	0.15
w/o trigger stem	41.81	0.15
w/o voice of sentence	41.85	0.19
w/o relative position	42.21	0.55
w/o parse tree path	42.67	1.01

Table 3: Effect of each argument feature type on Infectious Disease development data.

Within the argument feature set system, the parse tree path feature had a notable positive effect of 1.01. The features providing the greatest contribution were argument terms and argument type with effects of -5.50 and -2.16, respectively. Within the trigger feature set system, the lexical context and syntactic dependencies features showed the highest negative effect signifying positive contribution to the system. The text and known trigger types features showed a negative contribution to the system.

Event Class	Using Gold Triggers gold/ans./match	recall	prec.	fscore	Using Predicted Triggers gold/ans./match	recall	prec.	fscore
Gene_expression	134 / 110 / 100	74.63	90.00	81.60	134 / 132 / 85	64.18	64.39	64.29
Transcription	35 / 26 / 23	65.71	88.46	75.41	25 / 0 / 0	0.00	0.00	0.00
Protein_catabolism	0 / 0 / 0	0.00	0.00	0.00	0 / 0 / 0	0.00	0.00	0.00
Phosphorylation	13 / 13 / 13	100.00	100.00	100.00	13 / 14 / 13	100.00	92.86	96.30
Localization	1 / 1 / 0	0.00	0.00	0.00	1 / 10 / 0	0.00	0.00	0.00
Binding	17 / 6 / 0	0.00	0.00	0.00	17 / 3 / 0	0.00	0.00	0.00
Process	206 / 180 / 122	59.22	67.78	63.21	207 / 184 / 108	52.17	58.70	55.24
Regulation	81 / 61 / 20	24.69	32.79	28.17	80 / 0 / 0	0.00	0.00	0.00
Positive_regulation	113 / 91 / 36	31.86	39.56	35.29	113 / 42 / 13	11.50	30.95	16.77
Negative_regulation	90 / 71 / 32	35.56	45.07	39.75	90 / 42 / 11	12.22	26.19	16.67
TOTAL	690 / 559 / 346	50.14	61.72	55.33	680 / 427 / 230	33.97	53.86	41.66

Table 2: Results on Infectious Disease development data. The system is compared to a system using gold standard triggers to isolate performance of argument identification.

Features	fscore	effect
all features	41.66	
w/o lexical context	40.14	-1.52
w/o syntactic dependencies	40.28	-1.38
w/o ngrams	40.88	-0.78
w/o part of speech	41.48	-0.18
w/o capitalization	41.51	-0.15
w/o numerics	41.51	-0.15
w/o semantic group	41.55	-0.11
w/o punctuation	41.59	-0.07
w/o stem	41.74	0.08
w/o semantic type	41.82	0.16
w/o known trigger types	42.11	0.45
w/o text	42.31	0.65

Table 4: Effect of each trigger feature type on Infectious Disease development data.

3.2 Transcription and Regulation events

Lastly, we present representative examples of errors (e.g., false positive, false negative, poor recall) produced by our system in the Infectious Disease track core tasks. The discussion herein will cover evaluations where our system did not correctly predict (transcription and regulation) any events or partially predicted (binding and +/- regulation) event triggers and arguments. In the text examples that follow, triggers are underlined and arguments are italicized.

The following are transcription events from the document PMC1804205-02-Results-03 in the development data.

- In contrast to the phenotype of the pta ackA double mutant, *pbgP* <u>*transcription*</u> was reduced in the pmrD mutant (Fig. 3).
- Growth at pH 5.8 resulted in *pmrD* <u>*transcript levels*</u> that were approximately3.5-fold higher than in organisms grown at pH 7.7 (Fig. 4A).

In both the development and test data evaluations, our system did not predict any transcription events, resulting in a 0.0 fscore; however, the system achieved 75.41 fscore when the gold-standard triggers were provided to the evaluation. Because argument prediction performed well, the system will benefit most by improving transcription event trigger prediction.

The following are regulation events from the document PMC1804205-02-Results-01in the development data.

- ... we grew Salmonella cells harbouring chromosomal lacZYA transcriptional fusions to the *PmrA-<u>regulated</u>* genes *pbgP*, *pmrC* and *ugd* (Wosten and Groisman, 1999) in N-minimal media buffered at pH 5.8 or 7.7.
- We determined that Chelex 100 was effective at chelating iron because expression of the *pmrA-<u>independent</u>* iron-repressed iroA gene ...

Similar to the transcription task, our system did not predict any regulation events, resulting in a 0.0 fscore. Unlike transcription events though, our system performed poorly on both argument identification and trigger prediction. The system achieved a 28.17 fscore when gold-standard triggers were used

Event Class	gold	(match)	answer	(match)	recall	prec.	fscore
Gene_expression	152	80	148	80	52.63	54.05	53.33
Transcription	50	0	0	0	0.00	0.00	0.00
Protein_catabolism	5	1	12	1	20.00	8.33	11.76
Phosphorylation	16	10	12	10	62.50	83.33	71.43
Localization	7	4	22	4	57.14	18.18	27.59
Binding	56	7	14	7	12.50	50.00	20.00
Regulation	193	0	0	0	0.00	0.00	0.00
Positive_regulation	193	34	87	34	17.62	39.08	24.29
Negative_regulation	181	32	68	32	17.68	47.06	25.70
Process	516	234	401	234	45.35	58.35	51.04
TOTAL	1369	402	764	402	29.36	52.62	37.69

Table 5: Official results on Infectious Disease test data

in the evaluation. Hypotheses for poor performance on candidate argument prediction are addressed in the following sections.

We posit that false negative trigger identifications are due to the limited full text training data (i.e. transcription events) and the inability of our system to predict non-verb triggers (i.e. second transcription example above). The SVM classifier was unable to distinguish between true transcription event triggers and transcription-related terms and ultimately, did not predict any transcription event in the development or test evaluations. To improve transcription event prediction, immediate effort should focus on 1) providing additional training data (e.g., BioCreativečiteBioCreative) and 2) introduce a trigger word filter that defines a subset of event triggers that have the best hit rate in the corpus. The hit rate is the number of occurrences of the word in a sentence per event type, divided by the total count in the gold standard (Nguyen et al., 2010).

3.3 +/-Regulation and Binding

The following positive regulation event is from document PMC1874608-03-RESULTS-03 in the development data.

- Invasiveness for HEp-2 cells was reduced to 39.1% of the wild-type *level* by *mlc* mutation, whereas it was *increased* by 1.57-fold by *hilE* mutation (Figure 3B).

In the preceding example, our system correctly predicted the +regulation trigger and the theme hilE; however, the correct argument was a gene expression event, not the entity. Many errors in the positive and negative regulation events were of this type; the predicted argument was a theme and not an event.

Evaluation of our system's binding event predictions resulted in low recall (12.50 or 0.0) in the test and development evaluations. The proceeding binding events are from document PMC1874608-03-RESULTS-05 in the development data. In both of the examples, our system correctly predicted the trigger *binding*; however, no arguments were predicted. Evaluation on the development data with gold standard triggers also resulted in an fscore of 0.0; thus, further algorithm refinement is needed to improve binding scores.

- *Mlc* directly represses *hilE* by *binding* to the *P3 promoter*
- These results clearly demonstrate that *Mlc* can regulate directly the *hilE* P3 promoter by *binding* to the *promoter*.

The following binding event is from document PMC1874608-01-INTRODUCTION in the development data and is representative of errors across many of the tasks. Here, the trigger is correctly predicted; however, the candidate arguments did not match with the reference data. Upon closer look, the arguments were drawn from the entire sentence, rather than an independent clause. The syntactic parse feature was not sufficient to prevent over-predicting arguments for the trigger, a potential solution is to add the arguments syntactic dependency to the trigger as a feature to the candidate argument selection.

- Using two-hybrid analysis, it has been shown that *HilE* *interacts* with *HilD*, which suggests that HilE represses **hilA** expression by inhibiting the activity of **HilD** through a protein-protein interaction (19,20).

4 Summary

This article reports Pacific Northwest National Laboratory's entry to the BioNLP Shared Task 2011 Infectious Disease track competition. Our system uses a signature-based machine-learning approach incorporating traditional linguistic features and shallow semantic concepts from NIH's METAMAP Thesaurus. We examine the contribution of each of the linguistic and semantic features to the overall fscore for our system. This approach performs well on gene expression, process and phosphorylation event prediction. Transcription, regulation and binding events each achieve low fscores and warrant further research to improve their effectiveness. Lastly, we present a performance analysis of the transcription, regulation and binding tasks. Future work to improve our system's performance could include pre-processing using simple patterns (Nguyen et al., 2010), information extraction from figure captions (Kim and Yu, 2011) and text-to-text event extraction. The last suggested improvement is to add semantic features to the candidate argument prediction algorithm in addition to using rich features, such as semantic roles (Torii et al., 2011).

Acknowledgements

The authors thank the Signature Discovery Initiative, part of the Laboratory Directed Research and Development Program at Pacific Northwest National Laboratory (PNNL). PNNL is operated by Battelle for the U.S. Department of Energy under contract DE-ACO5-76RLO 1830.

References

Alan R Aronson and François-Michel Lang. 2010. An overview of metamap: historical perspective and recent advances. *J Am Med Inform Assoc*, 17(3):229–36, May.

J Bjorne, F Ginter, S Pyysalo, J Tsujii, and T Salakoski. 2010. Complex event extraction at pubmed scale. *Bioinformatics*, 26(12):i382–i390, Jun.

O. Bodenreider. 2004. The unified medical language system (UMLS): integrating biomedical terminology. *Nucleic acids research*, 32(suppl 1):D267.

M.C. deMarneffe, B. MacCartney, and C.D. Manning. 2006. Generating typed dependency parses from phrase structure parses. In *Proceedings of LREC 2006*.

D. Gildea and D. Jurafsky. 2002. Automatic labeling of semantic roles. *Computational Linguistics*, 28(3):245–288.

T. Joachims. 1999. Making large scale SVM learning practical. *Advances in Kernel Methods – Support Vector Learning*.

Daehyun Kim and Hong Yu. 2011. Figure text extraction in biomedical literature. *PLoS ONE*, 6(1):e15338, Jan.

JD Kim, T Ohta, S Pyysalo, Y Kano, and J Tsujii. 2009. Overview of bionlp'09 shared task on event extraction. *Proceedings of the Workshop on BioNLP: Shared Task*, pages 1–9.

Jin-Dong Kim, Sampo Pyysalo, Tomoko Ohta, Robert Bossy, and Jun'ichi Tsujii. 2011. Overview of BioNLP Shared Task 2011. In *Proceedings of the BioNLP 2011 Workshop Companion Volume for Shared Task*, Portland, Oregon, June. Association for Computational Linguistics.

D. Klein and C.D. Manning. 2003. Accurate unlexicalized parsing. In *Proceedings of the 41st Annual Meeting on Association for Computational Linguistics-Volume 1*, pages 423–430. Association for Computational Linguistics.

A MacKinlay, D Martinez, and T Baldwin. 2009. Biomedical event annotation with crfs and precision grammars. *Proceedings of the Workshop on BioNLP: Shared Task*, pages 77–85.

Makoto Miwa, Rune Saetre, Jin-Dong Kim, and Jun'ichi Tsujii. 2010. Event extraction with complex event classification using rich features. *J. Bioinform. Comput. Biol.*, 8(1):131–46, Feb.

Quang Long Nguyen, Domonkos Tikk, and Ulf Leser. 2010. Simple tricks for improving pattern-based information extraction from the biomedical literature. *J Biomed Semantics*, 1(1):9, Jan.

S. Pyysalo, T. Ohta, H.C. Cho, D. Sullivan, C. Mao, B. Sobral, J. Tsujii, and S. Ananiadou. 2010. Towards event extraction from full texts on infectious diseases. In *Proceedings of the 2010 Workshop on Biomedical Natural Language Processing*, pages 132–140. Association for Computational Linguistics.

Sampo Pyysalo, Tomoko Ohta, Rafal Rak, Dan Sullivan, Chunhong Mao, Chunxia Wang, Bruno Sobral, Jun'ichi Tsujii, and Sophia Ananiadou. 2011.

Overview of the Infectious Diseases (ID) task of BioNLP Shared Task 2011. In *Proceedings of the BioNLP 2011 Workshop Companion Volume for Shared Task*, Portland, Oregon, June. Association for Computational Linguistics.

Pontus Stenetorp, Goran Topić, Sampo Pyysalo, Tomoko Ohta, Jin-Dong Kim, and Jun'ichi Tsujii. 2011. BioNLP Shared Task 2011: Supporting Resources. In *Proceedings of the BioNLP 2011 Workshop Companion Volume for Shared Task*, Portland, Oregon, June. Association for Computational Linguistics.

Manabu Torii, Lanlan Yin, Thang Nguyen, Chand T Mazumdar, Hongfang Liu, David M Hartley, and Noele P Nelson. 2011. An exploratory study of a text classification framework for internet-based surveillance of emerging epidemics. *International Journal of Medical Informatics*, 80(1):56–66, Jan.

C.J. van Rijsbergen, S.E. Robertson, and M.F. Porter. 1980. New models in probabilistic information retrieval.

V. Vapnik. 1995. *The Nature of Statistical Learning Theory*. Springer, New York.

Using Kybots for Extracting Events in Biomedical Texts

Arantza Casillas (*) arantza.casillas@ehu.es
Arantza Díaz de Ilarraza (‡) a.diazdeilarraza@ehu.es
Koldo Gojenola (‡) koldo.gojenola@ehu.es

Maite Oronoz (‡) maite.oronoz@ehu.es
German Rigau (‡) german.rigau@ehu.es

IXA Taldea UPV/EHU
(*) Department of Electricity and Electronics
(‡) Department of Computer Languages and Systems

Abstract

In this paper we describe a rule-based system developed for the BioNLP 2011 GENIA event detection task. The system applies Kybots (Knowledge Yielding Robots) on annotated texts to extract bio-events involving proteins or genes. The main goal of this work is to verify the usefulness and portability of the Kybot technology to the domain of biomedicine.

1 Introduction

The aim of the BioNLP'11 Genia Shared Task (Kim *et al.*, 2011b) concerns the detection of molecular biology events in biomedical texts using NLP tools and methods. It requires the identification of events together with their gene or protein arguments. Nine event types are considered: localization, binding, gene expression, transcription, protein catabolism, phosphorylation, regulation, positive regulation and negative regulation.

When identifying the events related to the given proteins, it is mandatory to detect also the event triggers, together with its associated event-type, and recognize their primary arguments. There are "simple" events, concerning an event together with its arguments (Theme, Site, ...) and also "complex" events, or events that have other events as secundary arguments. Our system did not participate in the optional tasks of recognizing negation and speculation.

The training dataset contained 909 texts together with a development dataset of 259 texts. 347 texts were used for testing the system.

The main objective of the present work was to verify the applicability of a new Information Extraction (IE) technology developed in the KYOTO project[1] (Vossen *et al.*, 2008), to a new specific domain. The KYOTO system comprises a general and extensible multilingual architecture for the extraction of conceptual and factual knowledge from texts, which has already been applied to the environmental domain.

Currently, our system follows a rule-based approach (i.e. (Kim *et al.*, 2009), (Kim *et al.*, 2011a), (Cohen *et al.*, 2011) or (Vlachos, 2009)), using a set of manually developed rules.

2 System Description

Our system proceeds in two phases. Firstly, text documents are tokenized and structured using an XML layered structure called *KYOTO Annotation Format* (KAF) (Bosma et al., 2009). Secondly, a set of *Kybots* (Knowledge Yielding Robots) are applied to detect the biological events of interest occurring in the KAF documents. Kybots form a collection of general morpho-syntactic and semantic patterns on sequences of KAF terms. These patterns are defined in a declarative format using Kybot profiles.

2.1 KAF

Firstly, basic linguistic processors apply segmentation and tokenization to the text. Additionally, the offset positions of the proteins given by the task organizers are also considered. The output of this basic processing is stored in KAF, where words, terms, syntactic and semantic information can be stored in separate layers with references across them.

Currently, our system only considers a minimal amount of linguistic information. We are only using

[1] http://www.kyoto-project.eu/

Proceedings of BioNLP Shared Task 2011 Workshop, pages 138–142,

the word form and term layers. Figure 1 shows an example of a KAF document where proteins have been annotated using a special POS tag (`PRT`). Note that our approach did not use any external resource apart of the basic linguistic processing.

```
<?xml version="1.0" encoding="UTF-8" standalone="no"?>
<KAF xml:lang="en">
<text>
...
<wf wid="w210" sent="10">phosphorylation</wf>
<wf wid="w211" sent="10">of</wf>
<wf wid="w212" sent="10">I</wf>
<wf wid="w213" sent="10">kappaB</wf>
<wf wid="w214" sent="10">alpha<...
</text>
<term tid="t210" type="open" lemma="phosphorylation"
   start="1195" end="1210" pos="W">
<span><target id="w210"/></span>
</term>
<term tid="t211" type="open" lemma="of"
   start="1211" end="1213" pos="W">
<span><target id="w211"/></span>
</term>
<term tid="T5" type="open" lemma="I kappaB alpha"
   start="1214" end="1228" pos="PRT">
<span><target id="w212"/></span>
   <target id="w213"/>
   <target id="w214"/></span>
</term>...
</terms>
</KAF>
```

Figure 1: Example of a document in KAF format.

2.2 Kybots

Kybots (Knowledge Yielding Robots) are abstract patterns that detect actual concept instances and relations in KAF. The extraction of factual knowledge by the mining module is done by processing these abstract patterns on the KAF documents. These patterns are defined in a declarative format using Kybot profiles, which describe general morpho-syntactic and semantic conditions on sequences of terms. Kybot profiles are compiled to XQueries to efficiently scan over KAF documents uploaded into an XML database. These patterns extract and rank the relevant information from each match.

Kybot profiles are described using XML syntax and each one consists of three main declarative parts:

- *Variables*: In this part, the entities and its properties are defined
- *Relations*: This part specifies the positional relations among the previously defined variables
- *Events*: describes the output to be produced for every matching

Variables (see the Kybot section *variables* in figure 2) describe the term variables used by the Kybot. They have been designed with the aim of being flexible enough to deal with many different information associated with the KAF terms including semantic and ontological statements.

Relations (see the Kybot section *relations* in figure 2) define the sequence of variables the Kybot is looking for. For example, in the Kybot in figure 2, the variable named `Phosphorylation` is the main pivot, the variable `Of` must follow `Phosphorylation`(`immediate` is `true` indicating that it must be the next term in the sequence), and a variable representing a `Protein` must follow `Of`. Proteins and genes are identified with the `PRT` tag.

Events (expressions marked as *events* in figure 2) describes the output template of the Kybot. For every matched pattern, the kybot produces a new event filling the template structure with the selected pieces of information. For example, the Kybot in figure 2 selects some features of the event represented with the variable called `Phosphorylation`: its term-identification (*@tid*), its lemma, part of speech and offset. The expression also describes that the variable `Protein` plays the role of being the "Theme" of the event. The output obtained when aplying the Kybot in figure 2 is shown in figure 3. Comparing the examples in table 1 and in figure 3 we observe that all the features needed for generating the files for describing the results are also produced by the Kybot.

```
<doc shortname="PMID-9032271.kaf">
  <event eid="e1" target="t210" kybot="phosphorylation_of_P"
  type="Phosphorylation"
  lemma="phosphorylation" start="1195" end="1210" />
  <role target="T5" rtype="Theme"
  lemma="I kappaB alpha" start="1214" end="1228" />
</doc>
```

Figure 3: Output obtained after the application of the Kybot in figure 2.

3 GENIA Event Extraction Task and Results

We developed a set of basic auxiliary programs to extract event patterns from the training corpus. These programs obtain the struc-

```
<?xml version="1.0" encoding="utf-8"?>
<!-- Sentence:  phosphorylation of Protein
       Event1:  phosphorylation
       Role:  Theme Protein -->
<Kybot id="bionlp">
<variables>
    <var name="Phosphorylation" type="term" lemma="phosphorylat*">
    <var name="Of" type="term" lemma="of"/>
    <var name="Protein" type="term" pos="PRT"/>
</variables>
<relations>
    <root span="Phosphorylation"/>
    <rel span="Of" pivot="Phosphorylation" direction="following" immediate="true"/>
    <rel span="Protein" pivot="Of" direction="following" immediate="true"/>
</relations>
<events>
    <event eid="" target="$Phosphorylation/@tid" kybot="phosphorylation_of_P"
    type="Phosphorylation" lemma="$Phosphorylation/@lemma"
    pos="$Phosphorylation/@pos" start="$Phosphorylation/@start" end="$Phosphorylation/@end"/>
    <role target="$Protein/@tid" rtype="Theme" lemma="$Protein/@lemma" start="$Protein/@start"
    end="$Protein/@end"/>
</events>
</Kybot>
```

Figure 2: Example of a Kybot for the pattern `Event of Protein`.

.a1 file
T5 Protein **1214 1228 I kappaB alpha**
.a2 file
T20 **Phosphorylation 1195 1210** phosphorylation
E7 Phosphorylation:T20 **Theme:T5**

Table 1: Results in the format required in the GENIA shared task.

ture of the events, their associated trigger words and their frequency. For example, in the training corpus, a pattern of the type `Event of Protein` appears 35 times, where the *Event* is further described as `phosporylation`, `phosphorylated...`. Taking the most frequently occurring patterns in the training data into account, we manually developed the set of Kybots used to extract the events from the development and test corpora. For example, in this way we wrote the Kybot in figure 2 that fulfils the conditions of the pattern of interest.

The two phases mentioned in section 2, corresponding to the generation of the KAF documents and the application of Kybots, have different input files depending on the type of event we want to detect: *simple* or *complex* events. When extracting *simple* events (see figure 4), we used the input text and the files containing protein annotations (".a1" files in the task) to generate the KAF documents. These KAF documents and Kybots for simple events are provided to the mining module. In the case of *complex* events (events that have other events as arguments), the identifiers of the detected simple events are added to the KAF document in the first phase. A new set of Kybots describing complex events and KAF (now with annotations of the simple events) are used to obtain the final result (see figure 5).

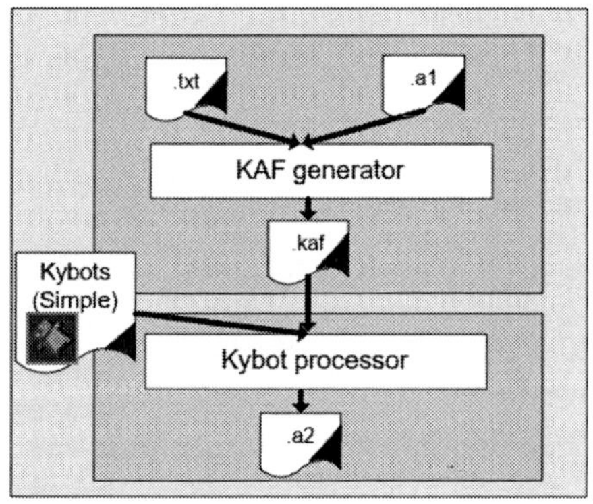

Figure 4: Application of Kybots. Simple events.

For the evaluation, we also developed some programs for adapting the output of the Kybots (see figure 3) to the required format (see table 1).

We used the development corpus to improve the Kybot performance. We developed 65 Kybots for detecting simple events. Table 2 shows the number of Kybots for each event type. Complex events relative to regulation (also including negative and positive regulations) were detected using a set of 24 Kybots.

The evaluation of the task was based on the output

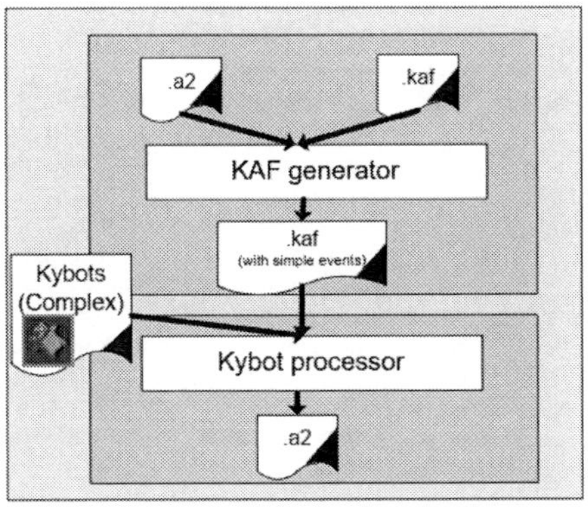

Figure 5: Application of Kybots. Complex events.

Event Class	Simple Kyb.	Complex Kyb.
Transcription	10	
Protein Catabolism	5	
Binding	5	
Regulation		3
Negative Regulation	5	4
Positive Regulation	3	17
Localization	7	
Phosphorylation	18	
Gene Exrpesion	12	
Total	65	24

Table 2: Number of Kybots generated for each event.

Event Class	Gold	R	P	F-score
Localization	191	19.90	97.44	33.04
Binding	491	5.30	50.00	9.58
Gene Expression	1002	54.19	42.22	47.47
Transcription	174	13.22	62.16	21.80
Protein catabolism	15	26.67	44.44	33.33
Phosphorylation	185	41.62	95.06	57.89
Non-reg total	2058	34.55	47.27	39.92
Regulation	385	7.53	9.63	8.45
Positive regulation	1443	6.38	62.16	11.57
Negative regulation	571	3.15	26.87	5.64
Regulatory total	2399	5.79	26.94	9.54
All total	4457	19.07	42.08	26.25

Table 3: Performance analysis on the test dataset.

of the system when applied to the test dataset of 347 previously unseen texts. Table 3 shows in the `Gold` column the number of instances for each event-type in the test corpus. `R`, `P` and `F-score` columns stand for the recall, precision and f-score the system obtained for each type of event. As a consequence of the characteristics of our system, precision is primed over recall. For example, the system obtains 95% and 97% precision on Phosphorylation an Localization events, respectively, although its recall is considerably lower (41% and 19%).

4 Conclusions and Future work

This work presents the first results of the application of the KYOTO text mining system for extracting events when ported to the biomedical domain. The KYOTO technology and data formats have shown to be flexible enough to be easily adapted to a new task and domain. Although the results are far from satisfactory, we must take into account the limited effort we dedicated to adapting the system and designing the kybots, which can be roughly estimated in two person/months.

After the final evaluation, our system obtained the thirteenth position out of 15 participating systems in the main task (processing PubMed abstracts and full paper articles), obtaining 19.07%, 42.08% and 26.25 recall, precision an f-score, respectively, far from the best competing system (49.41%, 64.75% and 56.04%). Although they are far from satisfactory, we must take into account the limited time we dedicated to adapting the system and designing the kybots. Apart from that, due to time restrictions, our system did not make use of the ample set of resources available, such as named entities, coreference resolution or syntactic parsing of the sentences. On the other hand, the system, based on manually developed rules, obtains reasonable accuracy in the task of processing full paper articles, obtaining 45% precision and 21% recall, compared to 59% and 47% for the best system, which means that the rule-based approach performs more robustly when dealing with long texts (5 full texts correspond to approximately 150 abstracts). As we have said before, our main objective was to evaluate the capabilities of the KYOTO technology without adding any additional information. The use of more linguistic information will probably facilitate our work and will benefit the system results. In the near future we will study the application of machine learning techniques for the automatic generation of Kybots from the training data. We also plan to include additional linguistic and semantic processing in the event extraction process to exploit the current semantic and ontological capabilities of the KYOTO technology.

Acknowledgments

This research was supported by the the KYOTO project (STREP European Community ICT-2007-211423) and the Basque Government (IT344-10).

References

Wauter Bosma, Piek Vossen, Aitor Soroa, German Rigau, Maurizio Tesconi, Andrea Marchetti, Monica Monachini and Carlo Aliprandi. *KAF: a generic semantic annotation format* Proceedings of the 5th International Conference on Generative Approaches to the Lexicon GL 2009 Pisa, Italy, September 17-19, 2009

Kevin Bretonnel Cohen, Karin Verspoor, Helen L. Johnson, Chris Roeder, Philip V. Ogren, Willian A. Baumgartner, Elizabeth White, Hannah Tipney, and Lawerence Hunter. High-precision biological event extraction: Effects of system and data. *Computational Intelligence*, to appear, 2011.

Jin-Dong Kim, Tomoko Ohta, Sampo Pyysalo, Yoshinobu Kano and Jun'ichi Tsujii. Overview of BioNLP'09 Shared Task on Event Extraction. *Proceedings of the BioNLP 2009 Workshop*. Association for Computational Linguistics. Boulder, Colorado, pp. 89–96., 2011

Jin-Dong Kim, Sampo Pyysalo, Tomoko Ohta, Robert Bossy, and Junichi Tsujii. 2011a. Overview of BioNLP Shared Task 2011. *Proceedings of the BioNLP 2011 Workshop Companion Volume for Shared Task.* Association for Computational Linguistics. Portland, Oregon, 2011.

Jin-Dong Kim, Yue Wang, Toshihisa Takagi, and Akinori Yonezawa. 2011b. Overview of the Genia Event task in BioNLP Shared Task 2011. *Proceedings of the BioNLP 2011 Workshop Companion Volume for Shared Task.* Association for Computational Linguistics. Portland, Oregon, 2011.

Andreas Vlachos. Two Strong Baselines for the BioNLP 2009 Event Extraction Task. *Proceedings of the 2010 Workshop on Biomedical Natural Language Processing.* Association for Computational Linguistics Uppsala, Sweden, pp. 1–9., 2010

Piek Vossen, Eneko Agirre, Nicoletta Calzolari, Christiane Fellbaum, Shu-kai Hsieh, Chu-Ren Huang, Hitoshi Isahara, Kyoko Kanzaki, Andrea Marchetti, Monica Monachini, Federico Neri, Remo Raffaelli, German Rigau, Maurizio Tescon, Joop VanGent. KYOTO: A System for Mining, Structuring, and Distributing Knowledge Across Languages and Cultures. *Proceedings of LREC 2008.* Marrakech, Morocco, 2008.

Extracting Biological Events from Text Using Simple Syntactic Patterns

Quoc-Chinh Bui, Peter M.A. Sloot
Computational Science, Informatics Institute
University of Amsterdam
Science Park 904, Amsterdam, The Netherlands
{c.buiquoc,p.m.a.sloot}@uva.nl

Abstract

This paper describes a novel approach presented to the BioNLP'11 Shared Task on GENIA event extraction. The approach consists of three steps. First, a dictionary is automatically constructed based on training datasets which is then used to detect candidate triggers and determine their event types. Second, we apply a set of heuristic algorithms which use syntactic patterns and candidate triggers detected in the first step to extract biological events. Finally, a post-processing is used to resolve regulatory events. We achieved an F-score of 43.94% using the online evaluation system.

1 Introduction

The explosive growth of biomedical scientific literature has attracted a significant interest on developing methods to automatically extract biological relations in texts. Until recently, most research was focused on extracting binary relations such as protein-protein interactions (PPIs), gene-disease, and drug-mutation relations. However, the extracted binary relations cannot fully represent the original biomedical data. Therefore, there is an increasing need to extract fine-grained and complex relations such as biological events (Miwa et al., 2010). The BioNLP'09 Shared Task (Kim et al., 2009) was the first shared task that provided a consistent data set and evaluation tools for extraction of such biological relations.

Several approaches to extract biological events have been proposed for this shared task. Based on their characteristics, these approaches can be divided into 3 groups. The first group uses a rule-based approach which implements a set of manually defined rules developed by experts or automatically learned from training data. These rules are then applied on dependency parse trees to extract biological events (Kaljurand et al., 2009; Kilicoglu and Bergler, 2009). The second group uses a machine learning (ML)-based approach which exploits various specific features and learning algorithms to extract events (Björne at al., 2009; Miwa et al., 2010). The third group uses hybrid methods that combine both rule- and ML-based approaches to solve the problem (Ahmed et al., 2009; Móra et al., 2009). Among these proposed approaches, the ML achieved the best results, however, it is non-trivial to apply.

In this paper, we propose a rule-based approach which uses two syntactic patterns derived from a parse tree. The proposed approach consists of the following components: a dictionary to detect triggers, text pre-processing, and event extraction.

2 System and method

2.1 Dictionary for event trigger detection

The construction of the dictionary consists of the following steps: grouping annotated triggers, filtering out irrelevant triggers, and calculating supportive scores. First, we collect all annotated triggers in the training and development datasets, convert them to lowercase format and group them based on their texture values and event types. For each trigger in a group, we count its frequency being annotated as trigger and its frequency being found in the training datasets to compute a confident score.

Next, we create a list of non-trigger words from the training dataset which consists of a list of prepositions (e.g. *to*, *by*), and a list of adjectives (e.g. *high*, *low*). We then filter out triggers that belong to the non-trigger list as well as triggers that consist of more than two words as suggested in the previous studies (Kilicoglu and Bergler, 2009). We further filter out more triggers by setting a frequency threshold for each event type. Triggers that

Proceedings of BioNLP Shared Task 2011 Workshop, pages 143–146,
Portland, Oregon, USA, 24 June, 2011. ©2011 Association for Computational Linguistics

have a frequency lower than a given threshold (which is empirically determined for each event type) are excluded.

In addition, for each binding trigger (i.e. trigger of binding event) we compute a *t2score* which is the ratio of having a second argument. For each regulatory trigger we compute an *escore* which is the ratio of having an event as the first argument (theme) and a *cscore* is the ratio of having a second argument (cause).

2.2 Text preprocessing

Text preprocessing includes splitting sentences, replacing protein names with place-holders, and parsing sentences using the Stanford Lexical Parser[1]. First, we split the input text (e.g. title, abstract, paragraph) into single sentences using LingPipe sentence splitter[2]. Sentences that do not contain protein names are dropped. Second, we replace protein names with their given annotated IDs in order to prevent the parser from segmenting multiple word protein names. Finally, the sentences are parsed with the Stanford parser to produce syntactic parse trees. All parse trees are stored in a local database for later use.

Detection of event trigger and event type: For each input sentence, we split the sentence into tokens and use the dictionary to detect a candidate trigger and determine its event type (hereafter we referred to as 'trigger' type). After this step, we obtain a list of candidate triggers and their related scores for each event type.

2.3 Event extraction

To extract the biological events from a parse tree after obtaining a list of candidate triggers, we adapt two syntactic patterns based on our previous work on extracting PPIs (Bui et al., 2011). These patterns are applied for triggers in noun, verb, and adjective form. In the following sections we describe the rules to extract events in more detail.

***Rule 1**: Extracting events from a noun phrase (NP)*
If the candidate trigger is a noun, we find a NP which is a joined node of this trigger and at least one protein from the parse tree. There are two NP patterns that can satisfy the given condition which are shown in Figure 1. In the first case (form1), NP does not contain a PP tag, and in the second case (form2), the trigger is the head of this NP. Depending on the trigger type (simple, binding or regulatory event), candidate events are extracted by the following rules as shown in Table 1.

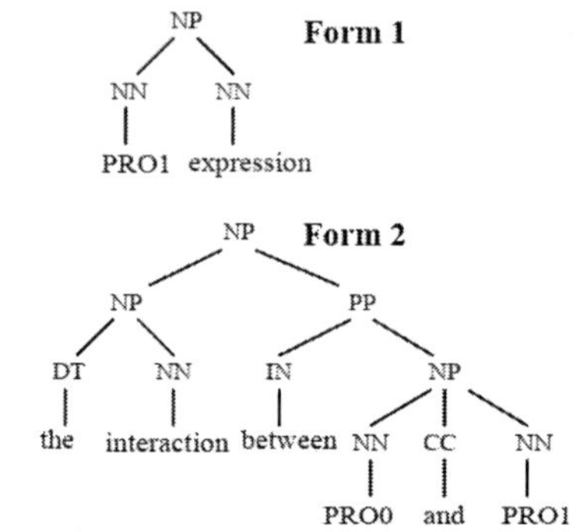

Figure 1: NP patterns containing trigger

Event type	Conditions and Actions
Simple or Regulatory	***NP in form1***: extract all proteins on the left of the trigger from NP. Form event pairs <trigger, protein>. ***NP in form2***: extract all proteins on the right of the trigger from NP. Form event pairs <trigger, protein>.
Binding	***NP in form1***: If proteins are in compound form i.e. PRO1/PRO2, PRO1-PRO2 then form an event triple <trigger, protein1, protein2>. Otherwise, form events pairs <trigger, protein>. ***NP in form2***: If NP contains one of the following preposition pairs: *between/and, of/with, of/to*, and the trigger' *t2score* >0.2 then split the proteins from NP into two lists: list1 and list2 based on the second PP (preposition phrase) or CC (conjunction). Form triples <trigger, protein1, protein2>, in which protein1 from list1 and protein2 from list2. Otherwise, form events the same way as simple event case.

Table 1: Conditions and actions to extract events from a NP. Simple and regulatory events use the same rules.

***Rule 2**: Extracting events from a verb phrase (VP)*
If the candidate trigger is a verb, we find a VP which is a direct parent of this trigger from the parse tree and find a sister NP immediately preceding this VP. Next, candidate events are extracted by the following rules as shown in Table 2.

[1] http://nlp.stanford.edu/software/lex-parser.shtml
[2] http://alias-i.com/lingpipe/

The event trigger is an adjective: For a candidate trigger which is an adjective, if the trigger is in a compound form (e.g. PRO1-mediated), we apply *rule1* to extract events. In this case, the compound protein (e.g. PRO1) is used as *cause* argument. Otherwise, we apply *rule 2* to extract.

2.4 Post-processing

Post-processing includes determination of an event type for a shared trigger and checking cross-references of regulatory events. For each extracted event which has a shared trigger[3], this event is verified using a list of modified words (e.g. *gene*, *mRNA*) to determine final event type. For regulatory events, the post-processing is used to find cross reference events. The post-processing is shown in Algorithm 1.

Event type	Conditions and Actions
Simple	If VP contains at least one protein then extract all proteins which have a position on the right of the trigger from the VP to create a protein list. Otherwise, extract all proteins that belong to the NP. Form event pairs <trigger, protein> with the obtained protein list.
Binding	If VP contains at least one protein then extract all proteins which have a position on the right of the trigger from VP to create a protein list1. Extracting all proteins that belong to the NP to create protein list2. If both list1 and list2 are not empty then form triples <trigger, protein1, protein2>, in which protein1 from list1 and protein2 from list2. Otherwise, form event pairs <trigger, protein> from the non-empty protein list.
Regulatory	If trigger' *cscore*>0.3 then extract the same way as for the binding event, in which protein from list1 is used for cause argument. Otherwise follows the rule of the simple event.

Table 2: Conditions and actions to extract events from a VP

2.5 Algorithm to extract events

The whole process of extracting biological event is shown in Algorithm 1

[3] A shared trigger is a trigger that appears in more than one group, see section 2.1.

Algorithm 1. *// Algorithm to extract biological events from sentence.*

Input: pre-processing sentence, parse tree, and lists of candidate triggers for each event type

Output: lists of candidate events of corresponding event type

Init: *found_list* = *null* // store extracted events for reference later

```
Step 1: Extracting events
  For each event type
    For each trigger of the current event type
      Extract candidate events using extraction rules
      If candidate event found
          Store this event to the found_list
      End if
    End for
  End for

Step 2: Post-preprocessing
  For each extracted event from found_list
    If event has a shared trigger
      Verify this event with the modified words
      If not satisfy
          Remove this event from found_list
      End if
    End if
    If event is a regulatory event and escore>0.3
      Check its argument (protein) for cross-reference
      If found
          Replace current protein with found event
      End if
    End if
  End for
```

3 Results and discussion

Table 3 shows the latest results of our system obtained from the online evaluation system (the official evaluation results are 38.19%). The results show that our method performs well on simple and binding events with an F-score of 63.03%. It outperforms previously proposed rule-based systems on these event types despite the fact that part of the test set consists of full text sentences. In addition, our system adapts two syntactic patterns which were previously developed for PPIs extraction. This means that the application of syntactic information is still relevant to extract biological events. In other words, there are some properties these extraction tasks share. However, the performance significantly decreases on regulatory events with an F-score of 26.61%.

Analyzing the performance of our system on regulatory events reveals that in most of false positive cases, the errors are caused by not resolving reference events properly. These errors can be reduced if we have a better implementation of the post-processing phase. Another source of errors is that the proposed method did not take into account the dependency among events. For example, most transcription events occurred when the regulatory events occurred (more than 50% cases). If association rules are applied here then the precision of both event types will increase.

Event Class	Recall	Precision	Fscore
Gene_expression	67.27	75.82	71.29
Transcription	46.55	79.41	58.70
Protein_catabolism	40.00	85.71	54.55
Phosphorylation	74.05	80.59	77.18
Localization	44.50	81.73	57.63
Binding	35.23	51.18	41.74
EVT-TOTAL	56.17	71.80	63.03
Regulation	19.22	27.11	22.49
Positive_regulation	22.52	33.89	27.06
Negative_regulation	24.34	33.74	28.28
REG-TOTAL	22.43	32.73	26.61
ALL-TOTAL	38.01	52.06	43.94

Table 3: Evaluation results on test set

To improve the overall performance of the system, there are many issues one should take into account. The first issue is related to the distance or the path length from the joined node between an event trigger and its arguments. By setting a threshold for the distance for each event type we increase the precision of the system. The second issue is related to setting thresholds for the extraction rules (e.g. *t2score, cscore*) which is done by using empirical data. Many interesting challenges remain to be solved, among which are the coreference, anaphora resolution, and cross sentence events. Furthermore, the trade-off between recall and precision needs to be taken into account, setting high thresholds for a dictionary might increase the precision, but could however drop the recall significantly.

4 Conclusion

In this paper we have proposed a novel system which uses syntactic patterns to extract biological events from a text. Our method achieves promising results on simple and binding events. The results also indicate that syntactic patterns for extracting PPIs and biological events share some common properties. Therefore systems developed for extracting PPIs can potentially be used to extract biological events.

Acknowledgements

The authors sincerely thank Dr. Sophia Katrenko and Rick Quax for their useful comments. This work was supported by the European Union through the DynaNets project, EU grant agreement no: 233847, and the Vietnamese Oversea Training Program.

References

S. Ahmed et al. 2009. BioEve: Bio-Molecular Event Extraction from Text Using Semantic Classification and Dependency Parsing. In *Proceedings of the Workshop on BioNLP'09 Shared Task*, pp. 99-102.

G. Móra et al. 2009. Exploring ways beyond the simple supervised learning approach for biological event extraction. In *Proceedings of the Workshop on BioNLP'09 Shared Task*, pp.137-140.

J. Kim et al. 2009. Overview of BioNLP'09 shared task on event extraction. *Proceedings of the Workshop on BioNLP'09 Shared Task*, pp. 1-9.

K. Kaljurand et al. 2009. UZurich in the BioNLP 2009 shared task. In *Proceedings of the Workshop on BioNLP'09 Shared Task*, pp. 28-36.

H. Kilicoglu and S. Bergler. 2009. Syntactic dependency based heuristics for biological event extraction. 2009. In *Proceedings of the Workshop on BioNLP'09 Shared Task*, pp. 119-127.

Q.C. Bui, S. Katrenko, and P.M.A. Sloot. 2011. A hybrid approach to extract protein-protein interactions. *Bioinformatics*. **27**(2), pp. 259-265.

M. Miwa, R. Sætre, J.-D. Kim, and J. Tsujii. 2010. Event Extraction with Complex Event Classification Using Rich Features. *Journal of Bioinformatics and Computational Biology*, **8**, pp. 131-146.

J. Björne, F. Ginter, S. Pyysalo, J. Tsujii, T. Salakoski. 2010. Complex event extraction at PubMed scale. *Bioinformatics*, **26**, pp. i382-390.

Detecting Entity Relations as a Supporting Task for Bio-Molecular Event Extraction

Sofie Van Landeghem[1,2], Thomas Abeel[1,2,3], Bernard De Baets[4] and Yves Van de Peer[1,2]
1. Dept. of Plant Systems Biology, VIB, Belgium
2. Dept. of Plant Biotechnology and Genetics, Ghent University, Belgium
3. Broad Institute of MIT and Harvard, Cambridge, MA, USA
4. Dept. of Applied Mathematics, Biometrics and Process Control, Ghent University, Belgium
yves.vandepeer@psb.ugent.be

Abstract

Recently, the focus in the BioNLP domain has shifted from binary relations to more expressive event representations, largely owing to the international popularity of the BioNLP Shared Task (ST) of 2009. This year, the ST'11 provides a further generalization on three key aspects: text type, subject domain, and targeted event types. One of the supporting tasks established to provide more fine-grained text predictions is the extraction of entity relations. We have implemented an extraction system for such non-causal relations between named entities and domain terms, applying semantic spaces and machine learning techniques. Our system ranks second of four participating teams, achieving 37.04% precision, 47.48% recall and 41.62% F-score.

1 Introduction

Understanding complex noun phrases with embedded gene symbols is crucial for a correct interpretation of text mining results (Van Landeghem et al., 2010). Such non-causal relations between a noun phrase and its embedded gene symbol are referred to as *entity relations*. As a supporting task for the BioNLP ST'11, we have studied two types of such entity relations: Subunit-Complex and Protein-Component. These relationships may occur within a single noun phrase, but also between two different noun phrases. A few examples are listed in Table 1; more details on the datasets and definitions of entity relations can be found in (Pyysalo et al., 2011).

Valid entity relations involve one GGP (gene or gene product) and one domain term (e.g. "promoter") and they always occur within a single sentence. In the first step towards classification of entity relations, we have calculated the semantic similarity between domain terms (Section 2). Supervised learning techniques are then applied to select sentences likely to contain entity relations (Section 3). Finally, domain terms are identified with a novel rule-based system and linked to the corresponding GGP in the sentence (Section 4).

2 Semantic analysis

To fully understand the relationship between a GGP and a domain term, it is necessary to account for synonyms and lexical variants. We have implemented two strategies to capture this textual variation, grouping semantically similar words together.

The first method takes advantage of manual annotations of semantic categories in the GENIA event corpus. This corpus contains manual annotation of various domain terms such as promoters, complexes and other biological entities in 1000 PubMed articles (Kim et al., 2008).

The second method relies on statistical properties of nearly 15.000 articles, collected by searching PubMed articles involving *human transcription factor blood cells*. From these articles, we have then calculated a semantic space using latent semantic analysis (LSA) as implemented by the S-Space Package (Jurgens and Stevens, 2010). The algorithm results in high-dimensional vectors that represent word contexts, and similar vectors then refer to semantically similar words. We have applied the Markov Cluster algorithm (MCL) (van Dongen, 2000) to group semantically similar terms together.

Proceedings of BioNLP Shared Task 2011 Workshop, pages 147–148,
Portland, Oregon, USA, 24 June, 2011. ©2011 Association for Computational Linguistics

Type of relation	Examples
Subunit-Complex	"the c-fos content of [AP-1]" / "c-jun, a component of the transcription factor [AP-1]"
Protein-Component	"the [IL-3 promoter]" / "the activating [ARRE-1 site] in the IL-2 promoter"

Table 1: Examples of entity relations. GGPs are underlined and domain terms are delimited by square brackets.

3 Machine learning framework

Our framework tries to define for each GGP in the data whether it is part of any of the two entity relations, by analysing the sentence context. To capture the lexical information for each sentence, we have derived bag-of-word features. In addition, 2- and 3-grams were extracted from the sentence. Finally, the content of the gene symbol was also used as lexical information. All lexical information in the feature vectors has undergone generalization by blinding the gene symbol with "protx" and all other co-occurring gene symbols with "exprotx". Furthermore, terms occurring in the semantic lexicons described in Section 2 were mapped to the corresponding cluster number or category. For each generalization, a blinded and a non-blinded variant is included in the feature vector.

Dependency graphs were further analysed for the extraction of grammatical patterns consisting of two nodes (word tokens) and their intermediate edge (grammatical relation). For the nodes, the same generalization rules as in the previous paragraph are applied. Finally, similar patterns are generated with the nodes represented by their part-of-speech tag.

The final feature vectors, representing sentences with exactly one tagged gene symbol, are classified using an SVM with a radial basis function as kernel. An optimal parameter setting (C and $gamma$) for this kernel was obtained by 5-fold cross-validation on the training data.

4 Entity detection

Once a sentence with a gene symbol is classified as containing a certain type of entity relation, it is necessary to find the exact domain term that is related to that gene symbol. To this end, we have designed a pattern matching algorithm that searches within a given window (number of tokens) around the gene symbol. The window size is increased to a predefined maximum as long as a maximal number of domain terms was not found.

Within the search window, a rule-based algorithm decides whether a given token qualifies as a relevant domain term, employing first a high-precision dictionary and then high-recall dictionaries.

5 Results

Our system achieves a global performance of 37.04% precision, 47.48% recall and 41.62% F-score, coming in second place after the university of Turku who obtained an F-score of 57.71%, and ranking before Concordia University who scores 32.04%. It remains an open question why the final results of the top ranked systems differ so much.

Acknowledgments

SVL and TA would like to thank the Research Foundation Flanders (FWO) for funding their research. TA is a post doctoral fellow of the Belgian American Education Foundation. The authors thank Jari Björne for his help with the manuscript.

References

David Jurgens and Keith Stevens. 2010. The S-Space package: an open source package for word space models. In *Proceedings of the ACL 2010 System Demonstrations*, ACLDemos '10, pages 30–35.

Jin-Dong Kim, Tomoko Ohta, and Jun'ichi Tsujii. 2008. Corpus annotation for mining biomedical events from literature. *BMC Bioinformatics*, 9(1):10.

Sampo Pyysalo, Tomoko Ohta, and Jun'ichi Tsujii. 2011. Overview of the Entity Relations (REL) supporting task of BioNLP Shared Task 2011. In *Proceedings of the BioNLP 2011 Workshop Companion Volume for Shared Task*, June.

Stijn van Dongen. 2000. *Graph Clustering by Flow Simulation*. Ph.D. thesis, University of Utrecht.

Sofie Van Landeghem, Sampo Pyysalo, Tomoko Ohta, and Yves Van de Peer. 2010. Integration of static relations to enhance event extraction from text. In *Proceedings of the 2010 Workshop on Biomedical Natural Language Processing*, BioNLP '10, pages 144–152.

A Pattern Approach for Biomedical Event Annotation

Quang Le Minh
Faculty of Information Technology
University of Science
Ho Chi Minh City, Vietnam
leeminhquang@gmail.com

Son Nguyen Truong
Faculty of Information Technology
University of Science
Ho Chi Minh City, Vietnam
ntson@fit.hcmus.edu.vn

Quoc Ho Bao
Faculty of Information Technology
University of Science
Ho Chi Minh City, Vietnam
hbquoc@fit.hcmus.edu.vn

Abstract

We describe our approach for the GENIA Event Extraction in the Main Task of BioNLP Shared Task 2011. There are two important parts in our method: Event Trigger Annotation and Event Extraction. We use rules and dictionary to annotate event triggers. Event extraction is based on patterns created from dependent graphs. We apply UIMA Framework to support all stages in our system.

1 Introduction

BioNLP Shared Task 2011 has been the latest event following the first attracted event in 2009-2010. We enrolled and submitted the results of Entity Relations Supporting Task and GENIA Event Extraction. In brief, the GENIA task requires the recognition of 9 biological events on genes or gene products described in the biomedical literature. Participants are required to extract and classify 9 kinds of event with appropriate arguments.

First time joining biomedical domain, we aim to learn current problems and approaches in biomedical research. Therefore, we have chosen simple approaches such as rule-based and pattern-based. In the following section, we will explain our work on GENIA Event Extraction Task (GENIA) in details. Finally, we will analyze and discuss results.

2 Our approach

The project uses UIMA Framework[1], an open source framework for analyzing unstructured information, to develop all analysis components. Events bounded in a sentence are 94.4% in training corpus. Consequently, sentences are processed in succession at each step. We divide the whole system into 3 parts: Preprocessing, Event Trigger annotation and Event annotation.

2.1 Preprocessing

At this step, the input documents are converted into objects of the framework. All analysis components will process objects and put results into them. Then we go through natural language processes that include sentence splitting, tokenizing and POS tagging by OpenNLP library. Lastly, the given Protein concepts are annotated.

2.2 Event Trigger annotation

According to our statistics in the training corpus, the percentage of single token trigger is 91.8%. To simplify it, we focus on triggers which span on one token. At this stage, rule-based and dictionary-based approaches are combined.

We choose tokens which are near a protein and have appropriate POS tags. Heuristic rules extracted from training corpus are used to identify candidate triggers. Those rules are, for instance, NN/NNS + of + PROTEIN, VBN + PROTEIN and so on.

Event triggers are diverse in lexical and ambiguous in classification (Björne et al. (2009) and Buyko et al. (2009)). Candidate triggers are classified by a dictionary. The dictionary containing words of triggers with their corresponding classes is built from training corpus. For ambiguous trigger classes, the class that has the highest rate of appearance is chosen.

2.3 Event annotation

Basing on the number of arguments and type of arguments, we categorize 9 event classes into 3 groups. The first group including Gene expression,

[1] Available at http://uima.apache.org/

Proceedings of BioNLP Shared Task 2011 Workshop, pages 149–150,
Portland, Oregon, USA, 24 June, 2011. ©2011 Association for Computational Linguistics

Transcription and Protein catabolism has only one Protein as the argument. The second group contains events with Protein and Entity as argument. Phosphorylation, Localization and Binding belong to that group. The third group has the most complex types, i.e. Regulation, Positive regulation and Negative regulation. These events can have other events as their argument.

Our method of event detection is using dependency graph as results of deep syntactic parsing. We prune parse tree and assign concept to nodes. Next, sub-trees which contains only conceptual node as patterns are extracted and represented as string form. We travel breadth-first and write conceptual labels to the string pattern. The pattern list is built from training data.

Firstly, for each sentence contains at least one trigger, we get the parse tree of the sentence. We prune nodes which contain only one child and that child node has zero or one descendant. It reduces the complexity and retains important and general parts of the parse tree.

Secondly, candidate arguments of events are identified by combining Protein, Entity and Event Trigger in that sentence. The number of combination can be huge, so we restrict it by the following conditions. Each combination has at least one Event Trigger with one Protein or Event. The number of argument depends on types of events and is usually less than 5. In addition, the difference of depth on tree between arguments has to be under a threshold.

Thirdly, concepts of arguments in each combination are assigned to parse tree nodes. The assignment bases on the span of argument and content of nodes. The pattern is extracted from the parse tree and examined whether it belongs to the pattern list. In order to increase the precision, we discard patterns having the depth of the tree greater than a threshold. The threshold is chosen by counting on the training corpus.

Finally, we classify events and determine role of arguments for each event. The type of the event is chosen by the type of the trigger of that event. We still simply assign roles of arguments in a fixed order of arguments.

3 Results and conclusions

Our fully official result in GENIA main task is described in Table 1. The F-score is only 14,75% and we were ranked 13th among 14 participants. It reflects many shortcomings in our system. We obtain a lot of experience.

In general, the patterns which we built are still generic. Besides, the OpenNLP library still encountered errors when processing documents, thus affected our result. For example, there are some sentences that OpenNLP parsed or tokenized wrongly and raised errors. In the step of Event Trigger annotation, there are a few rules to cover cases. The result of Regulation, Positive regulation and Negative regulation has the lowest result because we only process recursion with simple events.

Approach	**recall**	**precision**	**f-score**
Gene expression	26.45	39.73	31.76
Transcription	16.09	14.58	15.30
Protein catabolism	33.33	50.00	40.00
Phosphorylation	32.43	47.62	38.59
Localization	16.23	27.68	20.46
Binding	4.68	12.92	6.88
Regulation	0.26	1.35	0.44
Positive regulation	2.08	13.04	3.59
Negative regulation	1.40	11.27	2.49
All Total	**10.12**	**27.17**	**14.75**

Table 1: Our final result in GENIA BioNLP'11 Shared Task with approximately span and recursive matching

For future work, we intend to apply hybrid approach. We combine other methods such as machine learning in Event Trigger and Event annotation parts. We consider other NLP library to improve the performance of all steps relating to NLP processing. Rules from domain professions will be added to existent heuristic rules. We will try to add more features to improve the patterns.

References

Ekaterina Buyko, Erik Faessler, Joachim Wermter and Udo Hahn, "Event Extraction from Trimmed Dependency Graphs," in *Proceedings of the Workshop on BioNLP: Shared Task*, 2009, pp. 19-27.

Jari Bjorne, Juho Heimonen, Filip Ginter, Antti Airola, Tapio Pahikkala and Tapio Salakoski, "Extracting Complex Biological Events with Rich Graph-Based Feature Sets," in *Proceedings of the Workshop on BioNLP: Shared Task*, 2009, pp. 10-18.

An Incremental Model for the Coreference Resolution Task of BioNLP 2011

Don Tuggener, Manfred Klenner, Gerold Schneider, Simon Clematide, Fabio Rinaldi
Institute of Computational Linguistics, University of Zurich, Switzerland
{tuggener,klenner,gschneid,siclemat,rinaldi}@cl.uzh.ch

Abstract

We introduce our incremental coreference resolution system for the BioNLP 2011 Shared Task on Protein/Gene interaction. The benefits of an incremental architecture over a mention-pair model are: a reduction of the number of candidate pairs, a means to overcome the problem of underspecified items in pair-wise classification and the natural integration of global constraints such as transitivity. A filtering system takes into account specific features of different anaphora types. We do not apply Machine Learning, instead the system classifies with an empirically derived salience measure based on the dependency labels of the true mentions. The OntoGene pipeline is used for preprocessing.

1 Introduction

The Coreference Resolution task of BioNLP focused on finding anaphoric references to proteins and genes. Only antecedent-anaphora pairs are considered in evaluation and not full coreference sets. Although it might not seem to be necessary to generate full coreference sets, anaphora resolution still benefits from their establishment. Our incremental approach (Klenner et al., 2010) naturally enforces transitivity constraints and thereby reduces the number of potential antecedent candidates. The system achieved good results in the BioNLP 2011 shared task (Fig. 1)

Team	R	P	F1
A	22.18	73.26	34.05
Our model	21.48	55.45	30.96
B	19.37	63.22	29.65
C	14.44	67.21	23.77
D	3.17	3.47	3.31
E	0.70	0.25	0.37

Figure 1: Protein/Gene Coreference Task

2 Preprocessing: The OntoGene Pipeline

OntoGene's text mining system is based on an internally-developed fast, broad-coverage, deep-syntactic parsing system (Schneider, 2008). The parser is wrapped into a pipeline which uses a number of other NLP tools. The parser is a key component in a pipeline of NLP tools (Rinaldi et al., 2010), used to process input documents. First, in a pre-processing stage, the input text is transformed into a custom XML format, and sentences and tokens boundaries are identified. The OntoGene pipeline also includes a step of term annotation and disambiguation, which are not used for the BioNLP shared task, since relevant terms are already provided in both the training and test corpora. The pipeline also includes part-of-speech taggers, a lemmatizer and a syntactic chunker.

When the pipeline finishes, each input sentence has been annotated with additional information, which can be briefly summarized as follows: sentences are tokenized and their borders are detected; each sentence and each token has been assigned an ID; each token is lemmatized; tokens which belong to terms are grouped; each term is assigned a normal-form and a semantic type; tokens and terms are then grouped into chunks; each chunk has a type (NP or VP) and a head token; each sentence is described as a syntactic dependency structure. All this information is represented as a set of predicates and stored into the Knowledge Base of the system, which can then be used by different applications, such as the OntoGene Relation Miner (Rinaldi et al., 2006) and the OntoGene Protein-Protein Interaction discovery tool (Rinaldi et al., 2008).

3 Our Incremental Model for Coreference Resolution

for i=1 to length(I)
  for j=1 to length(C)
    r_j := virtual prototype of coreference set C_j
    Cand := Cand $\oplus$ r_j if compatible(r_j, m_i)
  for k= length(B) to 1
    b_k:= the k-th licensed buffer element
    Cand := Cand $\oplus$ b_k if compatible(b_k, m_i)
  if Cand = {} then B := B $\oplus$ m_i
  if Cand $\neq$ {} then
    $ante_i$:= most salient element of Cand
    C := augment(C,$ante_i$,m_i)

Figure 2: Incremental model: base algorithm

Proceedings of BioNLP Shared Task 2011 Workshop, pages 151–152,
Portland, Oregon, USA, 24 June, 2011. ©2011 Association for Computational Linguistics

Fig. 2 shows the base algorithm. Let I be the chronologically ordered list of NPs, C be the set of coreference sets and B a buffer, where NPs are stored, if they are not anaphoric (but might be valid antecedents). Furthermore m_i is the current NP and $\oplus$ means concatenation of a list and a single item. The algorithm proceeds as follows: a set of antecedent candidates is determined for each NP m_i (steps 1 to 7) from the coreference sets (r_j) and the buffer (b_k). A valid candidate r_j or b_k must be compatible with m_i. The definition of compatibility depends on the POS tags of the anaphor-antecedent pair. The most salient available candidate is selected as antecedent for m_i.

3.1 Restricted Accessibility of Antecedent Candidates

In order to reduce underspecification, m_i is compared to a virtual prototype of each coreference set (similar to e.g. (Luo et al., 2004; Yang et al., 2004; Rahman and Ng, 2009)). The virtual prototype bears morphologic and semantic information accumulated from all elements of the coreference set. Access to coreference sets is restricted to the virtual prototype. This reduces the number of considered pairs (from the cardinality of a set to 1).

3.2 Filtering based on Anaphora Type

Potentionally co-refering NPs are extracted from the OntoGene pipeline based on POS tags. We then apply filtering based on anaphora type: Reflexive pronouns must be bound to a NP that is governed by the same verb. Relative pronouns are bound to the closest NP in the left context. Personal and possessive pronouns are licensed to bind to morphologically compatible antecedent candidates within a window of two sentences. Demonstrative NPs containing the lemmata 'protein' or 'gene' are licensed to bind to name containing mentions. Demonstrative NPs not containing the trigger lemmata can be resolved to string matching NPs preceding them[1].

3.3 Binding Theory as a Filter

We know through binding theory that 'modulator' and 'it' cannot be coreferent in the sentence *"Overexpression of protein inhibited stimulus-mediated transcription, whereas modulator enhanced it"*. Thus, the pair 'modulator'-'it' need not be considered at all. We have not yet implemented a full-blown binding theory. Instead, we check if the antecedent and the anaphor are governed by the same verb.

[1]As we do not perform anaphoricity determination of nominal NPs, we do not consider bridging anaphora (anaphoric nouns that are connected to their antecedents through semantic relations and cannot be identified by string matching).

4 An Empirically-based Salience Measure

Our salience measure is a partial adaption of the measure from (Lappin and Leass, 1994). The salience of a NP is solely defined by the salience of the dependency label it bears. The salience of a dependency label, D, is estimated by the number of true mentions (i.e. co-refering NPs) that bear D (i.e. are connected to their heads with D), divided by the total number of true mentions (bearing any D). The salience of the label *subject* is thus calculated by:

$$\frac{Number\,of\,true\,mentions\,bearing\,subject}{Total\,number\,of\,true\,mentions}$$

We get a hierarchical ordering of the dependency labels (subject > object > pobject > ...) according to which antecedents are ranked and selected.

References

Manfred Klenner, Don Tuggener, and Angela Fahrni. 2010. Inkrementelle koreferenzanalyse für das deutsche. In *Proceedings der 10. Konferenz zur Verarbeitung Natürlicher Sprache*.

Shalom Lappin and Herbert J Leass. 1994. An algorithm for pronominal anaphora resolution. *Computational Linguistics*, 20:P. 535–561.

Xiaoqiang Luo, Abe Ittycheriah, Hongyan Jing, Nanda Kambhatla, and Salim Roukos. 2004. A mention-synchronous coreference resolution algorithm based on the bell tree. In *Proceedings of the 42nd Annual Meeting on Association for Computational Linguistics*.

Altaf Rahman and Vincent Ng. 2009. Supervised models for coreference resolution. In *Proceedings of the 2009 Conference on Empirical Methods in Natural Language Processing: Volume 2 - Volume 2*, EMNLP '09, pages 968–977, Stroudsburg, PA, USA. Association for Computational Linguistics.

Fabio Rinaldi, Gerold Schneider, Kaarel Kaljurand, Michael Hess, and Martin Romacker. 2006. An Environment for Relation Mining over Richly Annotated Corpora: the case of GENIA. *BMC Bioinformatics*, 7(Suppl 3):S3.

Fabio Rinaldi, Thomas Kappeler, Kaarel Kaljurand, Gerold Schneider, Manfred Klenner, Simon Clematide, Michael Hess, Jean-Marc von Allmen, Pierre Parisot, Martin Romacker, and Therese Vachon. 2008. OntoGene in BioCreative II. *Genome Biology*, 9(Suppl 2):S13.

Fabio Rinaldi, Gerold Schneider, Kaarel Kaljurand, Simon Clematide, Therese Vachon, and Martin Romacker. 2010. OntoGene in BioCreative II.5. *IEEE/ACM Transactions on Computational Biology and Bioinformatics*, 7(3):472–480.

Gerold Schneider. 2008. *Hybrid Long-Distance Functional Dependency Parsing*. Doctoral Thesis, Institute of Computational Linguistics, University of Zurich.

Xiaofeng Yang, Jian Su, Guodong Zhou, and Chew Lim Tan. 2004. An np-cluster based approach to coreference resolution. In *Proceedings of the 20th international conference on Computational Linguistics*.

Double Layered Learning for Biological Event Extraction from Text

Ehsan Emadzadeh, Azadeh Nikfarjam, Graciela Gonzalez
Arizona State University / Tempe, AZ 85283, USA
ehsan.emadzadeh@asu.edu, azadeh.nikfarjam@asu.edu
graciela.gonzalez@asu.edu

Abstract

This paper presents our approach (referred to as BioEvent) for protein-level complex event extraction, developed for the GENIA task (Kim et al., 2011b) of the BioNLP Shared Task 2011 (Kim et al., 2011a). We developed a double layered machine learning approach which utilizes a state-of-the-art minimized feature set for each of the event types. We improved the best performing system of BioNLP 2009 overall, and ranked first amongst 15 teams in finding "Localization" events in 2011[1][2]. BioEvent is available at http://bioevent.sourceforge.net/

1 Introduction

A biological event refers to a specific kind of interaction between biological entities. Events consist of two parts: event triggers and event arguments. Event extraction can be very challenging when dealing with complex events with multiple or nested arguments; for example, events themselves can be an argument for other events.

2 Methods

In general, to detect an event mentioned in text, the event trigger should be identified first, then complemented with event arguments. We divided the training and testing tasks into two phases: trigger detection and argument detection.

[1]Using the "Approximate Span without Event Trigger Matching/Approximate Recursive" metric

[2]http://www-tsujii.is.s.u-tokyo.ac.jp/GENIA/-SharedTask/evaluation.shtml

2.1 Event Trigger Detection

The trigger detection problem can be modeled as a multi-class classification of a word or combination of words (phrase). Instead of using all possible phrases in the training text as examples for the classifier, we only included those that were known triggers in the training set. For the official shared task submission we used SVM^{light} (Joachims, 1999). Detailed explanation of the trigger detection process includes three main steps: pre-processing, training of the SVM models, and combining SVM results.

Pre-processing. All tokenized documents provided by the shared task organizers (Stenetorp et al., 2011) were converted to database records. Then different sets of attributes were defined and calculated for words, sentences and documents.

Training SVM models and Combining Results. We trained 9 different binary SVM models using one-vs-many approach. One of the challenging tasks was to compare the results of different SVM models, given that each had different feature sets and their confidence values were not directly comparable and needed to be calibrated properly before comparing. We tried three approaches: 1) selecting the SVM result with highest positive distance to hyperplane, 2) using a trained decision tree and 3) using another SVM trained for voting. Model J48 from the WEKA library (Hall et al., 2009) was trained based on SVM distances for the training set examples and expected outputs. In the third approach, we tried SVM for voting, which generated better results than the decision tree. Last two approaches consist of two layers of classifiers which first layer includes event types classifiers and second layer generates final decision

Proceedings of BioNLP Shared Task 2011 Workshop, pages 153–154,
Portland, Oregon, USA, 24 June, 2011. ©2011 Association for Computational Linguistics

Event type	Bioevent	Turku09
Gene expression	**71.88**	70.84
Transcription	**47.62**	47.14
Protein catabolism	60.87	60.87
Phosphorylation	**75.14**	73.39
Localization	**61.49**	59.68
Binding	34.42	**35.97**
Regulation	**24.03**	22.26
Positive regulation	**33.41**	31.84
Negative regulation	**18.89**	18.58
ALL-TOTAL	**44.69**	43.54

Table 1: F-Value from our BioEvent system compared to Turku09 (Bjorne et al., 2009) results, using Approximate Span/Approximate Recursive matching

based on first layer outputs.

2.2 Arguments detection and Post-processing

Similar to trigger detection, argument detection can be modeled for a classification task by assigning an argument type label to each possible combination of an event trigger and a biological entity in a sentence. We obtained entities from a1 files, as well as the supportive analysis data provided by the shared task organizers (Bjorne et al., 2009). After generating events using SVM classification, we merged them with the output from the Turku system to generate the final result. For common events (detected by both systems) we used the arguments detected by the Turku system.

3 Results

Since we tried to improve upon the best performing system in the 2009 competition (Turku09), we compare the results of our system and Turku09's on the 2011 test set. Table 1 shows the performance of our proposed system and that of Turku09. We see that Binding was our worst event (negative change), Localization the most improved, no change for Protein Catabolism, and only a slight improvement in Negative Regulation.

4 Conclusion and future work

In this research we focused on event trigger detection by applying a SVM-based model. SVM is very sensitive to parameters and further tuning of parameters can improve the overall result. Furthermore, we want to evaluate our method independently and find the contribution of each modification to the final result. Our method is generalizable to other domains by using proper train-set and finding useful attributes for new event types.

Acknowledgments

The authors would like to thank Ryan Sullivan for his helps during this research. EE and GG acknowledge partial funding from NLM Contract HHSN276201000031C.

References

Jari Bjorne, Juho Heimonen, Filip Ginter, Antti Airola, Tapio Pahikkala, and Tapio Salakoski. 2009. Extracting Complex Biological Events with Rich Graph-Based Feature Sets. *Computational Linguistics*, (June):10–18.

M. Hall, E. Frank, G. Holmes, B. Pfahringer, P. Reutemann, and I.H. Witten. 2009. The WEKA data mining software: an update. *ACM SIGKDD Explorations Newsletter*, 11(1):10–18.

T. Joachims. 1999. Making large scale SVM learning practical. *Advances in Kernel Methods - Support Vector Learnin*, (B. Schölkopf and C. Burges and A. Smola (ed.)).

Jin-Dong Kim, Sampo Pyysalo, Tomoko Ohta, Robert Bossy, and Jun'ichi Tsujii. 2011a. Overview of BioNLP Shared Task 2011. In *Proceedings of the BioNLP 2011 Workshop Companion Volume for Shared Task*, Portland, Oregon, June. Association for Computational Linguistics.

Jin-Dong Kim, Yue Wang, Toshihisa Takagi, and Akinori Yonezawa. 2011b. Overview of the Genia Event task in BioNLP Shared Task 2011. In *Proceedings of the BioNLP 2011 Workshop Companion Volume for Shared Task*, Portland, Oregon, June. Association for Computational Linguistics.

Pontus Stenetorp, Goran Topić, Sampo Pyysalo, Tomoko Ohta, Jin-Dong Kim, and Jun'ichi Tsujii. 2011. BioNLP Shared Task 2011: Supporting Resources. In *Proceedings of the BioNLP 2011 Workshop Companion Volume for Shared Task*, Portland, Oregon, June. Association for Computational Linguistics.

MSR-NLP Entry in BioNLP Shared Task 2011

Chris Quirk, Pallavi Choudhury, Michael Gamon, and Lucy Vanderwende
Microsoft Research
One Microsoft Way
Redmond, WA 98052 USA
{chrisq,pallavic,mgamon,lucyv}@microsoft.com

Abstract

We describe the system from the Natural Language Processing group at Microsoft Research for the BioNLP 2011 Shared Task. The task focuses on event extraction, identifying structured and potentially nested events from unannotated text. Our approach follows a pipeline, first decorating text with syntactic information, then identifying the trigger words of complex events, and finally identifying the arguments of those events. The resulting system depends heavily on lexical and syntactic features. Therefore, we explored methods of maintaining ambiguities and improving the syntactic representations, making the lexical information less brittle through clustering, and of exploring novel feature combinations and feature reduction. The system ranked 4th in the GENIA task with an F-measure of 51.5%, and 3rd in the EPI task with an F-measure of 64.9%.

1 Introduction

We describe a system for extracting complex events and their arguments as applied to the BioNLP-2011 shared task. Our goal is to explore general methods for fine-grained information extraction, to which the data in this shared task is very well suited. We developed our system using only the data provided for the GENIA task, but then submitted output for two of the tasks, GENIA and EPI, training models on each dataset separately, with the goal of exploring how general the overall system design is with respect to text domain and event types. We used no external knowledge resources except a text corpus used to train cluster features. We further describe several system variations that we explored but which did not contribute to the final system submitted. We note that the MSR-NLP system consistently is among those with the highest recall, but needs additional work to improve precision.

2 System Description

Our event extraction system is a pipelined approach, closely following the structure used by the best performing system in 2009 (Björne et al., 2009). Given an input sentence along with tokenization information and a set of parses, we first attempt to identify the words that trigger complex events using a multiclass classifier. Next we identify edges between triggers and proteins, or between triggers and other triggers. Finally, given a graph of proteins and triggers, we use a rule-based post-processing component to produce events in the format of the shared task.

2.1 Preprocessing and Linguistic Analysis

We began with the articles as provided, with an included tokenization of the input and identification of the proteins in the input. However, we did modify the token text and the part-of-speech tags of the annotated proteins in the input to be PROT after tagging and parsing, as we found that it led to better trigger detection.

The next major step in preprocessing was to produce labeled dependency parses for the input. Note that the dependencies may not form a tree: there may be cycles and some words may not be connected. During feature construction, this parsing graph was used to find paths between

Proceedings of BioNLP Shared Task 2011 Workshop, pages 155–163,
Portland, Oregon, USA, 24 June, 2011.

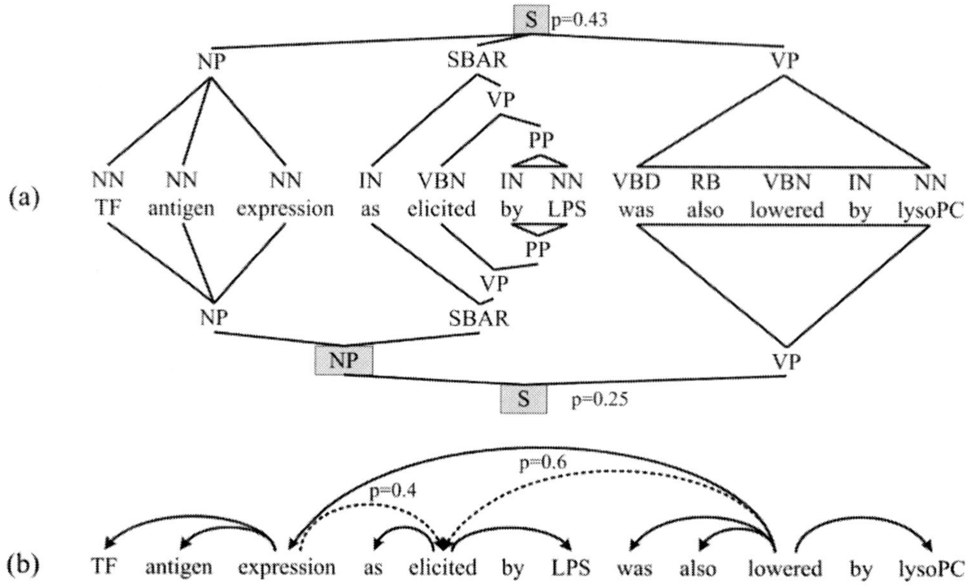

Figure 1: Example sentence from the GENIA corpus. (a) Two of the top 50 constituency parses from the MCCC-I parser; the first had a total probability mass of 0.43 and the second 0.25 after renormalization. Nodes that differ between parses are shaded and outlined. (b) The dependency posteriors (labels omitted due to space) after conversion of 50-best parses. Solid lines indicate edges with posterior > 0.95; edges with posterior < 0.05 were omitted. Most of the ambiguity is in the attachment of "*elicited*".

words in the sentence. Since proteins may consist of multiple words, for paths we picked a single representative word for each protein to act as its starting point and ending point. Generally this was the token inside the protein that is closest to the root of the dependency parse. In the case of ties, we picked the rightmost such node.

2.1.1 McClosky-Charniak-Stanford parses

The organizers provide parses from a version of the McClosky-Charniak parser, MCCC (McClosky and Charniak, 2008), which is a two-stage parser/reranker trained on the GENIA corpus. In addition, we used an improved set of parsing models that leverage unsupervised data, MCCC-I (McClosky, 2010). In both cases, the Stanford Parser was used to convert constituency trees in the Penn Treebank format into labeled dependency parses: we used the collapsed dependency format.

2.1.2 Dependency posteriors

Effectively maintaining and leveraging the ambiguity present in the underlying parser has improved task accuracy in some downstream tasks (e.g., Mi et al. 2008). McClosky-Charniak parses in two passes: the first pass is a generative model that produces a set of n-best candidates, and the second pass is a discriminative reranker that uses a rich set of features including non-local information. We renormalized the outputs from this log-linear discriminative model to get a posterior distribution over the 50-best parses. This set of parses preserved some of the syntactic ambiguity present in the sentence.

The Stanford parser deterministically converts phrase-structure trees into labeled dependency graphs (de Marneffe et al., 2006). We converted each constituency tree into a dependency graph separately and retained the probability computed above on each graph.

One possibility was to run feature extraction on each of these 50 parses, and weight the resulting features in some manner. However, this caused a significant increase in feature count. Instead, we gathered a posterior distribution over dependency edges: the posterior probability of a labeled dependency edge was estimated by the sum of the probability of all parses containing that edge. Gathering all such edges produced a single labeled graph that retained much of the ambiguity of the input sentence. Figure 1 demonstrates this process on a simple example. We applied a threshold of 0.5 and retained all edges above that threshold, although there are many alternative ways to exploit this structure.

As above, the resulting graph is likely no longer a connected tree, though it now may also be cyclic and rather strange in structure. Most of the dependency features were built on shortest paths between words. We used the algorithm in Cormen et al. (2002, pp.595) to find shortest paths in a cyclic graph with non-negative edge weights. The shortest path algorithm used in feature finding was supplied uniform positive edge weights. We could also weight edges by the negative log probability to find the shortest, most likely path.

2.1.3 ENJU

We also experimented with the ENJU parses (Miyao and Tsujii, 2008) provided by the shared task organizers. The distribution contained the output of the ENJU parser in a format consistent with the Stanford Typed Dependency representation .

2.1.4 Multiple parsers

We know that even the best modern parsers are prone to errors. Including features from multiple parsers helps mitigate these errors. When different parsers agree, they can reinforce certain classification decisions. The features that were extracted from a dependency parse have names that include an identifier for the parser that produced them. In this way, the machine learning algorithm can assign different weights to features from different parsers. For finding heads of multi-word entities, we preferred the ENJU parser if present in that experimental condition, then fell back to MCCC parses, and finally MCCC-I.

2.1.5 Dependency conversion rules

We computed our set of dependency features (see 2.2.1) from the collapsed, propagated Stanford Typed Dependency representation (see http://nlp.stanford.edu/software/dependencies_manual.pdf and de Marneffe et al., 2006), made available by the organizers. We chose this form of representation since we are primarily interested in computing features that hold between content words. Consider, for example, the noun phrase "phosphorylation of TRAF2". A dependency representation would specify *head-modifier* relations for the tuples (*phosphorylation*, *of*) and (*of*, *TRAF2*). Instead of *head-modifier*, a typed dependency representation specifies **PREP** and **PPOBJ** as the two grammatical relations: **PREP**(*phosphorylation*-1, *of*-2) and **PPOBJ**(*of*-2, *TRAF2*-3). A collapsed representation has a single triplet specifying the relation between the content words directly, **PREP_OF**(*phosphorylation*-1, *TRAF2*-3); we considered this representation to be the most informative.

We experimented with a representation that further normalized over syntactic variation. The system submitted for the GENIA subtask does not use these conversion rules, while the system submitted for the EPI subtask does use these rules. See Table 2 for further details. While for some applications it may be useful to distinguish whether a given relation was expressed in the active or passive voice, or in a main or a relative clause, we believe that for this application it is beneficial to normalize over these types of syntactic variation. Accordingly, we had a set of simple renaming conversion rules, followed by a rule for expansion; this list was our first effort and could likely be improved. We modeled this normalized level of representation on the logical form, described in Jensen (1993), though we were unable to explore NP-or VP-anaphora

Renaming conversion rules:

1. **ABBREV** -> **APPOS**
2. **NSUBJPASS** -> **DOBJ**
3. **AGENT** -> **NSUBJ**
4. **XSUBJ** -> **NSUBJ**
5. **PARTMOD**(head, modifier where last 3 characters are "ing") -> **NSUBJ**(modifier, head)
6. **PARTMOD**(head, modifier where last 3 characters are "ed") -> **DOBJ**(modifier, head)

Expansion:

1. For **APPOS**, find all edges that point to the head (*gene*-20) and duplicate those edges, but replacing the modifier with the modifier of the **APPOS** relation (*kinase*-26).

Thus, in the 2nd sentence in PMC-1310901-01-introduction, "*... leading to expression of a bcr-abl fusion gene, an aberrant activated tyrosine kinase,* ....", there are two existing grammatical relations:

PREP_OF(*expression*-15, *gene*-20)
APPOS(*gene*-20, *kinase*-26)

to which this rule adds:

PREP_OF(*expression*-15, *kinase*-26)

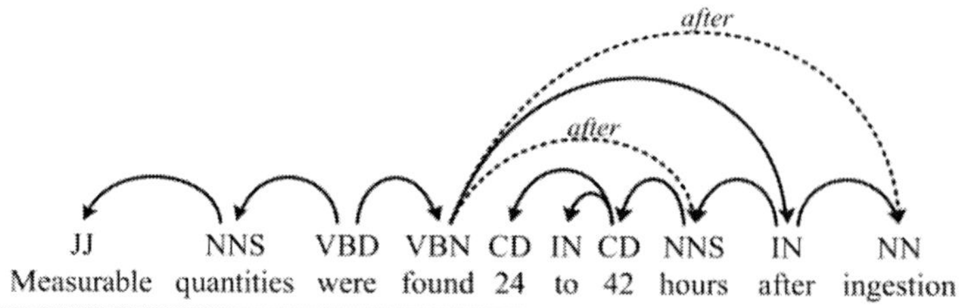

Key	Relation	Value	Key	Relation	Value
quantities	child(left, NNS→JJ)	measurable	measurable	child^{-1}(left, NNS→JJ)	quantities
found	child(after, VBN→NNS)	hours	hours	child^{-1}(after, VBN→NNS)	found
found	child(after, VBN→NN)	ingestion	ingestion	child^{-1}(after, VBN→NN)	found

Figure 2: A sample PubMed sentence along with its dependency parse, and some key/relation/value triples extracted from that parse for computation of distributional similarity. Keys with a similar distribution of values under the same relation are likely semantically related. Inverse relations are indicated with a superscript -1. Prepositions are handled specially: we add edges labeled with the preposition from its parent to each child (indicated by dotted edges).

2.2 Trigger Detection

We treated trigger detection as a multi-class classification problem: each token should be annotated with its trigger type or with NONE if it was not a trigger. When using the feature set detailed below, we found that an SVM (Tsochantaridis et al., 2004) outperformed a maximum entropy model by a fair margin, though the SVM was sensitive to its free parameters. A large value of C, the penalty incurred during training for misclassifying a data point, was necessary to achieve good results.

2.2.1 Features for Trigger Detection

Our initial feature set for trigger detection was strongly influenced by features that were successful in Björne et al., (2009).

Token Features. We included stems of single tokens from the Porter stemmer (Porter, 1980), character bigrams and trigrams, a binary indicator feature if the token has upper case letters, another indicator for the presence of punctuation, and a final indicator for the presence of a number. We gathered these features for both the current token as well as the three immediate neighbors on both the left and right hand sides.

We constructed a gazetteer of possible trigger lemmas in the following manner. First we used a rule-based morphological analyzer (Heidorn, 2000) to identify the lemma of all words in the training, development, and test corpora. Next, for each word in the training and development sets, we mapped it to its lemma. We then computed the number of times that each lemma occurred as a trigger for each type of event (and none). Lemmas that acted as a trigger more than 50% of the time were added to the gazetteer.

During feature extraction for a given token, we found the lemma of the token, and then look up that lemma in the gazetteer. If found, we included a binary feature to indicate its trigger type.

Frequency Features. We included as features the number of entities in the sentence, a bag of words from the current sentence, and a bag of entities in the current sentence.

Dependency Features. We used primarily a set of dependency chain features that were helpful in the past (Björne et al., 2009); these features walk the Stanford Typed Dependency edges up to a distance of 3.

We also found it helpful to have features about the path to the nearest protein, regardless of distance. In cases of multiple shortest paths, we took only one, exploring the dependency tree generally in left to right order. For each potential trigger, we looked at the dependency edge labels leading to that nearest protein. In addition we had a feature including both the dependency edge labels and the token text (lowercased) along that path. Finally, we had a feature indicating whether some token along that path was also in the trigger gazetteer. The formulation of this set of features is still not optimal especially for the "binding" events as the training data will include paths to more than one protein argument. Nevertheless, in Table 3, we can see that this set of features contributed to improved precision.

Cluster Features. Lexical and stem features were crucial for accuracy, but were unfortunately sparse and did not generalize well. To mitigate this, we incorporated word cluster features. In addition to the lexical item and the stem, we added another feature indicating the cluster to which each word belongs. To train clusters, we downloaded all the PubMed abstracts (http://pubmed.gov), parsed them with a simple dependency parser (a reimplementation of McDonald, 2006 trained on the GENIA corpus), and extracted dependency relations to use in clustering: words that occur in similar contexts should fall into the same cluster. An example sentence and the relations that were extracted for distributional similarity computation are presented in Figure 2. We ran a distributional similarity clustering algorithm (Pantel et al., 2009) to group words into clusters.

Tfidf features. This set of features was intended to capture the salience of a term in the medical and "general" domain, with the aim of being able to distinguish domain-specific terms from more ambiguous terms. We calculated the tf.idf score for each term in the set of all PubMed abstracts and did the same for each term in Wikipedia. For each token in the input data, we then produced three features: (i) the tf.idf value of the token in PubMed abstracts, (ii) the tf.idf value of the token in Wikipedia, and (iii) the delta between the two values. Feature values were rounded to the closest integer. We found, however, that adding these features did not improve results.

2.2.2 Feature combination and reduction

We experimented with feature reduction and feature combination within the set of features described here. For feature reduction we tried a number of simple approaches that typically work well in text classification. The latter is similar to the task at hand, in that there is a very large but sparse feature set. We tried two feature reduction methods: a simple count cutoff, and selection of the top n features in terms of log likelihood ratio (Dunning, 1993) with the target values. For a count cutoff, we used cutoffs from 3 to 10, but we failed to observe any consistent gains. Only low cutoffs (3 and occasionally 5) would ever produce any small improvements on the development set. Using log likelihood ratio (as determined on the training set), we reduced the total number of features to between 10,000 and 75,000. None of these experiments improved results, however. One potential reason for this negative result may be that there were a lot of features in our set that capture the same phenomenon in different ways, i.e. which correlate highly. By retaining a subset of the original feature set using a count cutoff or log likelihood ratio we did not reduce this feature overlap in any way. Alternative feature reduction methods such as Principal Component Analysis, on the other hand, would target the feature overlap directly. For reasons of time we did not experiment with other feature reduction techniques but we believe that there may well be a gain still to be had.

For our feature combination experiments the idea was to find highly predictive Boolean combinations of features. For example, while the features *a* and *b* may be weak indicators for a particular trigger, the cases where both *a* and *b* are present may be a much stronger indicator. A linear classifier such as the one we used in our experiments by definition is not able to take such Boolean combinations into account. Some classifiers such as SVMs with non-linear kernels do consider Boolean feature combinations, but we found the training times on our data prohibitive when using these kernels. As an alternative, we decided to pre-identify feature combinations that are predictive and then add those combination features to our feature inventory. In order to pre-identify feature combinations, we trained decision tree classifiers on the training set, and treated each path from the root to a leaf through the decision tree classifier as a feature combination. We also experimented with adding all partial paths through the tree (as long as they started from the root) in addition to adding all full paths. Finally, we tried to increase the diversity of our combination features by using a "bagging" approach, where we trained a multitude of decision trees on random subsets of the data. Again, unfortunately, we did not find any consistent improvements. Two observations that held relatively consistently across our experiments with combination features and different feature sets were: (i) only adding full paths as combination features sometimes helped, while adding partial paths did not, and (ii) bagging hardly ever led to improvements.

	Development Set				Test Set			
Event Class	**Count**	**Recall**	**Precision**	**F1**	**Count**	**Recall**	**Precision**	**F1**
Gene_expression	749	76.37	81.46	78.83	1002	73.95	73.22	73.58
Transcription	158	49.37	73.58	59.09	174	41.95	65.18	51.05
Protein_catabolism	23	69.57	80.00	74.42	15	46.67	87.50	60.87
Phosphorylation	111	73.87	84.54	78.85	185	87.57	81.41	84.37
Localization	67	74.63	75.76	75.19	191	51.31	79.03	62.22
=[SVT-TOTAL]=	1108	72.02	80.51	76.03	1567	68.99	74.03	71.54
Binding	373	47.99	50.85	49.38	491	42.36	40.47	41.39
=[EVT-TOTAL]=	1481	65.97	72.73	69.18	2058	62.63	65.46	64.02
Regulation	292	32.53	47.05	38.62	385	24.42	42.92	31.13
Positive_Regulation	999	38.74	51.67	44.28	1443	37.98	44.92	41.16
Negative_Regulation	471	35.88	54.87	43.39	571	41.51	42.70	42.10
=[REG-TOTAL]=	1762	36.95	51.79	43.13	2399	36.64	44.08	40.02
ALL-Total	3243	50.20	62.60	55.72	4457	48.64	54.71	51.50

Table 1: Approximate span matching/approximate recursive matching on development and test data sets for GENIA Shared Task -1 with our system.

2.3 Edge Detection

This phase of the pipeline was again modeled as multi-class classification. There could be an edge originating from any trigger word and ending in any trigger word or protein. Looking at the set of all such edges, we trained a classifier to predict the label of this edge, or NONE if the edge was not present. Here we found that a maximum entropy classifier performed somewhat better than an SVM, so we used an in-house implementation of a maximum entropy trainer to produce the models.

2.3.1 Features for Edge Detection

As with trigger detection, our initial feature set for edge detection was strongly influenced by features that were successful in Björne et al. (2009). Additionally, we included the same dependency path features to the nearest protein that we used for trigger detection, described in 2.2.1. Further, for a prospective edge between two entities, where the entities are either a trigger and a protein, or a trigger and a second trigger, we added a feature that indicates (i) if the second entity is in the path to the nearest protein, (ii) if the head of the second entity is in the path to the nearest protein, (iii) the type of the second entity.

2.4 Post-processing

Given the set of edges, we used a simple deterministic procedure to produce a set of events. This step is not substantially different from that used in prior systems (Björne et al., 2009).

2.4.1 Balancing Precision and Recall

As in Björne et al. (2009), we found that the trigger detector had quite low recall. Presumably this is due to the severe class imbalance in the training data: less than 5% of the input tokens are triggers. Thus, our classifier had a tendency to overpredict NONE. We tuned a single free parameter $\beta \in \mathbb{R}^+$ (the "recall booster") to scale back the score associated with the NONE class before selecting the optimal class. The value was tuned for whole-system F-measure; optimal values tended to fall in the range 0.6 to 0.8, indicating that only a small shift toward recall led to the best results.

Trigger Detection Features	**Trigger Loss**	**Recall**	**Prec.**	**F1**
B	2.14	48.44	64.08	55.18
B + TI	2.14	48.17	62.49	54.40
B + TI + C	2.14	50.32	60.90	55.11
B + TI + C + PI	2.03	50.20	62.60	55.72
B + TI + C + PI +D	2.02	49.21	62.75	55.16

Table 2: Recall/Precision/F1 on the GENIA development set using MCCC-I + Enju parse; adding different features for Trigger Detection. B = Base set Features, TI = Trigger inflect forms.

Parser	SVT-Total			Binding			REG-Total			All-Total		
	Recall	**Prec.**	**F1**	**Recall**	**Prec.**	**F1**	**Recall**	**Prec.**	**F1**	**Recall**	**Prec.**	**F1**
MCCC	70.94	**82.72**	76.38	45.04	55.26	**49.63**	34.39	51.88	41.37	48.10	64.39	55.07
MCCC-I	68.59	82.59	74.94	42.63	**58.67**	49.38	32.58	52.76	40.28	46.06	**65.50**	54.07
Enju	71.66	82.18	**76.56**	40.75	51.01	45.31	32.24	49.39	39.01	46.69	62.70	53.52
MCCC-I + Posteriors	70.49	78.87	74.44	47.72	51.59	49.58	35.64	50.40	41.76	48.94	61.47	54.49
MCCC + Enju	71.84	82.04	76.60	44.77	53.02	48.55	34.96	**53.15**	42.18	48.69	64.59	55.52
MCCC-I + Enju	**72.02**	80.51	76.03	**47.99**	50.85	49.38	**36.95**	51.79	**43.13**	**50.20**	62.60	**55.72**

Table 3: Comparison of Recall/Precision/F1 on the GENIA Task-1 development set using various combinations of parsers: Enju, MCCC (Mc-Closky Charniak), and MCCC-I (Mc-Closky Charniak Improved self-trained biomedical parsing model) with Stanford collapsed dependencies were used for evaluation. Results on Simple, Binding and Regulation and all events are shown.

	Development Set				Test Set			
Event Class	**Count**	**Recall**	**Precision**	**F1**	**Count**	**Recall**	**Precision**	**F1**
Hydroxylation	31	25.81	61.54	36.36	69	30.43	84.00	44.68
Dehydroxylation	0	100.00	100.00	100.00	0	100.00	100.00	100.00
Phosphorylation	32	71.88	85.19	77.97	65	72.31	85.45	78.33
Dephosphorylation	1	0.00	0.00	0.00	4	0.00	0.00	0.00
Ubiquitination	76	63.16	75.00	68.57	180	67.78	81.88	74.16
Deubiquitination	8	0.00	0.00	0.00	10	0.00	0.00	0.00
DNA_methylation	132	72.73	72.18	72.45	182	71.43	73.86	72.63
DNA_demethylation	9	0.00	0.00	0.00	6	0.00	0.00	0.00
Glycosylation	70	61.43	67.19	64.18	169	39.05	69.47	50.00
Deglycosylation	7	0.00	0.00	0.00	12	0.00	0.00	0.00
Acetylation	65	89.23	75.32	81.69	159	87.42	85.28	86.34
Deacetylation	19	68.42	92.86	78.79	24	62.50	93.75	75.00
Methylation	65	64.62	75.00	69.42	193	62.18	73.62	67.42
Demethylation	7	0.00	0.00	0.00	10	0.00	0.00	0.00
Catalysis	60	3.33	15.38	5.48	111	4.50	33.33	7.94
====[TOTAL]====	582	57.22	72.23	63.85	1194	55.70	77.60	64.85

Table 4: Approximate span matching/approximate recursive matching on development and test data sets for EPI CORE Task with our system

3 Results

Of the five evaluation tracks in the shared task, we participated in two: the GENIA core task, and the EPI (Epigenetics and Post-translational modifications) task. The systems used in each track were substantially similar; differences are called out below. Rather than building a system customized for a single trigger and event set, our goal was to build a more generalizable framework for event detection.

3.1 GENIA Task

Using F-measure performance on the development set as our objective function, we trained the final system for the GENIA task with all the features described in section 2, but without the conversion rules and without either feature combination or reduction. Furthermore, we trained the cluster features using the full set of PubMed documents (as of January 2011). The results of our final submission are summarized in Table 1. Overall, we saw a substantial degradation in F-measure when moving from the development set to the test set, though this was in line with past experience from our and other systems.

We compared the results for different parsers in Table 3. MCCC-I is not better in isolation but does produce higher F-measures in combination with other parsers. Although posteriors were not particularly helpful on the development set, we ran a system consisting of MCCC-I with posteriors (MCCC-I + Posteriors) on the test set after the final results were submitted, and found that it was competitive with our submitted system (MCCC-I + ENJU). We believe that ambiguity preservation has merit, and hope to explore more of this area in the future. Diversity is important: although the ENJU parser alone was not the best, combining it with other parsers led to consistently strong results.

Table 2 explores feature ablation: TI appears to degrade performance, but clusters regain that loss. Protein depth information was helpful, but dependency rule conversion was not. Therefore the B+TI+C+PI combination was our final submission on GENIA.

3.2 EPI Task

We trained the final system for the Epigenetics task with all the features described in section 2. Further, we produced the clusters for the Epigenetics task using only the set of GENIA documents provided in the shared task.

In contrast to GENIA, we found that the dependency rule conversions had a positive impact on development set performance. Therefore, we included them in the final system. Otherwise the system was identical to the GENIA task system.

4 Discussion

After two rounds of the BioNLP shared task, in 2009 and 2011, we wonder whether it might be possible to establish an upper-bound on recall and precision. There is considerable diversity among the participating systems, so it would be interesting to consider whether there are some annotations in the development set that cannot be predicted by any of the participating systems[1]. If this is the case, then those triggers and edges would present an interesting topic for discussion. This might result either in a modification of the annotation protocols, or an opportunity for all systems to learn more.

After a certain amount of feature engineering, we found it difficult to achieve further improvements in F1. Perhaps we need a significant shift in architecture, such as a shift to joint inference (Poon and Vanderwende, 2010). Our system may be limited by the pipeline architecture.

MWEs (multi-word entities) are a challenge. Better multi-word triggers accuracy may improve system performance. Multi-word proteins often led to incorrect part-of-speech tags and parse trees.

Cursory inspection of the Epigenetics task shows that some domain-specific knowledge would have been beneficial. Our system had significant difficulties with the rare inverse event types, e.g. "demethylation" (e.g., there are 319 examples for "methylation" in the combined training/development set, but only 12 examples for "demethylation"). Each trigger type was treated independently, thus we did not share information between an event and its related inverse event type. Furthermore, our system also failed to identify edges for these rare events. One approach would be to share parameters between types that differ only in a prefix, e.g., "de". In general, some knowledge about the hierarchy of events may let the learner generalize among related events.

5 Conclusion and Future Work

We have described a system designed for fine-grained information extraction, which we show to be general enough to achieve good performance across different sets of event types and domains. The only domain-specific characteristic is the pre-annotation of proteins as a special class of entities. We formulated some features based on this knowledge, for instance the path to the nearest protein. This would likely have analogues in other domains, given that there is often a special class of target items for any Information Extraction task.

As the various systems participating in the shared task mature, it will be viable to apply the automatic annotations in an end-user setting. Given a more specific application, we may have clearer criteria for balancing the trade-off between recall and precision. We expect that fully-automated systems coupled with reasoning components will need very high precision, while semi-automated systems, designed for information visualization or for assistance in curating knowledge bases, could benefit from high recall. We believe that the data provided for the shared tasks will support system development in either direction. As mentioned in our discussion, though, we find that improving recall continues to be a major challenge. We seek to better understand the data annotations provided.

[1] Our system output for the 2011development set can be downloaded from http://research.microsoft.com/bionlp/

Our immediate plans to improve our system include semi-supervised learning and system combination. We will also continue to explore new levels of linguistic representation to understand where they might provide further benefit. Finally, we plan to explore models of joint inference to overcome the limitations of pipelining and deterministic post-processing.

Acknowledgments

We thank the shared task organizers for providing this interesting task and many resources, the Turku BioNLP group for generously providing their system and intermediate data output, and Patrick Pantel and the MSR NLP group for their help and support.

References

Jari Björne, Juho Heimonen, Filip Ginter, Antti Airola, Tapio Pahikkala and Tapio Salakoski. 2009. Extracting Complex Biological Events with Rich Graph-Based Feature Sets. In *Proceedings of the Workshop on BioNLP: Shared Task*.

Thomas Cormen, Charles Leiserson, and Ronald Rivest. 2002. *Introduction to Algorithms*. MIT Press.

Ted Dunning. 1993. Accurate methods for the statistics of surprise and coincidence. *Computational Linguistics*, 19(1), pp. 61-74.

George E. Heidorn, 2000. Intelligent Writing Assistance. In Handbook of Natural Language Processing, ed. Robert Dale, Hermann Moisl, and Harold Somers. Marcel Dekker Publishers.

Karen Jensen. 1993. PEGASUS: Deriving Argument Structures after Syntax. In *Natural Language Processing: the PLNLP approach*, ed. Jensen, K., Heidorn, G.E., and Richardson, S.D. Kluwer Academic Publishers.

Marie-Catherine de Marneffe, Bill MacCartney and Christopher D. Manning. 2006. Generating Typed Dependency Parses from Phrase Structure Parses. In *LREC 2006*.

Yusuke Miyao and Jun'ichi Tsujii. 2008. Feature forest models for probabilistic HPSG parsing. *Computational Linguistics* 34(1): 35-80.

David McClosky and Eugene Charniak. 2008. Self-Training for Biomedical Parsing. In *Proceedings of the Association for Computational Linguistics 2008*.

David McClosky. 2010. Any Domain Parsing: Automatic Domain Adaptation for Natural Language Parsing. Ph.D. thesis, Department of Computer Science, Brown University.

Ryan McDonald. 2006. Discriminative training and spanning tree algorithms for dependency parsing. Ph. D. Thesis. University of Pennsylvania.

Haitao Mi, Liang Huang, and Qun Liu. 2008. Forest-based Translation. In *Proceedings of ACL 2008*, Columbus, OH.

Patrick Pantel, Eric Crestan, Arkady Borkovsky, Ana-Maria Popescu and Vishnu Vyas. 2009. Web-Scale Distributional Similarity and Entity Set Expansion. In *Proceedings of EMNLP 2009*.

Hoifung Poon and Lucy Vanderwende. 2010. Joint inference for knowledge extraction from biomedical literature. In *Proceedings of NAACL-HLT 2010*.

Martin.F. Porter, 1980, An algorithm for suffix stripping, *Program*, **14**(3):130–137.

Ioannis Tsochantaridis, Thomas Hofmann, Thorsten Joachims, and Yasemin Alton. 2004. Support vector machine learning for interdependent and structured output spaces. In *ICML 2004*.

From Graphs to Events: A Subgraph Matching Approach for Information Eextraction from Biomedical Text

Haibin Liu, Ravikumar Komandur, Karin Verspoor
Center for Computational Pharmacology
University of Colorado School of Medicine
PO Box 6511, MS 8303, Aurora, CO, 80045 USA

Abstract

We participated in the BioNLP Shared Task 2011, addressing the GENIA event extraction (GE) and the Epigenetics and Post-translational Modifications (EPI) tasks. A graph-based approach is employed to automatically learn rules for detecting biological events in the life-science literature. The event rules are learned by identifying the key contextual dependencies from full syntactic parsing of annotated text. Event recognition is performed by searching for an isomorphism between event rules and the dependency graphs of sentences in the input texts. While we explored methods such as performance-based rule ranking to improve precision, we merged rules across multiple event types in order to increase recall.

We achieved a 41.13% F-score in detecting events of nine types in the Task 1 of the GE task, and a 52.67% F-score in identifying events across fifteen types in the core task of the EPI task. Our performance on both tasks is comparable to the state-of-the-art systems. Our approach does not require any external domain-specific resources. The consistent performance on the two tasks supports the claim that the method generalizes well to extract events from different domains where training data is available.

1 Introduction

Recent research in information extraction in the biological domain has focused on extracting semantic events involving genes or proteins, such as binding events or post-translational modifications. To date, most of the biological knowledge about these events has only been available in the form of unstructured text in scientific articles (Abulaish and Dey, 2007; Ananiadou et al., 2010).

When a biological event is described in text, it can be analyzed by recognizing its type, the trigger that signals the event, and one or more event arguments. The BioNLP-ST 2009 (Kim et al., 2009) focused on the recognition of semantically typed, complex events in the biological literature. Although the best-performing system achieved a 51.95% F-score in identifying events across nine types, only 4 of the rest 23 participating teams obtained an F-score in the 40% range. This suggests that the problem of biological event extraction is difficult and far from solved.

Graphs provide a powerful primitive for modeling biological data such as pathways and protein interaction networks (Tian et al., 2007; Yan et al., 2006). More recently, the dependency representations obtained from full syntactic parsing, with its ability to reveal long-range dependencies, has shown an advantage in biological relation extraction over the traditional Penn Treebank-style phrase structure trees (Miyao et al., 2009). Since the dependency representation maps straightforwardly onto a directed graph, operations on graphs can be naturally applied to the problem of biological event extraction.

We participated in the BioNLP-ST 2011 (Kim et al., 2011a), and applied a graph matching-based approach (Liu et al., 2010) to tackling the Task 1 of the GENIA event extraction (GE) task (Kim et al., 2011b), and the core task of the Epigenetics and Post-translational Modifications (EPI) task (Ohta et al., 2011), two main tasks of the BioNLP-ST 2011. Event recognition is performed by searching for an isomorphism between dependency representations of automatically learned event rules and complete sentences in the input texts. This process is treated as a subgraph matching problem, which corresponds to the search for a subgraph isomorphic to a rule graph within a sentence graph. While we explored methods such as performance-based rule ranking to improve the precision of the GE and EPI tasks, we merged rules across multiple event types in order to increase the recall of the EPI task.

The rest of the paper is organized as follows: In Section 2, we introduce the BioNLP Shared Task 2011. Section 3 describes the subgraph matching-based event extraction method. Section 4 and Section 5 elabo-

Proceedings of BioNLP Shared Task 2011 Workshop, pages 164–172,
Portland, Oregon, USA, 24 June, 2011.

rate the implementation details and our performance respectively. Finally, Section 6 summarizes the paper and introduces future work.

2 BioNLP Shared Task 2011

The BioNLP-ST 2011 is the extension of the BioNLP-ST 2009 that focused on the recognition of events in the biological literature. The BioNLP-ST 2011 extends the previous task in three directions: the type of the investigated text, the domain of the subject, and the targeted event types. As a result, the shared task was organized into four independent tasks: GENIA Event Extraction Task (GE), Epigenetics and Post-translational Modifications Task (EPI), Infectious Diseases Task (ID) and Bacteria Track.

The definition of the GE task remained the same as the BioNLP-ST 2009. However, additional annotated texts that come from full papers were provided together with the dataset of the 2009 task to generalize the task from PubMed abstracts to full text articles. The primary task of the GE task was to detect biological events of nine types such as protein binding and regulation, given the annotation of protein names. It was required to extract type, trigger, and primary arguments of each event. This task is an example of extraction of semantically typed, complex events for which the arguments can also be other events. Such embedding results in a nested structure that captures the underlying biological statements more accurately.

Different from the subject domain of the GE task on transcription factors in human blood cells, the EPI task focused on events related to epigenetic change, including DNA methylation and histone modification, as well as other common post-translational protein modifications. The core task followed the definition for Phosphorylation event extraction in the 2009 task, and extended that basic event type to a total of fifteen types including both positive and negative variants, for example *Acetylation* and *Deacetylation*. The task dataset was prepared from relevant PubMed abstracts, with additional evidence sentences from databases such as PubMeth (Ongenaert et al., 2007). Given the annotation of protein names, the core task required to extract type, trigger, and primary arguments of each event.

We focused on the primary task of GE and the core task of EPI, and tackled the event extraction problem in both cases using a graph matching-based method.

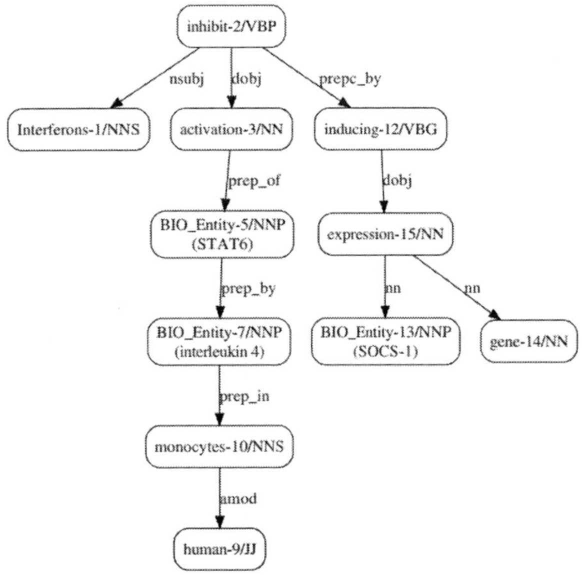

Figure 1: Dependency Graph Example

3 Subgraph Matching-based Event Extraction

3.1 Dependency Representation

The dependency representation of a sentence is formed by tokens in the sentence and binary relations between them. A single dependency relation is represented as *relation*(*governor*, *dependent*), where *governor* and *dependent* are tokens, and *relation* is a type of the grammatical dependency relation. This representation is essentially a labeled directed graph, which is named *dependency graph* and defined as follows:

Definition 1. A dependency graph is a pair of sets $G = (V, E)$, where V is a set of nodes that correspond to the tokens in a sentence, and E is a set of directed edges, for which the edge labels are types of dependency relations between the tokens, and the edge direction is from *governor* to *dependent* node.

Figure 1 illustrates the dependency graph for the sentence: "Interferons inhibit activation of STAT6 by interleukin 4 in human monocytes by inducing SOCS-1 gene expression." (MEDLINE: 10485906). The token number in the sentence is appended to each token in order to differentiate identical tokens that co-occur in a sentence. All the protein names in the sentence have been replaced with a unified tag "BIO_Entity". The POS tag of each token is noted. "BIO_Entity" tokens are uniformly tagged as proper nouns.

3.2 Event Rule Induction

The premise of our work is that there is a set of frequently occurring event rules that match a majority of stated events about protein biology. We consider that an event rule encodes the detailed description and characterizes the typical contextual structure of a group of biological events. The rules are learned from labeled training sentences using a graph-based rule induction method (Liu et al., 2010), and we briefly describe the algorithm as follows.

Starting with the dependency graph of each training sentence, edge directions are first removed so that the directed graph is transformed into an undirected graph, where a path must exist between any two nodes since the graph is always connected. For each gold event, the shortest dependency path in the undirected graph connecting the event trigger nodes to each event argument node is selected. The union of all shortest dependency paths is then computed, and the original directed dependency representation of the path union is retrieved and used as the graph representation of the event.

For multi-token event triggers, the shortest dependency path connecting the node of every trigger token to the node of each event argument is selected, and the union of the paths is then computed for each trigger. For regulation events, when a sub-event is used as an argument, only the type and the trigger of the sub-event are preserved as the argument of the main events. The shortest dependency path is extracted so as to connect the trigger nodes of the main event to the trigger nodes of the sub-event. In case that there exists more than one shortest path, all of the paths are considered. As a result, each gold event is transformed into the form of a biological event rule. The algorithm is elaborated in more detail in (Liu et al., 2010). The obtained rules are categorized in terms of the event types of the tasks.

3.3 Sentence Matching

We attempted to match event rules to each testing sentence to extract events from the sentence using a sentence matching approach. Since the event rules and the sentences all possess a dependency graph, the matching process is a subgraph matching problem, which corresponds to the search for a subgraph isomorphic to an event rule graph within the graph of a testing sentence. The subgraph matching problem is also called *subgraph isomorphism*, defined in this work as follows:

Definition 2. An event rule graph $G_r = (V_r, E_r)$ is isomorphic to a subgraph of a sentence graph $G_s = (V_s, E_s)$, denoted by $G_r \cong S_s \subseteq G_s$, if there is an injective mapping $f : V_r \rightarrow V_s$ such that, for every directed pair of nodes $v_i, v_j \in V_r$, if $(v_i, v_j) \in E_r$ then $(f(v_i), f(v_j)) \in E_s$, and the edge label of (v_i, v_j) is the same as the edge label of $(f(v_i), f(v_j))$.

The subgraph isomorphism problem is NP-complete (Cormen et al., 2001). A number of algorithms have been designed to tackle the problem of subgraph isomorphism in different applications (Ullmann, 1976; Cordella et al., 2004; Pelillo et al., 1999). Considering that the graphs of rules and sentences involved in the matching process are small, a simple subgraph matching algorithm using a backtracking approach (Liu et al., 2010) was used in this work. It is named "Injective Graph Embedding Algorithm" and designed based on the Huet's graph unification algorithm (Huet, 1975). The formalized algorithm and the detailed description are given in (Liu et al., 2010).

When matching between graphs, different combinations of matching features can be applied, resulting in different matching criteria. The features include edge features (E) which are edge label and edge direction, and node features which are POS tags (P), trigger tokens (T), and all tokens (A), ranging from the least specific matching criterion, E, to the much stricter criterion, A. For each sentence, the algorithm returns all the matched rules together with the corresponding injective mappings from rule nodes to sentence tokens. Biological events are then extracted by applying the event descriptions of tokens in each matched rule consisting of the type, the trigger and the arguments to the corresponding tokens of the sentence.

4 Implementation

4.1 Preprocessing

The same preprocessing steps as in (Liu et al., 2010) are completed on the datasets of the GE and the EPI tasks before performing text mining strategies. These include sentence segmentation and tokenization, Part-of-Speech tagging, and sentence parsing.

The Stanford unlexicalized natural language parser (version 1.6.5), which includes Genia Treebank 1.0 (Ohta et al., 2005) as training material, is used to analyze the syntactic structure of the sentences. The parser returns a dependency graph for each sentence.

4.2 Rule Induction and Sentence Matching

For each gold event, the shortest path in the undirected graph connecting the event trigger to each event argument is extracted using Dijkstra's algorithm (Cormen et al., 2001) with equal weight for edges.

Sentence matching is performed and the raw matching results are then postprocessed based on the specifications of the shared task, such as event trigger cannot be a protein name or another event.

5 Results and Evaluation

This section presents our results on the GE and the EPI tasks (Kim et al., 2011b; Ohta et al., 2011) respectively. Different experimental methods in processing the obtained event rules are described for the purpose of improving the precision of both tasks and increasing the recall of the EPI task.

5.1 GE task

5.1.1 Preprocessing Results

For training data, only sentences that contain at least one protein and one event are considered candidates for further processing. For testing data, candidate sentences contain at least one protein. Our event recognition method focuses on extracting events from sentences. Therefore, only sentence-based events are considered in this work. Table 1 presents some statistics of the preprocessed datasets.

Attributes Counted	Training	Dev.	Testing
Abstracts&Full articles	908	259	347
Total sentences	8,759	2,954	3,437
Candidate sentences	3,615	1,989	2,353
Total events	10,287	3,243	4,457
Sentence-based events	9,583	3,058	hidden

Table 1: Statistics of GE dataset

We were able to build event rules for 9,414 gold events. Gold events in which the event trigger and an event argument are not connected by a path in the undirected dependency graph of the sentence could not be transformed into a biological event rule. After removing duplicate rules, we obtained 8,677 event rules, which are distributed over nine event types. The rules that are isomorphic to each other in terms of their graph representation are not filtered at this stage as the duplicate events they produce will be removed eventually to prepare the annotations for the shared task.

5.1.2 Probability-based rule refining

We observed that some event rules of an event type overlap with rules of other event types. For instance, a *Transcription* rule is isomorphic to a *Gene_expression* rule in terms of the graph representation and they also share a same event trigger token. In fact, tokens like "gene expression" and "induction" are used as event trigger of both *Transcription* and *Gene_expression* in training data. Therefore, the detection of some *Gene_expression* events is always accompanied by certain *Transcription* events. This will have detrimental effects on the precision of both *Transcription* and *Gene_expression* event types.

As transcription is the first step leading to gene expression (Ananiadou and Mcnaught, 2005), there exist some correlations or associations between the two event types. In tackling this problem, we processed the overlapping rules based on a conditional probability $P(t|E)$, where t stands for an event trigger and E represents one of the event types. Eq.(1) is used to estimate the value of $P(t_i|E)$.

$$P(t_i|E) = \frac{f(t_i, E)}{\sum_i f(t_i, E)}, \quad (1)$$

where $f(t_i, E)$ is the frequency of the event trigger t_i of the event type E in the training data, and $\sum_i f(t_i, E)$ calculates the total frequency of all event triggers of the event type E in the training data.

$P(t_i|E)$ evaluates the degree of the importance of a trigger to an event type. When the dependency graphs of two rules of different event types are isomorphic to each other, and two rules share a same event trigger, we examine the $P(t_i|E)$ of each event type, and only retain the rule for which the $P(t_i|E)$ is higher.

Compared to the "once a trigger, always a trigger" method employed in other work (Buyko et al., 2009; Kilicoglu and Bergler, 2009), triggers are treated in a more flexible way in our work. A token is not necessarily always a trigger unless it appears in the appropriate context. Also, the same token can serve as trigger for different event types as long as it appears in the different context. A trigger will only be classified into a fixed event type when it could serve as trigger for different event types in the same context.

5.1.3 Performance-based rule ranking

In addition to the process of refining rules across event types, we proposed a performance-based rule ranking method to evaluate each rule under one event type. We matched each rule to sentences in the development set using the subgraph matching approach. For rules that produce at least one event prediction, we ranked them by $PRC(r_i)$, the precision of each rule r_i, which is computed via Eq.(2).

$$PRC(r_i) = \frac{\#correctly\_predicted\_events\_by\_r_i}{\#predicted\_events\_by\_r_i} \quad (2)$$

We manually examined the rules with low rank. In our experiments, the $PRC(r_i)$ ratio of these rules is bigger than 4:1. We removed the ones that are either incorrect or ambiguous in semantics and syntactics based on our domain knowledge. Our assumption is that these rules will keep producing false positive events on the testing data if they are retained in the rule set. For rules that do not make any predictions on the development data, we keep them in the set in the hope that they may contribute to the event recognition from the testing data. Without affecting much on the recall, this process helps to improve the precision of the events extracted from the development data.

5.1.4 GE Results on Development Set

In our previous work (Liu et al., 2010), the matching criteria, "E+P+T" and and "E+P+A", achieved the highest F-score and the highest precision respectively among all the investigated matching criteria. "E+P+T" requires that edge directions and labels of all edges (E) be identical, POS tags (P) of all tokens be identical, and tokens of only event triggers (T) be identical for the edges and the nodes of a rule and a sentence to match with each other. "E+P+A" requires that edges (E), POS tags (P) and all tokens (A) be exactly the same. In this work, we focused on these two criteria and explored to extend them for graph matching between event rules and sentences.

We attempted to relax the matching criterion of POS tags for nouns and verbs. For nouns, the plural form of nouns is allowed to match with the singular form, and proper nouns are allowed to match with regular nouns. For verbs, past tense, present tense and base present form are allowed to match with each other.

Next, letters of each token are transformed into lower case, and tokens containing hyphens are normalized into non-hyphenated forms. Lemmatization is then performed on every pair of tokens to be matched using WordNet (Fellbaum, 1998) as the lemmatizer to allow tokens that share a same lemma to match. Since WordNet is a lexical database only for the general English language, the lemma of a fair amount of domain-specific vocabulary cannot be found in WordNet, such as "Phosphorylation" and "Methylation". In this case, a backup process is invoked to stem the tokens to their root forms using the Porter's stemming algorithm (Porter, 1997) allowing the tokens derived from a same root word to match.

To further generalize event rules, we extended the matching criteria "E+P*+A*" to "E+P*+A*S" to allow tokens to match if their lemmatized forms have a common synonym in terms of the synsets of WordNet. Since WordNet will relate verbs such as "induce" and "receive" together as they share a synonym "have", and allow nouns like "expression" and "aspect" to match as they share a synonym "face", we limited this extension to only adjective tokens to avoid too many false positive events and allow tokens like "crucial" and "critical" to match.

Table 2 shows the event extraction results on the development data based on different matching criteria. The performance is evaluated by "Approximate Span Matching/Approximate Recursive Matching", the primary evaluation measure of the shared task. "E+P*+T*", "E+P*+A*" and "E+P*+A*S" demonstrate the performance of the extended criteria.

Feature	Recall(%)	Prec.(%)	F-score(%)
E+P+A	28.03	66.74	39.48
E+P+T	31.17	52.38	39.09
E+P*+A*	31.45	63.51	42.07
E+P*+T*	35.71	46.26	40.31
E+P*+A*S	31.51	63.32	42.08

Table 2: GE results on development set using different matching criteria

As the strictest matching criteria, "E+P+A" performs better than "E+P+T" in both precision and F-score. Although "E+P+T" achieves a better recall, when relaxing the matching criteria from all tokens being the same to only event trigger tokens having to be identical, the precision of "E+P+T" is decreased by a large margin, nearly 14%. This indicates that a certain number of biological events are described in very similar ways in the literature, involving same grammatical structures and identical contextual contents. While producing more incorrect events, "E+P*+A*" and "E+P*+T*" significantly improve the recall, leading to a better F-score over "E+P+A" and "E+P+T". This confirms the effectiveness of the POS relaxation and the token lemmatization on the generalization of event rules. "E+P*+A*S" obtains a comparable performance with "E+P*+A*" with only a 0.06% increase in recall and a 0.2% drop in precision.

5.1.5 GE Results on Testing Set

Table 3 shows our results of "E+P*+A*" on the testing data using the official metric. We are listed as team "CCP-BTMG". Ranked by F-score, our performance ranked 10th out of 15 participating groups. It is worth noting that our result on the event type "Protein_catabolism" ranked 1st.

Event type	Rec.(%)	Prec.(%)	F(%)
Gene_expression	58.68	75.77	66.14
Transcription	39.08	51.91	44.59
Protein_catabolism	66.67	83.33	74.07
Phosphorylation	63.78	85.51	73.07
Localization	29.32	91.80	44.44
Binding	22.61	49.12	30.96
Regulation	12.99	46.73	20.33
Positive_regulation	21.90	44.51	29.35
Negative_regulation	15.76	40.18	22.64
All total	31.57	58.99	41.13

Table 3: GE results of "E+P*+A*" on testing set by "Approximate Span /Approximate Recursive Matching"

The performance of our system on the testing set is consistent with that of the development set. We achieved a comparable precision with the top systems and ranked 6th by precision. However, our recall was lower, ranking 11th. This adversely impacted the overall F-score. The lower recall is not surprising because the graph matching criteria "E+P*+A*" strictly demand that every lemmatized token in the patterns, other than protein names represented as"BIO_Entity", has to find its exact match in the input sentences. The detailed analysis on the recall problem is presented in the "Error Classification" section.

While examining the false positives, we found that for many cases our result matched the gold annotation but for the trigger word. We believe that event type and their arguments are more important biologically than the trigger. We consulted some domain experts who reinforced our intuition in many cases that different words could be considered as trigger for the event in question. Following this we contacted organizers and they agreed to release a new evaluation scheme to ignore the trigger match requirement in order to support evaluation of the event extraction itself.

Table 4 shows our results of "E+P*+A*" evaluated by other official evaluation metrics of the task. The strict matching scheme requires exact trigger span as well as all its nested events to be recursively correct for an event to be considered correctly extracted. Our F-score in terms of the strict matching is only 2.65% lower than the relaxed, primary measure, indicating that most of the detected triggers are captured with correct text span. The organizers also provided the evaluation results on PubMed abstracts and PMC full text articles separately. Our system performs consistently on both abstracts and full papers and the difference between F-scores is less than 1% (41.39% vs. 40.47%) mostly due to the small recall loss on full texts.

Measures	R(%)	P(%)	F(%)
Strict Matching	29.55	55.13	38.48
Appr. SpanNoTrigger/Recur.	33.68	62.17	43.69
Appr. Span/Recur./Decomp.	32.56	66.20	43.65
Appr. Sp. No T./Recur./Decomp.	34.96	69.87	46.60
Appr. Span/Recur. (Abstract)	31.87	59.02	41.39
Appr. Span/Recur. (Full paper)	30.82	58.92	40.47

Table 4: GE results on testing set by other evaluation measures

5.2 EPI task

5.2.1 Preprocessing Results

Table 5 presents some statistics of the datasets. We were able to build event rules for 1598 gold events. After removing duplicate rules, we obtained 1,562 event rules distributed over fifteen event types.

Attributes Counted	Training	Dev.	Testing
Abstracts	600	200	440
Total sentences	6,411	2,218	4,640
Candidate sentences	1,054	1,241	2,839
Total events	1,738	582	1,194
Sentence-based events	1,643	536	hidden

Table 5: Statistics of EPI dataset

We processed the obtained rules following the same rule refining and ranking processes of the GE task. We experimented with two graph matching criteria for extracting EPI events, "E+P*+T*" and "E+P*+A*". From the preliminary results, we observed that "E+P*+A*" achieves a high precision over 80% but a lower recall around 33%. Compared to the GE task results, "E+P*+T*" achieves a better recall against a small tradeoff for precision. We consider that this is because the event triggers themselves for the EPI task such as "acetylation", "deglycosylation" and "demethylation" are powerful enough to differentiate among event types without the need to resort to more contextual content of the patterns. Therefore, we focused on using "E+P*+T*" to extract events.

5.2.2 Recall-oriented rule merging

Since all the event types except *Catalysis*, *DNA_methylation* and *DNA_demethylation* in the

EPI task involve addition or removal of biochemical functional groups at a particular amino acid residue of a protein (Hunter, 2009), common syntactic structures of expressing the protein PTM events might be shared across event types. To further improve the recall, we proposed a rule merging strategy to take advantage of the syntactic structures of rules across event types.

We first experimented with a "pairwise flip" approach which combines rules of the pairwise, positive and negative event types by flipping the type and the trigger of event rules. For instance, the event rules of *Phosphorylation* and *Dephosphorylation* are merged together and then used to detect events of the two types respectively.

Next, the "pairwise flip" approach was extended to an "all in one" method. For one event type, the rules of all other PTM event types are processed and merged into the rules of the current type if the trigger of rules of other types contains one of these 12 morphemes: "acetyl", "glycosyl", "hydroxyl", "methyl", "phosphoryl", "ubiqui", "deacetyl", "deglycosyl", "dehydroxyl", "demethyl", "dephosphoryl", "deubiqui". We consider that event rules involving these morphemes in trigger are more likely to discuss representative protein post-translational modifications.

5.2.3 EPI Results on Development Set

Table 6 shows the event extraction results on the development data using different matching criteria and rule merging methods. The performance is evaluated by the primary evaluation measure.

Feature	Recall(%)	Prec.(%)	F(%)
E+P*+A*	32.65	79.83	46.34
E+P*+T*	38.14	73.51	50.23
E+P*+A*(pairwise)	35.22	80.39	48.98
E+P*+T*(pairwise)	40.89	77.52	53.54
E+P*+T*(all in one)	46.39	63.08	53.47

Table 6: EPI results on development set

The two rule merging methods using "E+P*+T*" outperform others in terms of F-score. The "pairwise flip" method achieves higher precision as the syntactic structures of rules to describe the pairwise, positive and negative events tend to be highly similar. However, when merging all the rules across PTM event types, although more events are captured, rules that involve syntactic structures for expressing very specific events of certain types may not generalize well on some other types, resulting in incorrect events. Thus, the "all in one" approach significantly improves the recall while producing many false positive events, leading to a F-score comparable with the "pairwise flip" method.

5.2.4 EPI Results on Testing Set

We conducted two runs on the testing data in terms of "E+P*+T*(pairwise)" and "E+P*+T*(all in one)". Since the two rule merging methods achieve comparable F-scores, we decided to submit a run with higher recall. Table 7 shows our results of "E+P*+T*" using the "all in one" approach on the official metrics. Only 7 teams participated in this task. For the core task, our performance ranked 7th, only 0.16% lower in F-score than the 6th team. When evaluating our results in terms of the full task, we ranked 6th.

Feature	Recall(%)	Prec.(%)	F(%)
E+P*+T*(core task)	45.06	63.37	52.67
E+P*+T*(full task)	23.44	37.93	28.97

Table 7: EPI results on testing set

Compared to the top teams, our F-score is mostly affected by the lower recall. Although the run we submitted achieves the highest recall among all our runs, our recall is about 20% less than the best performing system. Considering that most of the event types of the EPI task tend to use tokens containing only a small fixed set of domain-specific morphemes as triggers, the recall deficit is assumed to be lack of event rules that describe syntactic structures of expressing a fair amount of EPI events.

5.3 Error Classification

Since the gold event annotation of the testing data is hidden, we examined the event extraction results of the development data to analyze the underlying errors. The detailed analysis is reported in terms of false negative and false positive events.

5.3.1 False negatives

It is shown that false negative events have a substantial impact on the performance of all 15 participating teams of the GE task. The best recall, 49.56%, captures less than half of the gold events in the testing set. In our work, three major causes of false negatives are determined for both tasks.

(1) **Low coverage of rule set**: For the GE task, the graph matching criteria "E+P*+A*" strictly asks every lemmatized token in the patterns to find its exact match in the input sentences. Although maintaining the precision at a high level, this directly limits the contextual

structure and content around the proteins and thus prevents the recall from being higher.

Lemmatization helps to detect more events, however, further generalization needs to be performed on the existing rules to relax the token matching requirement. For instance, when "lysine" appears in an event rule, knowing that "lysine" is an amino acid, the rule might be further generalized to allow all amino acids to match with each other in order to recognize more events.

For the EPI task, although "E+P*+T*" requires tokens of only event triggers to be identical, we captured less than half of the gold events. We noticed that many trigger tokens in the development sentences do not appear as triggers in the training set. This leads to the failure of extracting the corresponding events. Since the training data is the only source of triggers in our work, the coverage of triggers limits the generalization power of event rules.

For both tasks, we found that many gold events are described in grammatical structures that are not covered by the existing rules induced from the training sentences. These structures tend to be more complex, involving a long dependency path from the trigger to arguments in the graphs of sentences. Events that consist of these structures are not recognized as no matched rules will be returned from the subgraph matching.

In order to further improve the recall, some postprocessing steps are necessary to be performed on the raw dependency graphs of both rules and sentences instead of using them in the graph matching directly. By eliminating semantically unimportant nodes and grouping lexically connected nodes together, the rules can be generalized to retain only their skeleton structures while complex sentences can be syntactically simplified to allow event rules to match them.

(2) **Compound error effect**: In both tasks, regulation and catalysis event types can take sub-events as arguments. Therefore, if the nested sub-events are not correctly identified, the main events will not be extracted due to the compound error effect.

(3) **Anaphora and coreference**: Since our system focuses on extracting events from sentences, events that contain protein names spanning multiple sentences will not be captured. Recognition of these events requires the ability to do anaphora and coreference resolution in biological text (Gasperin and Briscoe, 2008).

5.4 False positives

Three major causes of false positives are generalized from our analysis.

(1) **Assignment of overlapping event rules**: The conditional probability-based method to assign overlapped rules of different event types effectively reduces the number of event candidates but leads to errors. For instance, "methylation" is used as the trigger for two overlapping rules of *DNA_methylation* and *Methylation*. Based on the $P(t_i|E)$, "methylation" is classified into *DNA_methylation*. An erroneous *DNA_methylation* event is then detected from a development sentence instead of the gold *Methylation* event. Although the trigger and the participant are all identified correctly, the event type is assigned wrongly.

In fact, the same contextual structure and content appear in both *DNA_methylation* and *Methylation* events in the training data. According to the EPI task (Ohta et al., 2011), *Methylation* is to abbreviate for "protein methylation" and thus is different from *DNA_methylation*. In this case, the only way to distinguish between the two types is to identify that the biological entity mentioned in the sentence is a gene for *DNA_methylation* and a protein for *Methylation*. Since genes and their products are uniformly annotated as "Protein" in the task, it is not possible to assign a correct event type in this case from the perspective of the event extraction itself.

(2) **Lack of postprocessing rules**: Some misidentified events require customized postprocessing rules. For instance, a *Gene_expression* event is detected from the phrase "Tax expression vector" of a development sentence. However, since "Tax expression" is only used as an adjective to describe "vector" in this context, the identified *Gene_expression* event is not appropriate. Likewise, "Sp1 transcription" should not be identified as an event in the context of "Sp1 transcription factors".

(4) **Inconsistencies in gold annotation**: Some extracted events are considered biologically meaningful but evaluated as false positives due to the inconsistencies in the gold annotation. In Table 4, the 3.2% increase in precision of the no-trigger evaluation measure over the primary evaluation scheme indicates that the inconsistent gold annotations of event triggers.

6 Conclusion and future work

We used dependency graphs to automatically induce biological event rules from annotated events. We explored methods such as performance-based rule ranking to improve the accuracy of the obtained rules, and we merged rules across multiple event types in order to increase the coverage of the rules. The event extraction process is treated as a subgraph matching problem to search for the graph of an event rule within the graph of a sentence. We tackled two main tasks of the BioNLP Shared Task 2011. We achieved a 41.13% F-score in detecting events across nine types in the Task 1 of the GE task, and a 52.67% F-score in identifying events across fifteen types in the core task of the EPI task.

In future work, we would like to explore the approaches of generalizing the raw dependency graphs of both event rules and sentences in order to improve the recall of our event extraction system. We also plan to extend our system to tackle the other sub-tasks in GE and EPI tasks, such as to extract events with additional arguments like site and location, and to recognize negations and speculations regarding the extracted events.

References

Muhammad Abulaish and Lipika Dey. 2007. Biological relation extraction and query answering from medline abstracts using ontology-based text mining. *Data & Knowledge Engineering*, 61(2):228–262.

Sophia Ananiadou and John Mcnaught. 2005. *Text Mining for Biology And Biomedicine*. Artech House Publishers.

Sophia Ananiadou, Sampo Pyysalo, Jun'ichi Tsujii, and Douglas B. Kell. 2010. Event extraction for systems biology by text mining the literature. *Trends in Biotechnology*, 28(7):381–390.

Ekaterina Buyko, Erik Faessler, Joachim Wermter, and Udo Hahn. 2009. Event extraction from trimmed dependency graphs. In *BioNLP '09: Proceedings of the Workshop on BioNLP*, pages 19–27, Morristown, NJ, USA. Association for Computational Linguistics.

Luigi P. Cordella, Pasquale Foggia, Carlo Sansone, and Mario Vento. 2004. A (sub)graph isomorphism algorithm for matching large graphs. *IEEE Trans. Pattern Anal. Mach. Intell.*, 26(10):1367–1372.

Thomas H. Cormen, Charles E. Leiserson, Ronald L. Rivest, and Clifford Stein. 2001. *Introduction to Algorithms*. The MIT Press.

Christiane Fellbaum. 1998. *WordNet: An Electronic Lexical Database*. Bradford Books.

Caroline Gasperin and Ted Briscoe. 2008. Statistical anaphora resolution in biomedical texts. In *COLING '08: Proceedings of the 22nd International Conference on Computational Linguistics*, pages 257–264, Morristown, NJ, USA. Association for Computational Linguistics.

Gérard P. Huet. 1975. A unification algorithm for typed lambda-calculus. *Theor. Comput. Sci.*, 1(1):27–57.

Lawrence Hunter. 2009. *The Processes of Life: An Introduction to Molecular Biology*. The MIT Press.

Halil Kilicoglu and Sabine Bergler. 2009. Syntactic dependency based heuristics for biological event extraction. In *Proceedings of the Workshop on Current Trends in Biomedical Natural Language Processing: Shared Task*, BioNLP '09, pages 119–127.

Jin-Dong Kim, Yoshinobu Kano Tomoko Ohta, Sampo Pyysalo, and Jun'ichi Tsujii. 2009. Overview of bionlp'09 shared task on event extraction. In *Proceedings of the NAACL-HLT 2009 Workshop on Natural Language Processing in Biomedicine (BioNLP'09)*, pages 1–9. ACL.

Jin-Dong Kim, Sampo Pyysalo, Tomoko Ohta, Robert Bossy, and Jun'ichi Tsujii. 2011a. Overview of BioNLP Shared Task 2011. In *Proceedings of the BioNLP 2011 Workshop Companion Volume for Shared Task*, Portland, Oregon, June. Association for Computational Linguistics.

Jin-Dong Kim, Yue Wang, Toshihisa Takagi, and Akinori Yonezawa. 2011b. Overview of the Genia Event task in BioNLP Shared Task 2011. In *Proceedings of the BioNLP 2011 Workshop Companion Volume for Shared Task*, Portland, Oregon, June. Association for Computational Linguistics.

Haibin Liu, Vlado Keselj, and Christian Blouin. 2010. Biological event extraction using subgraph matching. In *Proceedings of the 4th International Symposium on Semantic Mining in Biomedicine (SMBM-2010)*, October.

Yusuke Miyao, Kenji Sagae, Rune Saetre, Takuya Matsuzaki, and Jun'ichi Tsujii. 2009. Evaluating contributions of natural language parsers to protein–protein interaction extraction. *Bioinformatics*, 25(3):394–400.

Tomoko Ohta, Yuka Tateisi, and Junichi Tsujii. 2005. Syntax annotation for the genia corpus. In *Proceedings of the IJCNLP 2005*, pages 222–227.

Tomoko Ohta, Sampo Pyysalo, and Jun'ichi Tsujii. 2011. Overview of the Epigenetics and Post-translational Modifications (EPI) task of BioNLP Shared Task 2011. In *Proceedings of the BioNLP 2011 Workshop Companion Volume for Shared Task*, Portland, Oregon, June. Association for Computational Linguistics.

Mate Ongenaert, Leander Van Neste, Tim De Meyer, Gerben Menschaert, Sofie Bekaert, and Wim Van Criekinge. 2007. Pubmeth: a cancer methylation database combining text-mining and expert annotation. *Nucleic Acids Research*, pages 1–5.

Marcello Pelillo, Kaleem Siddiqi, and Steven W. Zucker. 1999. Matching hierarchical structures using association graphs. *IEEE Trans. Pattern Anal. Mach. Intell.*, 21(11):1105–1120.

M. F. Porter. 1997. An algorithm for suffix stripping. pages 313–316.

Yuanyuan Tian, Richard C. Mceachin, Carlos Santos, David J. States, and Jignesh M. Patel. 2007. Saga: a subgraph matching tool for biological graphs. *Bioinformatics*, 23(2):232–239.

J. R. Ullmann. 1976. An algorithm for subgraph isomorphism. *J. ACM*, 23(1):31–42.

Xifeng Yan, Feida Zhu, Jiawei Han, and Philip S. Yu. 2006. Searching substructures with superimposed distance. In *ICDE '06: Proceedings of the 22nd International Conference on Data Engineering*, page 88, Washington, DC, USA. IEEE Computer Society.

Adapting a General Semantic Interpretation Approach to Biological Event Extraction

Halil Kilicoglu and Sabine Bergler
Department of Computer Science and Software Engineering
Concordia University
1455 de Maisonneuve Blvd. West
Montréal, Canada
{h_kilico,bergler}@cse.concordia.ca

Abstract

The second BioNLP Shared Task on Event Extraction (BioNLP-ST'11) follows up the previous shared task competition with a focus on *generalization* with respect to text types, event types and subject domains. In this spirit, we re-engineered and extended our event extraction system, emphasizing linguistic generalizations and avoiding domain-, event type- or text type-specific optimizations. Similar to our earlier system, syntactic dependencies form the basis of our approach. However, diverging from that system's more pragmatic nature, we more clearly distinguish the shared task concerns from a general semantic composition scheme, that is based on the notion of *embedding*. We apply our methodology to core bio-event extraction and speculation/negation detection tasks in three main tracks. Our results demonstrate that such a general approach is viable and pinpoint some of its shortcomings.

1 Introduction

In the past two years, largely due to the availability of GENIA event corpus (Kim et al., 2008) and the resulting shared task competition (BioNLP'09 Shared Task on Event Extraction (Kim et al., 2009)), event extraction in biological domain has been attracting greater attention. One of the criticisms towards this paradigm of corpus annotation/competition has been that they are concerned with narrow domains and specific representations, and that they may not generalize well. For instance, GENIA event corpus contains only Medline abstracts on transcription factors in human blood cells. Whether models trained on this corpus would perform well on full-text articles or on text focusing on other aspects of biomedicine (e.g., treatment or etiology of disease) remains largely unclear. Since annotated corpora are not available for every conceivable domain, it is desirable for automatic event extraction systems to be generally applicable to different types of text and domains without requiring much training data or customization.

	GENIA	EPI	ID	BB	BI
# core events	9	15	10	2	10
Triggers?	Y	Y	Y	N	N
Full-text?	Y	N	Y	N	N
Spec/Neg?	Y	Y	Y	N	N

Table 1: An overview of BioNLP-ST'11 tracks

In the follow-up event to BioNLP'09 Shared Task on Event Extraction, organizers of the second BioNLP Shared Task on Event Extraction (BioNLP-ST'11) (Kim et al., 2011a) address this challenge to some extent. The theme of BioNLP-ST'11 is *generalization* and the net is cast much wider. There are 4 event extraction tracks: in addition to the GENIA track that again focuses on transcription factors (Kim et al., 2011b), the epigenetics and post-translational modification track (EPI) focuses on events relating to epigenetic change, such as DNA methylation and histone modification, as well as other common post-translational protein modifications (Ohta et al., 2011), whereas the infectious diseases track (ID) focuses on bio-molecular mechanisms of infectious diseases (Pyysalo et al., 2011a). Both GENIA and ID tracks include data pertaining to full-text articles, as well. The fourth track, Bacteria, consists of two sub-tracks: Biotopes (BB) and Interactions (BI) (Bossy et al. (2011) and Jourde

Proceedings of BioNLP Shared Task 2011 Workshop, pages 173–182,
Portland, Oregon, USA, 24 June, 2011.

et al. (2011), respectively). A summary of the BioNLP-ST'11 tracks is given in Table (1).

We participated in three tracks: GENIA, EPI, and ID. In the spirit of the competition, our aim was to demonstrate a methodology that was general and required little, if any, customization or training for individual tracks. For this purpose, we used a two-phase approach: a syntax-driven *composition* phase that exploits linguistic generalizations to create a general semantic representation in a bottom-up manner and a *mapping* phase, which relies on the shared task event definitions and constraints to map relevant parts of this semantic representation to event instances. The composition phase takes as its input simple entities and syntactic dependency relations and is intended to be fully general. On the other hand, the second phase is more task-specific even though the kind of task-specific knowledge it requires is largely limited to event definitions and trigger expressions. In addition to extracting core biological events, our system also addresses speculation and negation detection within the same framework. Our results demonstrate the feasibility of a methodology that uses little training data or customization.

2 Methodology

In our general research, we are working towards a linguistically-grounded, bottom-up discourse interpretation scheme. In particular, we focus on lower level discourse phenomena, such as *causation*, *modality*, and *negation*, and investigate how they interact with each other, as well as their effect on basic propositional semantic content (who did what to who?) and higher discourse/pragmatics structure. In our model, we distinguish three layers of propositions: *atomic*, *embedding*, and *discourse*. An *atomic proposition* corresponds to the basic unit and lowest level of meaning: in other words, a semantic relation whose arguments correspond to ontologically simple entities. Atomic propositions form the basis for *embedding propositions*, that is, propositions taking as arguments other propositions (embedding them). In turn, embedding and atomic propositions act as arguments for *discourse relations*[1]. Our main motivation in casting the problem of discourse interpretation in this structural manner is two-fold: a) to explore the semantics of the embedding layer in a systematic way b) to allow a bottom-up semantic composition approach, which works its way from atomic propositions towards discourse relations in creating general semantic representations.

The first phase of our event extraction system (*composition*) is essentially an implementation of this semantic composition approach. Before delving into further details regarding our implementation for the shared task, however, it is necessary to briefly explain the embedding proposition categorization that our interpretation scheme is based on. With this categorization, our goal is to make explicit the kind of semantic information expressed at the embedding layer. We distinguish three basic classes of embedding propositions: MODAL, ATTRIBUTIVE, and RELATIONAL. We provide a brief summary below.

2.1 MODAL type

The embedding propositions of MODAL type *modify* the status of the embedded proposition with respect to its factuality, possibility, or necessity, and so on. They typically involve a) judgement about the status of the proposition, b) evidence for the proposition, c) ability or willingness, and d) obligations and permissions, corresponding roughly to EPISTEMIC, EVIDENTIAL, DYNAMIC and DEONTIC types (cf. Palmer (1986)), respectively. Further subdivisions are given in Figure (1). In the shared task context, the MODAL class is mostly relevant to the speculation and negation detection tasks.

2.2 ATTRIBUTIVE type

The ATTRIBUTIVE type of embedding serves to *specify* an attribute of an embedded proposition (semantic role of an argument). They typically involve a verbal predicate (*undergo* in Example (1) below), which takes a nominalized predicate (*degradation*) as one of its syntactic arguments. The other syntactic argument of the verbal predicate corresponds to a semantic argument of the embedded predicate. In Example (1), *p105* is a semantic argument of PATIENT type for the proposition indicated by *degradation*.

[1] Discourse relations, also referred to as *coherence* or *rhetorical relations* (Mann and Thompson, 1988), are not relevant to the shared task and, thus, we will not discuss them further in this paper.

(1) ... *p105* <u>*undergoes*</u> ***degradation*** ...

Verbs functioning in this way are plenty (e.g., *perform* for the AGENT role, *experience* for *experiencer* role). With respect to the shared task, we found that the usefulness of the ATTRIBUTIVE type of embedding was largely limited to verbal predicates *involve* and *require* and their nominal forms.

2.3 RELATIONAL type

The RELATIONAL type of embedding serves to semantically *link* two propositions, providing a discourse/pragmatic function. It is characterized by permeation of a limited set of discourse relations to the clausal level, often signalled lexically by "discourse verbs" (Danlos, 2006) (e.g., *cause*, *mediate*, *lead*, *correlate*), their nominal forms or other abstract nouns, such as *role*. We categorize the RELATIONAL class into CAUSAL, TEMPORAL, CORRELATIVE, COMPARATIVE, and SALIENCY types. In the example below, the verbal predicate *leads to* indicates a CAUSAL relation between the propositions whose predicates are highlighted.

(2) ***Stimulation*** *of cells* <u>*leads to*</u> *a rapid* ***phosphorylation*** *of* $I\kappa B\alpha$ ...

While not all the subtypes of this class were relevant to the shared task, we found that CAUSAL, CORRELATIVE, and SALIENCY subtypes play a role, particularly in complex regulatory events. The portions of the classification that pertain to the shared task are given in Figure (1).

3 Implementation

In the shared task setting, embedding propositions correspond to complex regulatory events (e.g., Regulation, Catalysis) as well as event modifications (Negation and Speculation), whereas atomic propositions correspond to simple event types (e.g., Phosphorylation). While the treatment of these two types differ in significant ways, they both require that simple entities are recognized, syntactic dependencies are identified and a dictionary of trigger expressions is available. We first briefly explain the construction of the trigger dictionary.

3.1 Dictionary of Trigger Expressions

In the previous shared task, we relied on training data and simple statistical measures to identify good

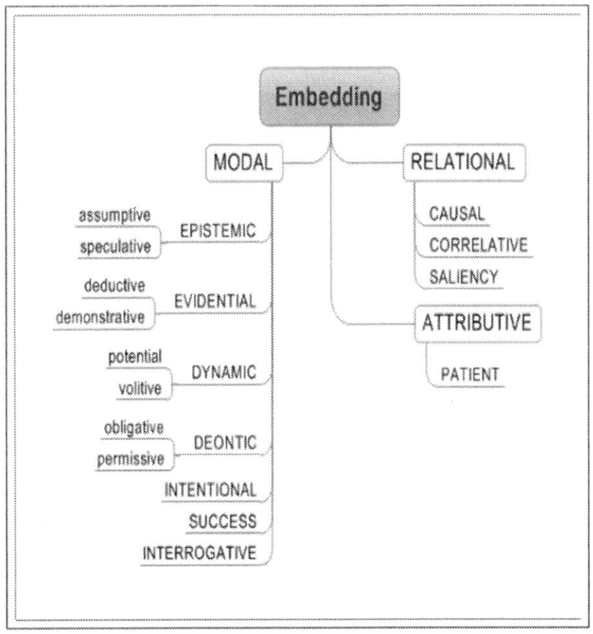

Figure 1: Embedding proposition categorization relevant to the shared task

trigger expressions for events and used a list of triggers that we manually compiled for speculation and negation detection (see Kilicoglu and Bergler (2009) for details). With respect to atomic propositions, our method of constructing a dictionary of trigger expressions remains essentially the same, including the use of statistical measures to distinguish good triggers. The only change we made was to consider affixal negation and set polarity of several atomic proposition triggers to *negative* (e.g., *nonexpression*, *unglycosylated*). On the other hand, we have been extending our manually compiled list of speculation/negation triggers to include other types of embedding triggers and to encode finer grained distinctions in terms of their categorization and trigger behaviors. The training data provided for the shared task also helped us expand this trigger dictionary, particularly with respect to RELATIONAL trigger expressions. It is worth noting that we used the same embedding trigger dictionary for all three tracks that we participated in. Several entries from the embedding trigger dictionary are summarized in Table (2).

Lexical polarity and *strength* values play a role in the composition phase in associating a context-dependent scalar value with propositions. Lexical polarity values are largely derived from a polarity lexicon (Wilson et al., 2005) and extended by us-

Trigger	POS	Semantic Type	Lexical Polarity	Strength
show	VB	DEMONSTRATIVE	*positive*	1.0
unknown	JJ	EPISTEMIC	*negative*	0.7
induce	VB	CAUSAL	*positive*	1.0
fail	VB	SUCCESS	*negative*	0.0
effect	NN	CAUSAL	*neutral*	0.5
weakly	RB	HEDGE	*neutral*	-
absence	NN	REVERSE	*negative*	-

Table 2: Several entries from the embedding dictionary

ing heuristics involving the event types associated with the trigger[2]. Some polarity values were assigned manually. Some strength values were based on prior work (Kilicoglu and Bergler, 2008), others were manually assigned. As Table (2) shows, in some cases, the semantic type (e.g., DEMONSTRATIVE, CAUSAL) is simply a mapping to the embedding categorization. In other cases, such as *weakly* or *absence*, the semantic type identifies the role that the trigger plays in the composition phase. The embedding trigger dictionary incorporates ambiguity; however, for the shared task, we limit ourselves to *one semantic type per trigger* to avoid the issue of disambiguation. For ambiguous triggers extracted from the training data, the semantic type with the maximum likelihood is used. On the other hand, we determined the semantic type to use manually for triggers that we compiled independent of the training data. In this way, we use 466 triggers for atomic propositions and 908 for embedding ones[3].

3.2 Composition

As mentioned above, the composition phase assumes simple entities, syntactic dependency relations and trigger expressions. Using these elements, we construct a semantic embedding graph of the document. To obtain syntactic dependency relations, we segment documents into sentences, parse them using the re-ranking parser of Charniak and Johnson (2005) adapted to the biomedical domain (McClosky and Charniak, 2008) and extract syntactic dependencies from parse trees using the Stanford dependency scheme (de Marneffe et al., 2006). In addition to syntactic dependencies, we also require information regarding individual tokens, including lemma, part-of-speech, and positional information, for which we also rely on Stanford parser tools. We present a high level description of the composition phase below.

3.2.1 From syntactic dependencies to embedding graphs

As the first step in composition, we convert syntactic dependencies into embedding relations. An embedding relation, in our definition, is very similar to a syntactic dependency; it is typed and holds between two textual elements. It diverges from a syntactic dependency in two ways: its elements can be multi-word expressions and it is aimed at better reflecting the direction of the semantic dependency between its elements. Take, for example, the sentence fragment in Example (3a). Syntactic dependencies are given in (3b) and the corresponding embedding relations in (3c). The fact that the adjectival predicate in modifier position (*possible*) semantically embeds its head (*involvement*) is captured with the first embedding relation. The second syntactic dependency already reflects the direction of the semantic dependency between its elements accurately and, thus, is unchanged as an embedding relation.

(3) (a) *...possible involvement of HCMV...*
(b) *amod(involvement,possible)*
prep_of(involvement,HCMV)
(c) *amod(possible,involvement)*
prep_of(involvement,HCMV)

To obtain the embedding relations in a sentence, we apply a series of transformations to its syntactic

[2] For example, if the most likely event type associated with the trigger is Negative_regulation, its polarity is considered negative.

[3] Note, however, that not all embedding propositions (or their triggers) were directly relevant to the shared task.

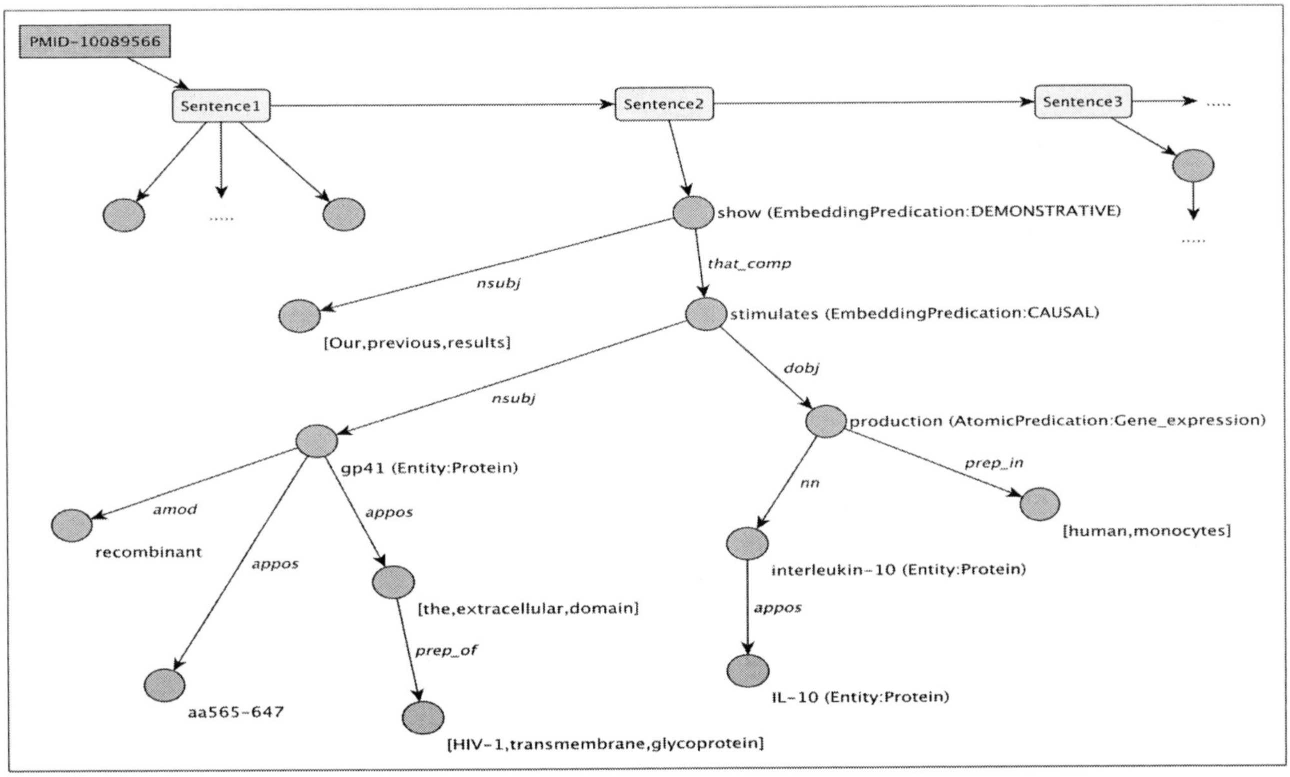

Figure 2: The embedding graph for the sentence *Our previous results show that recombinant gp41 (aa565-647), the extracellular domain of HIV-1 transmembrane glycoprotein, stimulates interleukin-10 (IL-10) production in human monocytes.* in the context of the document embedding graph for the Medline abstract with PMID 10089566.

dependencies. A transformation may not be necessary, as with the *prep_of* dependency in the example above. It may result in collapsing several syntactic dependencies into one, as well, or in splitting one into several embedding relations. In addition to capturing semantic dependency behavior explicitly, these transformations serve to incorporate semantic information (entities and triggers) into the embedding structure and to correct syntactic dependencies that are systemically misidentified, such as those that involve modifier coordination.

After these transformations, the resulting directed acyclic embedding graph is, in the simplest case, a tree, but more often a forest. An example graph is given in Figure (2). The edges are associated with the embedding relation types, and the nodes with textual elements.

3.2.2 Composing Propositions

After constructing the embedding graph, we traverse it in a bottom-up manner and compose semantic propositions. Before this procedure can take place, though, the embedding graph pertaining to each sentence is further linked to the document embedding graph in a way to reflect the proximity of sentences, as illustrated in Figure (2). This is done to enable discourse interpretation across sentences, including coreference resolution.

Traversal of the embedding structure is guided by *argument identification rules*, which apply to non-leaf nodes in the embedding graph. An argument identification rule is essentially a mapping from *the type of the embedding relation* holding between a parent node and its child node and *part-of-speech* of the parent node to a logical argument type (*logical subject*, *logical object* or *adjunct*). Constraints on and exclusions from a rule can be defined, as shown in Table (3). We currently use about 80 such rules, mostly adapted from our previous shared task system (Kilicoglu and Bergler, 2009).

After all the descendants of a non-leaf node are recursively processed for arguments, a semantic proposition can be composed. We define a semantic proposition as consisting of a trigger, a collection

Relation	Applies to	Argument	Constrained to	Exclusions
prep_on	NN	Object	*influence,impact,effect*	-
agent	VB	Subject	-	-
nsubjpass	VB	Object	-	-
whether_comp	VB	Object	INTERROGATIVE	-
prep_in	NN	Adjunct	-	*effect, role, influence, importance*

Table 3: Several argument identification rules. Note that constraints and exclusions may apply to trigger categories, as well as to lemmas.

of core and adjunct arguments as well as a polarity value and a scalar value. The polarity value can be *positive*, *negative* or *neutral*. The scalar value is in the (0,1) range. Atomic propositions are simply assigned polarity value of *neutral*[4] and the scalar value of 1.0. On the other hand, in the context of embedding propositions, the computation of these values, through which we attempt to capture some of the interactions occurring at the embedding layer, is more involved. For the sentence depicted in Figure (2), the relevant resulting embedding and atomic propositions are given below.

(4) DEMONSTRATIVE(em_1,Trigger=*show*, Object=em_2, Subject=*Our previous results*, Polarity=positive, Value=1.0)

(5) CAUSAL(em_2, Trigger=*stimulates*, Object=ap_1, Subject=*recombinant gp41*, Polarity=positive, Value=1.0)

(6) Gene_expression(ap_1, Trigger= *production*, Object= *interleukin-10*, Adjunct= *human monocytes*, Polarity=neutral, Value=1.0)

The composition phase also deals with coordination of entities and propositions as well as with propagation of arguments at the lower levels.

3.3 Mapping Propositions to Events

The goal of the *mapping* phase is to impose the shared task constraints on the partial interpretation achieved in the previous phase. We achieve this in three steps.

The first step is to map embedding proposition types to event (or event modification) types. We defined constraints that guide this mapping. Some of these mappings are presented in Table (4). In this way, Example (4) is pruned, since embedding propositions of DEMONSTRATIVE type satisfy the constraints only if they have negative polarity, as shown in Table (4).

We then apply constraints concerned with the semantic roles of the participants. For this step, we define a small number of *logical argument/semantic role mappings*. These are similar to argument identification rules, in that the mapping can be constrained to certain event types or event types can be excluded from it. We provide some of these mappings in Table (5). With these mappings, the Object and Subject arguments of the proposition in Example (5) are converted to Theme and Cause semantic roles, respectively.

As the final step, we prune event participants that do not conform to the event definition as well as the propositions whose types could not be mapped to a shared task event type. For example, a Cause participant for a Gene_expression event is pruned, since only Theme participants are relevant for the shared task. Further, a proposition with DEONTIC semantic type is pruned, because it cannot be mapped to a shared task type. The infectious diseases track (ID) event type Process is interesting, because it may take no participants at all, and we deal with this idiosyncrasy at this step, as well. This concludes the progressive transformation of the graph to event and event modification annotations.

4 Results and Discussion

With the two-phase methodology presented above, we participated in three tracks: GENIA (Tasks 1 and 3), ID, and EPI. The official evaluation results we obtained for the GENIA track are presented in Table (6) and the results for the EPI and ID tracks in

[4]Unless affixal negation is involved, in which case the assigned polarity value is *negative*.

Track	Prop. Type	Polarity	Value	Correspond. Event (Modification) Type
GENIA,ID	CAUSAL	neutral	-	Regulation
GENIA,ID,EPI	SUCCESS	negative	-	Negation
EPI	CAUSAL	positive	-	Catalysis
GENIA,ID,EPI	SPECULATIVE	-	> 0.0	Speculation
GENIA,ID,EPI	DEMONSTRATIVE	negative	-	Speculation

Table 4: Several event (and event modification) mappings

Logical Arg.	Semantic Role	Constraint	Exclusion
Object	Theme	-	Process
Subject	Cause	-	-
Subject	Theme	Binding	-
Object	Participant	Process	-
Object	Scope	Speculation, Negation	-

Table 5: Logical argument to semantic role mappings

Table (7). With the official evaluation criteria, we were ranked 5th in the GENIA track (5/15), 7th in the EPI track (7/7) and 4th in the ID track (4/7). There were only two submissions for the GENIA speculation/negation task (Task 3) and our results in this task were comparable to those of the other participating group: our system performed slightly better with speculation, and theirs with negation.

Our core module extracts adjunct arguments, using ABNER (Settles, 2005) as its source for additional named entities. We experimented with mapping these arguments to non-core event participants (Site, Contextgene, etc.); however, we did not include them in our official submission, because they seemed to require more work with respect to mapping to shared task specifications. Due to this shortcoming, the performance of our system suffered significantly in the EPI track.

A particularly encouraging outcome for our system is that our results on the GENIA development set versus on the test set were very close (an F-score of 51.03 vs. 50.32), indicating that our general approach avoided overfitting, while capturing the linguistic generalizations, as we intended. We observe similar trends with the other tracks, as well. In the EPI track, development/test F-score results were 29.10 vs. 27.88; while, in the ID track, interestingly, our test set performance was better (39.64 vs. 44.21). We also obtained the highest recall in the ID track, despite the fact that our system typically favors precision. We attribute this somewhat idiosyncratic performance in the ID track partly to the fact that we did not use a track-specific trigger dictionary. Most of the ID track event types are the same as those of GENIA track, which probably led to identification of some ID events with GENIA-only triggers[5].

Event Class	Recall	Precis.	F-score
Localization	39.27	90.36	54.74
Binding	29.33	49.66	36.88
Gene_expression	65.87	86.84	74.91
Transcription	32.18	58.95	41.64
Protein_catabolism	66.67	71.43	68.97
Phosphorylation	75.14	94.56	83.73
EVT-TOTAL	52.67	78.04	62.90
Regulation	33.77	42.48	37.63
Positive_regulation	35.97	47.66	41.00
Negative_regulation	36.43	43.88	39.81
REG-TOTAL	35.72	45.85	40.16
Negation	18.77	44.26	26.36
Speculation	21.10	38.46	27.25
MOD-TOTAL	19.97	40.89	26.83
ALL-TOTAL	43.55	59.58	50.32

Table 6: Official GENIA track results, with *approximate span matching/approximate recursive matching* evaluation criteria

One of the interesting aspects of the shared task was its inclusion of full-text articles in training and evaluation. Cohen et al. (2010) show that structure and content of biomedical abstracts and article bodies differ markedly and suggest that some of these

[5]This clearly also led to low precision particularly in complex regulatory events.

Track-Eval. Type	Recall	Precis.	F-score
EPI-FULL	20.83	42.14	27.88
EPI-CORE	40.28	76.71	52.83
ID-FULL	49.00	40.27	44.21
ID-CORE	50.77	43.25	46.71

Table 7: Official evaluation results for EPI and ID tracks. Primary evaluation criteria underlined.

differences may pose problems in processing full-text articles. Since one of our goals was to determine the generality of our system across text types, we did not perform any full text-specific optimization. Our results on article bodies are notable: our system had stable performance across text types (in fact, we had a very slight F-score improvement on full-text articles: 50.40 vs. 50.28). This contrasts with the drop of a few points that seems to occur with other well-performing systems. Taking only full-text articles into consideration, we would be ranked 4th in the GENIA track. Furthermore, a preliminary error analysis with full-text articles seems to indicate that parsing-related errors are more prevalent in the full-text article set than in the abstract set, consistent with Cohen et al.'s (2010) findings. At the same time, our results confirm that we were able to abstract away from this complexity to some degree with our approach.

We have a particular interest in speculation and negation detection. Therefore, we examined our results on the GENIA development set with respect to Task 3 more closely. Consistent with our previous shared task results, we determined that the majority of errors were due to misidentified or missed base events (70% of the precision errors and 83% of the recall errors)[6]. Task 3-specific precision errors included cases in which speculation or negation was debatable, as the examples below show. In Example (7a), our system detected a Speculation instance, due to the verbal predicate *suggesting*, which scopes over the event indicated by *role*. In Example (7b), our system detected a Negation instance, due to the nominal predicate *lack*, which scopes over the events indicated by *expression*. Neither were annotated as such in the shared task corpus.

(7) (a) *. . . suggesting a **role** of these 3' elements in beta-globin gene expression.*

(b) *. . . DT40 B cell lines that lack **expression** of either PKD1 or PKD3 . . .*

Another class of precision errors was due to argument propagation up the embedding graph. It seems the current algorithm may be too permissive in some cases and a more refined approach to argument propagation may be necessary. In the following example, while *suggest*, an epistemic trigger, does not embed *induction* directly (as shown in (8b)), the intermediate nodes simply propagate the proposition associated with the *induction* node up the graph, leading us to conclude that the proposition triggered by *induction* is speculated, leading to a precision error.

(8) (a) *. . . these findings suggest that PWM is able to initiate an intracytoplasmic signaling cascade and EGR-1 **induction** . . .*

(b) *suggest* $\rightarrow$ *able* $\rightarrow$ *initiate* $\rightarrow$ *induction*

Among the recall errors, some of them were due to shortcomings of the composition algorithm, as it is currently implemented. One recall problem involved the embedding status of and rules concerning copular constructions, which we had not yet addressed. Therefore, we miss the relatively straightforward Speculation instances in the following examples.

(9) (a) *. . . the A3G promoter appears constitutively **active**.*

(b) *. . . the precise factors that **mediate** this induction mechanism remain unknown.*

Similarly, the lack of a trigger expression in our dictionary may cause recall errors. The example below shows an instance where this occurs, in addition to lack of an appropriate argument identification rule:

(10) *mRNA was quantified by real-time PCR for FOXP3 and GATA3 **expression**.*

Our system also missed an interesting, domain-specific type of negation, in which the minus sign indicates negation of the event that the entity participates in.

(11) *. . . CD14- surface Ag **expression** . . .*

[6]Even a bigger percentage of speculation/negation-related errors in the EPI and ID tracks were due to the same problem, as the overall accuracy in those tracks is lower.

5 Conclusions and Future Work

We explored a two-phase approach to event extraction, distinguishing general linguistic principles from task-specific aspects, in accordance with the *generalization* theme of the shared task. Our results demonstrate the viability of this approach on both abstracts and article bodies, while also pinpointing some of its shortcomings. For example, our error analysis shows that some aspects of semantic composition algorithm (argument propagation, in particular) requires more refinement. Furthermore, using the same trigger expression dictionary for all tracks seems to have negative effect on the overall performance. The incremental nature of our system development ensures that some of these shortcomings will be addressed in future work.

We participated in three supporting tasks, two of which (Co-reference (CO) and Entity Relations (REL) tasks (Nguyen et al. (2011) and Pyysalo et al. (2011b), respectively) were relevant to the main portion of the shared task; however, due to time constraints, we were not able to fully incorporate these modules into our general framework, with the exception of the co-reference resolution of relative pronouns. Since our goal is to move towards discourse interpretation, we plan to incorporate these modules (inter-sentential co-reference resolution, in particular) into our framework. After applying the lessons we learned in the shared task and fully incorporating these modules, we plan to make our system available to the scientific community.

References

Robert Bossy, Julien Jourde, Philippe Bessières, Marteen van de Guchte, and Claire Nédellec. 2011. BioNLP Shared Task 2011 - Bacteria Biotope. In *Proceedings of the BioNLP 2011 Workshop Companion Volume for Shared Task*, Portland, Oregon, June. Association for Computational Linguistics.

Eugene Charniak and Mark Johnson. 2005. Coarse-to-fine n-best parsing and maxent discriminative reranking. In *Proceedings of the 43rd Meeting of the Association for Computational Linguistics*, pages 173–180.

K Bretonnel Cohen, Helen L Johnson, Karin Verspoor, Christophe Roeder, and Lawrence E Hunter. 2010. The structural and content aspects of abstracts versus bodies of full text journal articles are different. *BMC Bioinformatics*, 11:492.

Laurence Danlos. 2006. "Discourse verbs" and discourse periphrastic links. In C Sidner, J Harpur, A Benz, and P Kühnlein, editors, *Second Workshop on Constraints in Discourse (CID06)*, pages 59–65.

Marie-Catherine de Marneffe, Bill MacCartney, and Christopher D Manning. 2006. Generating typed dependency parses from phrase structure parses. In *Proceedings of the 5th International Conference on Language Resources and Evaluation*, pages 449–454.

Julien Jourde, Alain-Pierre Manine, Philippe Veber, Karën Fort, Robert Bossy, Erick Alphonse, and Philippe Bessières. 2011. BioNLP Shared Task 2011 - Bacteria Gene Interactions and Renaming. In *Proceedings of the BioNLP 2011 Workshop Companion Volume for Shared Task*, Portland, Oregon, June. Association for Computational Linguistics.

Halil Kilicoglu and Sabine Bergler. 2008. Recognizing speculative language in biomedical research articles: a linguistically motivated perspective. *BMC Bioinformatics*, 9 Suppl 11:s10.

Halil Kilicoglu and Sabine Bergler. 2009. Syntactic dependency based heuristics for biological event extraction. In *Proceedings of Natural Language Processing in Biomedicine (BioNLP) NAACL 2009 Workshop*, pages 119–127.

Jin-Dong Kim, Tomoko Ohta, and Jun'ichi Tsujii. 2008. Corpus annotation for mining biomedical events from literature. *BMC Bioinformatics*, 9:10.

Jin-Dong Kim, Tomoko Ohta, Sampo Pyysalo, Yoshinobu Kano, and Jun'ichi Tsujii. 2009. Overview of BioNLP'09 Shared Task on Event Extraction. In *Proceedings of Natural Language Processing in Biomedicine (BioNLP) NAACL 2009 Workshop*, pages 1–9.

Jin-Dong Kim, Sampo Pyysalo, Tomoko Ohta, Robert Bossy, and Jun'ichi Tsujii. 2011a. Overview of BioNLP Shared Task 2011. In *Proceedings of the BioNLP 2011 Workshop Companion Volume for Shared Task*, Portland, Oregon, June. Association for Computational Linguistics.

Jin-Dong Kim, Yue Wang, Toshihisa Takagi, and Akinori Yonezawa. 2011b. Overview of the Genia Event task in BioNLP Shared Task 2011. In *Proceedings of the BioNLP 2011 Workshop Companion Volume for Shared Task*, Portland, Oregon, June. Association for Computational Linguistics.

William C Mann and Sandra A Thompson. 1988. Rhetorical structure theory: Toward a functional theory of text organization. *Text*, 8(3):243–281.

David McClosky and Eugene Charniak. 2008. Self-training for biomedical parsing. In *Proceedings of the 46th Meeting of the Association for Computational Linguistics*, pages 101–104.

Ngan Nguyen, Jin-Dong Kim, and Jun'ichi Tsujii. 2011. Overview of the Protein Coreference task in BioNLP Shared Task 2011. In *Proceedings of the BioNLP 2011 Workshop Companion Volume for Shared Task*, Portland, Oregon, June. Association for Computational Linguistics.

Tomoko Ohta, Sampo Pyysalo, and Jun'ichi Tsujii. 2011. Overview of the Epigenetics and Post-translational Modifications (EPI) task of BioNLP Shared Task 2011. In *Proceedings of the BioNLP 2011 Workshop Companion Volume for Shared Task*, Portland, Oregon, June. Association for Computational Linguistics.

Frank R Palmer. 1986. *Mood and modality*. Cambridge University Press, Cambridge, UK.

Sampo Pyysalo, Tomoko Ohta, Rafal Rak, Dan Sullivan, Chunhong Mao, Chunxia Wang, Bruno Sobral, Jun'ichi Tsujii, and Sophia Ananiadou. 2011a. Overview of the Infectious Diseases (ID) task of BioNLP Shared Task 2011. In *Proceedings of the BioNLP 2011 Workshop Companion Volume for Shared Task*, Portland, Oregon, June. Association for Computational Linguistics.

Sampo Pyysalo, Tomoko Ohta, and Jun'ichi Tsujii. 2011b. Overview of the Entity Relations (REL) supporting task of BioNLP Shared Task 2011. In *Proceedings of the BioNLP 2011 Workshop Companion Volume for Shared Task*, Portland, Oregon, June. Association for Computational Linguistics.

Burr Settles. 2005. ABNER: An open source tool for automatically tagging genes, proteins, and other entity names in text. *Bioinformatics*, 21(14):3191–3192.

Theresa Wilson, Janyce Wiebe, and Paul Hoffmann. 2005. Recognizing contextual polarity in phrase-level sentiment analysis. In *Proceedings of the Human Language Technologies Conference/Conference on Empirical Methods in Natural Language Processing (HLT/EMNLP-2005)*, pages 347–354.

Generalizing Biomedical Event Extraction

Jari Björne and **Tapio Salakoski**
Department of Information Technology, University of Turku
Turku Centre for Computer Science (TUCS)
Joukahaisenkatu 3-5, 20520 Turku, Finland
firstname.lastname@utu.fi

Abstract

We present a system for extracting biomedical events (detailed descriptions of biomolecular interactions) from research articles. This system was developed for the BioNLP'11 Shared Task and extends our BioNLP'09 Shared Task winning Turku Event Extraction System. It uses support vector machines to first detect event-defining words, followed by detection of their relationships. The theme of the BioNLP'11 Shared Task is generalization, extending event extraction to varied biomedical domains. Our current system successfully predicts events for every domain case introduced in the BioNLP'11 Shared Task, being the only system to participate in all eight tasks and all of their subtasks, with best performance in four tasks.

1 Introduction

Biomedical event extraction is the process of automatically detecting statements of molecular interactions in research articles. Using natural language processing techniques, an event extraction system predicts relations between proteins/genes and the processes they take part in. Manually annotated corpora are used to evaluate event extraction techniques and to train machine-learning based systems.

Event extraction was popularised by the BioNLP'09 Shared Task on Event Extraction (Kim et al., 2009), providing a more detailed alternative for the older approach of binary interaction detection, where each pair of protein names co-occurring in the text is classified as interacting or not. Events extend this formalism by adding to the relations *direction*, *type* and *nesting*. Events define the type of interaction, such as *phosphorylation*, and commonly mark in the text a *trigger word* (e.g. "phosphorylates") describing the interaction. Directed events can define the role of their protein or gene arguments as e.g. *cause* or *theme*, the agent or the target of the biological process. Finally, events can act as arguments of other events, creating complex nested structures that accurately describe the biological interactions stated in the text. For example, in the case of a sentence stating "Stat3 phosphorylation is regulated by Vav", a *phosphorylation*-event would itself be the argument of a *regulation*-event.

We developed for the BioNLP'09 Shared Task the Turku Event Extraction System, achieving the best performance at 51.95% F-score (Björne et al., 2009). This system separated event extraction into multiple classification tasks, detecting individually the trigger words defining events, and the arguments that describe which proteins or genes take part in these events. Other approaches used in the Shared Task included e.g. joint inference (Riedel et al., 2009). An overall notable trend was the use of full dependency parsing (Buyko et al., 2009; Van Landeghem et al., 2009; Kilicoglu and Bergler, 2009).

In the following years, event extraction has been the subject of continuous development. In 2009, after the BioNLP'09 Shared Task, we extended our system and improved its performance to 52.85% (Björne et al., 2011). In 2010, the system introduced by Miwa et. al. reached a new record performance of 56.00% (Miwa et al., 2010a).

Proceedings of BioNLP Shared Task 2011 Workshop, pages 183–191,
Portland, Oregon, USA, 24 June, 2011.

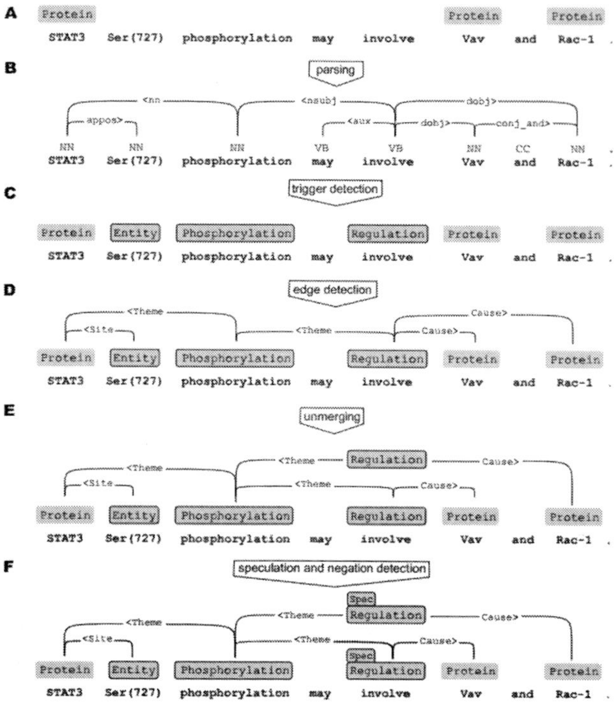

Figure 1: Event extraction. In most tasks named entities are given (A). Sentences are parsed (B) to produce a dependency parse. Entities not given are predicted through trigger detection (C). Edge detection predicts event arguments between entities (D) and unmerging creates events (E). Finally, event modality is predicted (F). When the graph is converted to the Shared Task format, site arguments are paired with core arguments that have the same target protein.

In 2010, we applied the Turku Event Extraction System to detecting events in all 18 million PubMed abstracts, showing its scalability and generalizability into real-world data beyond domain corpora (Björne et al., 2010). In the current BioNLP'11 Shared Task[1] (Kim et al., 2011), we demonstrate its generalizability to different event extraction tasks by applying what is, to a large extent, the same system to every single task and subtask.

2 System Overview

Our system divides event extraction into three main steps (Figure 1 C, D and E). First, entities are predicted for each word in a sentence. Then, arguments are predicted between entities. Finally, entity/argument sets are separated into individual events.

[1] http://sites.google.com/site/bionlpst/

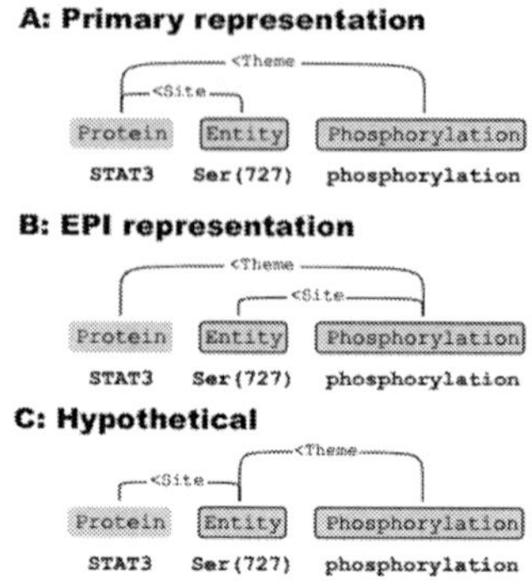

Figure 2: Site argument representation. Site arguments add detail to core arguments. (A) In most tasks we link both core and site arguments to given protein nodes. This minimizes the number of outgoing edges per trigger node, simplifying unmerging, but loses the connection between site and core arguments. (B) In the EPI task, all events with site-arguments have a single core argument, so linking sites to the trigger node preserves the site/core connection. (C) To both limit number of arguments in trigger nodes and preserve site information, event arguments using sites could be linked to protein nodes through the site entity. However, in this approach the core argument would remain undetected if the site wasn't detected.

2.1 Graph Representation

The BioNLP'11 Shared Task consists of eight separate tasks. Most of these follow the BioNLP'09 Shared Task annotation scheme, which defines events as having a trigger entity and one or more arguments that link to other events or protein/gene entities. This annotation can be represented as a graph, with trigger and protein/gene entities as nodes, and arguments (e.g. *theme*) as edges. In our graph representation, an event is defined implicitly as a trigger node and its outgoing edges (see Figure 1 F).

Most of the BioNLP'11 Shared Task tasks define task-specific annotation terminology, but largely follow the BioNLP'09 definition of events. Some new annotation schemes, such as the bracket notation in the CO-task can be viewed simply as alternative representations of arguments. The major new feature is *relations* or *triggerless events*, used in the REL, REN, BB and BI tasks. In our graph representation, this type of event is a single, directed edge.

Some event arguments have a matching *site* argument that determines the part of the protein the argument refers to (Figure 2). To allow detection of core arguments independently of site arguments, in most tasks we link site arguments directly to proteins (Figure 2 A). This maximises extraction performance on core events, but losing the connection between site and core arguments limits performance on site arguments.

To further simplify event extraction all sentences are processed in isolation, so events crossing sentence boundaries (intersentence events, Table 2) cannot be detected. This also limits the theoretical maximum performance of the system (see Figure 3).

In the provided data an event is annotated only once for a set of equivalent proteins. For example, in the sentence "Ubiquitination of caspase 8 (casp8)" a *ubiquitination* event would be annotated only for "caspase 8", "casp8" being marked as equivalent to "caspase 8". To improve training data consistency, our system fully resolves these equivalences into new events, also recursively when a duplicated event is nested in another event (Table 2). Resolved equivalences were used for event extraction in the BioNLP'11 GE, ID, EPI and BB tasks, although based on tests with the GE dataset their impact on performance was negligible.

2.2 Machine Learning

The machine learning based event detection components classify examples into one of the positive classes or as negatives, based on a feature vector representation of the data. To make these classifications, we use the SVM^{multiclass} support vector machine[2] (Tsochantaridis et al., 2005) with a linear kernel. An SVM must be optimized for each classification task by experimentally determining the regularization parameter C. This is done by training the system on a training dataset, and testing a number of C values on a development dataset. When producing predictions for the test set, the classifier is retrained with combined training and development sets, and the test data is classified with the previously determined optimal value of C.

Unlike in the BioNLP'09 Shared Task where the three main parameters (trigger-detector, recall-adjustment and edge-detector) were optimized in an exhaustive grid search against the final metric, in the new system only the recall-adjustment parameter (see Section 2.5) is optimized against the final metric, edge and trigger detector parameters being optimized in isolation to speed up experiments.

2.3 Syntactic Analyses

The machine learning features that are used in event detection are mostly derived from the syntactic parses of the sentences. Parsing links together related words that may be distant in their linear order, creating a parse tree (see Figure 1 B).

We used the Charniak-Johnson parser (Charniak and Johnson, 2005) with David McClosky's biomodel (McClosky, 2010) trained on the GENIA corpus and unlabeled PubMed articles. The parse trees produced by the Charniak-Johnson parser were further processed with the Stanford conversion tool (de Marneffe et al., 2006), creating a dependency parse (de Marneffe and Manning, 2008).

In the supporting tasks (REL, REN and CO) this parsing was done by us, but in the main tasks the organizers provided official parses which were used. All parses for tasks where named entities were given as gold data were further processed with a *protein name splitter* that divides at punctuation tokens which contain named entities, such as "p50/p65" or "GATA3-binding".

2.4 Feature Groups

To convert text into features understood by the classifier, a number of analyses are performed on the sentences, mostly resulting in binary features stating the presence or absence of some feature. Applicable combinations of these features are then used by the trigger detection, edge detection and unmerging steps of the event extraction system.

Token features can be generated for each word token, and they define the text of the token, its Porter-stem (Porter, 1980), its Penn treebank part-of-speech-tag, character bi- and trigrams, presence of punctuation or numeric characters etc.

Sentence features define the number of named entities in the sentence as well as bag-of-words counts for all words.

Dependency chains follow the syntactic dependencies up to a depth of three, starting from a token of interest. They are used to define the immediate context of these words.

[2] `http://svmlight.joachims.org/svm_multiclass.html`

Dependency path N-grams, are built from the shortest undirected path of tokens and dependencies linking together two entities, and are used in edge detection. N-grams join together a token with its two flanking dependencies as well as each dependency with its two flanking tokens. While these N-grams follow the direction of the entire path, the governor-dependent directions of individual dependencies are used to define token bigrams.

Trigger features can be built in cases where triggers are already present, such as edge detection and event construction. These features include the types and supertypes of the trigger nodes, and combinations thereof.

External features are additional features based on data external to the corpus being processed. Such features can include e.g. the presence of a word in a list of key terms, Wordnet hypernyms, or other resources that enhance performance on a particular task. These are described in detail in Section 3.

2.5 Trigger Detection

Trigger words are detected by classifying each token as negative or as one of the positive trigger classes. Sometimes several triggers overlap, in which case a merged class (e.g. *phosphorylation–regulation*) is used. After trigger prediction, triggers of merged classes are split into their component classes.

Most tasks evaluate trigger detection using approximate span, so detecting a single token is enough. However, this token must be chosen consistently for the classifier to be able to make accurate predictions. For multi-token triggers, we select as the trigger word the *syntactic head*, the root token of the dependency parse subtree covering the entity.

When optimizing the SVM C-parameter for trigger and edge detection, it is optimized in isolation, maximizing the F-score for that classification task. Edges can be predicted for an event only if its trigger has been detected, but often the C-parameter that maximizes trigger detection F-score has too low recall for optimal edge detection. A *recall adjustment* step is used to fit together the trigger and edge detectors. For each example, the classifier gives a confidence score for each potential class, and picks as the predicted class the one with the highest score. In recall adjustment, the confidence score of each negative example is multiplied with a multiplier, and if the result falls below the score of another class, that class becomes the new classification. This multiplier is determined experimentally by optimizing against overall system performance, using the official task metric for cases where a downloadable evaluator is available (GE and BB).

2.6 Edge Detection

Edge detection is used to predict event arguments or triggerless events and relations, all of which are defined as edges in the graph representation. The edge detector defines one example per direction for each pair of entities in the sentence, and uses the SVM classifier to classify the examples as negatives or as belonging to one of the positive classes. As with the trigger detector, overlapping positive classes are predicted through merged classes (e.g. *cause–theme*). Task-specific rules defining valid argument types for each entity type are used to considerably reduce the number of examples that can only be negatives.

2.7 Unmerging

In the graph representation, events are defined through their trigger word node, resulting in overlapping nodes for overlapping events. The trigger detector can however predict a maximum of one trigger node per type for each token. When edges are predicted between these nodes, the result is a *merged graph* where overlapping events are merged into a single node and its set of outgoing edges. Taking into account the limits of trigger prediction, the edge detector is also trained on a merged graph version of the gold data.

To produce the final events, these merged nodes need to be "pulled apart" into valid trigger and argument combinations. In the BioNLP'09 Shared Task, this was done with a rule-based system. Since then, further research has been done on machine learning approaches for this question (Miwa et al., 2010b; Heimonen et al., 2010). In our current system, unmerging is done as an SVM-classification step. An example is constructed for each argument edge combination of each predicted node, and classified as a true event or a false event to be removed. Tested on the BioNLP'09 Shared Task data, this system performs roughly on par with our earlier rule-based system, but has the advantage of being more general and thus applicable to all BioNLP'11 Shared Task

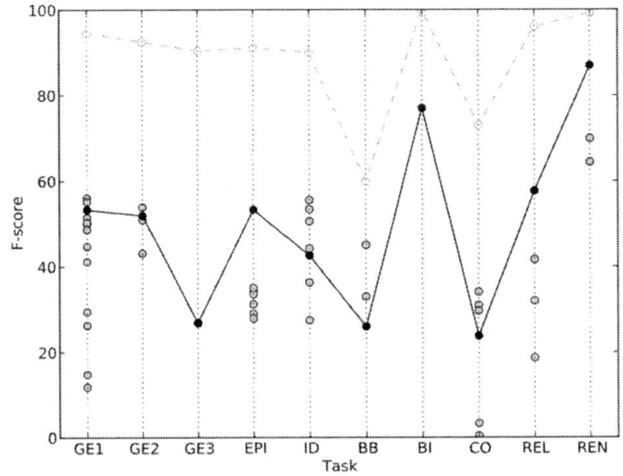

Figure 3: Ranking of the systems participating in the BioNLP'11 Shared Task. Our system is marked with black dots and the dotted line shows its theoretical maximum performance (see Section 2.1) with all correct classifications.

tasks. The unmerging step is not required for *triggerless events* which are defined by a single edge.

All of the tasks define varied, detailed limits on valid event type and argument combinations. A final validation step based on task-specific rules is used to remove structurally incorrect events left over from preceding machine learning steps.

2.8 Modality Detection

Speculation and negation are detected independently, with binary classification of trigger nodes. The features used are mostly the same as for trigger detection, with the addition of a list of speculation-related words based on the BioNLP'09 ST corpus.

3 Tasks and Results

The BioNLP'11 Shared Task consists of five main tasks and three supporting tasks. Additionally, many of these tasks specify separate subtasks. Except for the GE-task, which defines three main evaluation criteria, all tasks have a single primary evaluation criterion. All evaluations are based on F-score, the harmonic mean of precision and recall. Performance of all systems participating in the BioNLP'11 Shared Task is shown in Figure 3. Our system's performance on both development and test sets of all tasks is shown in Table 1.

Corpus	Devel F	Test F
GE'09 task 1	56.27	53.15
GE'09 task 2	54.25	50.68
GE task 1	55.78	53.30
GE task 2	53.39	51.97
GE task 3	38.34	26.86
EPI	56.41	53.33
ID	44.92	42.57
BB	27.01	26
BI	77.24	77
CO	36.22	23.77
REL	65.99	57.7
REN	84.62	87.0

Table 1: Devel and test results for all tasks. The performance of our new system on the BioNLP'09 ST GENIA dataset is shown for reference, with task 3 omitted due to a changed metric. For GE-tasks, the Approximate Span & Recursive matching criterion is used.

3.1 GENIA (GE)

The GENIA task is the direct continuation of the BioNLP'09 Shared Task. The BioNLP'09 ST corpus consisted only of abstracts. The new version extends this data by 30% with full text PubMed Central articles.

Our system applied to the GE task is the most similar to the one we developed for the BioNLP'09 Shared Task. The major difference is the replacement of the rule-based unmerging component with an SVM based one.

The GE task has three subtasks, task 1 is detection of events with their main arguments, task 2 extends this to detection of sites defining the exact molecular location of interactions, and task 3 adds the detection of whether events are stated in a negated or speculative context.

For task 3, speculation and negation detection, we considered the GE, EPI and ID task corpora similar enough to train a single model on. Compared to training on GE alone, example classification F-score decreased for negation by 8 pp and increased for speculation by 4 pp. Overall task 3 processing was considerably simplified.

Our system placed third in task 1, second in task 2 and first in task 3. Task 1 had the most participants, making it the most useful for evaluating overall performance. Our F-score of 53.30% was within three percentage points of the best performing system (by

Corpus	sentences	events	equiv events	nesting events	intersentence events	neg/spec events
GE'09	8906	11285	7.9%	38.8%	6.0%	12.1%
GE	11581	14496	6.6%	37.2%	6.0%	13.3%
EPI	7648	2684	9.1%	10.2%	9.3%	10.1%
ID	3193	2931	5.3%	21.3%	3.9%	4.9%
BB	1762	5843	79.4%	N/A	86.0%	0%
BI	120	458	0%	N/A	0%	0%
CO	8906	5284	0%	N/A	8.5%	N/A
REL	8906	2440	4.2%	N/A	0%	0%
REN	13235	373	0%	N/A	2.4%	0%

Table 2: Corpus statistics. Numbers are for all available annotated data, i.e. the merged training and development sets.

team FAUST), indicating that our chosen event detection approach still remains competitive. For reference, we ran our system also on the BioNLP'09 data, reaching an F-score of 53.15%, a slight increase over the 52.85% we previously reported in Björne et al. (2011).

3.2 Epigenetics and Post-translational Modifications (EPI)

All events in the EPI task that have additional arguments (comparable to the site-arguments in the GE-task) have a single core argument. We therefore use for this task a slightly modified graph representation, where all additional arguments are treated as core arguments, linking directly to the event node (Figure 2 B). The number of argument combinations per predicted event node remains manageable for the unmerging system and full recovery of additional arguments is possible.

Eight of the EPI event types have corresponding reverse events, such as *phosphorylation* and *dephosphorylation*. Many of these reverse events are quite rare, resulting in too little training data for the trigger detector to find them. Therefore we merge each reverse event type into its corresponding forward event type. After trigger detection, an additional rule-based step separates them again. Most of the reverse classes are characterized by a "de"-prefix in their trigger word. On the EPI training dataset, the rule-based step determined correctly whether an event was reversed in 99.6% of cases (1698 out of 1704 events). Using this approach, primary criterion F-score on the development set increased 1.33 percentage points from 55.08% to 56.41%. Several previously undetectable small reverse classes became detectable, with e.g. *deubiquitination* (8 instances in the development set) detected at 77.78% F-score.

Our system ranked first on the EPI task, outperforming the next-best system (team FAUST) by over 18 percentage points. On the alternative core metric our system was also the first, but the FAUST system was very close with only a 0.27 percentage point difference. Since the core metric disregards additional arguments, it may be that our alternative approach for representing these arguments (Figure 2 B) was important for the primary criterion difference.

3.3 Infectious Diseases (ID)

The annotation scheme for the ID task closely follows the GE task, except for an additional *process* event type that may have no arguments, and for five different entity types in place of the *protein* type. Our approach for the ID task was identical to the GE task, but performance relative to the other teams was considerably lower. Primary evaluation metric F-score was 42.57% vs. 43.44% for the core metric which disregards additional arguments, indicating that these are not the reason for low performance.

3.4 Bacteria Biotopes (BB)

The BB task considers detection of events describing bacteria and their habitats. The task defines only two event types but a large number of entity types which fall into five supertypes. All entities must be predicted and all events are triggerless.

Unlike in the other main tasks, in the BB task exact spans are required for *Bacterium*-type entities, which usually consist of more than one token (e.g. *B. subtilis*). After trigger detection, a rule-based step attempts to extend predicted trigger spans forwards and backwards to cover the correct span. When extending the spans of BB training set gold entity head tokens, this step produced the correct span for 91% (399 out of 440) of *Bacterium*-type entities.

To aid in detecting *Bacterium*-entities a list of bacteria names from the List of Prokaryotic names with Standing in Nomenclature[3] was used (Euzéby, 1997) as external features. To help in detecting the heterogeneous habitat-entities, synonyms and hypernyms from Wordnet were used (Fellbaum, 1998). The development set lacked some event classes, so we moved some documents from the training set to the development set to include these.

Our F-score was the lowest of the three participating systems, and detailed results show a consistently lower performance in detecting the entities. The large number of intersentence events (Table 2) also considerably limited performance (Figure 3).

3.5 Bacteria Gene Interactions (BI)

The BI-task considers events related to genetic processes of the bacterium *Bacillus subtilis*. This task defines a large number of both entity and event types, but all entities are given as gold-standard data, therefore we start from edge detection (Figure 1 D). All BI events are triggerless.

In this task manually curated syntactic parses are provided. As also automated parses were available, we tested them as an alternative. With the Charniak-Johnson/McClosky parses overall performance was only 0.65 percentage points lower (76.59% vs. 77.24%). As with the BB task, we moved some documents from the training set to the development set to include missing classes.

Despite this task being very straightforward compared to the other tasks we were the only participant. Therefore, too many conclusions shouldn't be drawn from the performance, except to note that a rather high F-score is to be expected with all the entities being given as gold data.

3.6 Protein/Gene Coreference (CO)

In the CO supporting task the goal is to extract anaphoric expressions. Even though our event extraction system was not developed with coreference resolution in mind, the graph representation can be used for the coreference annotation, making coreference detection possible. *Anaphoras* and *Antecedents* are both represented as *Exp*-type entities, with *Coref*-type edges linking *Anaphora*-entities to *Antecedent*-entities and *Target*-type edges linking *Protein*-type entities to *Antecedent*-entities.

In the CO-task, character spans for detected entities must be in the range of a full span and minimum span. Therefore in this task we used an alternative trigger detector. Instead of predicting one trigger per token, this component predicted one trigger per each syntactic phrase created by the Charniak-Johnson parser. Since these phrases don't cover most of the CO-task triggers, they were further subdivided into additional phrases, e.g. by cutting away determiners and creating an extra phrase for each noun-token, with the aim of maximizing the number of included triggers and minimizing the number of candidates.

Our system placed fourth out of six, reaching an F-score of 23.77%. Coreference resolution being a new subject for us and our system not being developed for this domain, we consider this an encouraging result, but conclude that in general dedicated systems should be used for coreference resolution.

3.7 Entity Relations (REL)

The REL supporting task concerns the detection of static relationships, *Subunit-Complex* relations between individual proteins and protein complexes and *Protein-Component* relations between a gene or protein and its component, such as a protein domain or gene promoter. In the graph representation these relations are defined as edges that link together given protein/gene names and *Entity*-type entities that are detected by the trigger detector.

To improve entity detection, additional features are used. Derived from the REL annotation, these features highlight structures typical for biomolecular components, such as aminoacids and their shorthand forms, domains, motifs, loci, termini and promoters. Many of the REL entities span multiple tokens. Since the trigger detector predicts one entity per token, additional features are defined to mark whether a token is part of a known multi-token name.

Our system had the best performance out of four participating systems with an F-score of 57.7%, over 16 percentage points higher than the next. Development set results show that performance for the two event classes was very close, 66.40% for Protein-Component and 65.23% for Subunit-Complex.

[3] http://www.bacterio.cict.fr/

3.8 Bacteria Gene Renaming (REN)

The REN supporting task is aimed at detecting statements of *B. Subtilis* gene renaming where a synonym is introduced for a gene. The REL task defines a single relation type, *Renaming*, and a single entity type, *Gene*. All entities are given, so only edge detection is required. Unlike the other tasks, the main evaluation criterion ignores the direction of the relations, so they are processed as *undirected edges* in the graph representation.

Edge detection performance was improved with external features based on two sources defining known *B. Subtilis* synonym pairs: The Uniprot *B. Subtilis* gene list "bacsu"[4] and *Subti*Wiki[5], the *B. Subtilis* research community annotation wiki.

For the 300 renaming relations in the REN training data, the synonym pair was found from the Uniprot list in 66% (199 cases), from *Subti*Wiki in 79% (237 cases) and from either resource in 81.3% (244 cases). For the corresponding negative edge examples, Uniprot or *Subti*Wiki synonym pairs appeared in only 2.1% (351 out of 16640 examples).

At 87.0% F-score our system had the highest performance out of the three participants, exceeding the next highest system by 17.1 percentage points. If Uniprot and *Subti*Wiki features are not used, performance on the development set is still 67.85%, close to the second highest performing system on the task.

4 Conclusions

We have developed a system that addresses all tasks and subtasks in the BioNLP'11 Shared Task, with top performance in several tasks. With the modular design of the system, all tasks could be implemented with relatively small modifications to the processing pipeline. The graph representation which covered naturally all different task annotations was a key feature in enabling fast system development and testing. As with the Turku Event Extraction System developed for the BioNLP'09 Shared Task, we release this improved system for the BioNLP community under an open source license at `bionlp.utu.fi`.

Of all the tasks, the GE-task, which extends the BioNLP'09 corpus, is best suited for evaluating advances in event extraction in the past two years. Comparing our system's performance on the GE'09 corpus with the current one, we can assume that the two corpora are of roughly equal difficulty. Therefore we can reason that overall event extraction performance has increased about three percentage points, the highest performance on the current GE-task being 56.04% by team FAUST. It appears that event extraction is a hard problem, and that the immediate easy performance increases have already been found. We hope the BioNLP'11 Shared Task has focused more interest in the field, hopefully eventually leading to breakthroughs in event extraction and bringing performance closer to established fields of BioNLP such as syntactic parsing or named entity recognition.

That our system could be generalized to work on all tasks and subtasks, indicates that the event extraction approach can offer working solutions for several biomedical domains. A potential limiting factor currently is that most task-specific corpora annotate a non-overlapping set of sentences, necessitating the development of task-specific machine learning models. Training on multiple datasets could mean that positives of one task would be unannotated on text from the other task, confusing the classifier. On the other hand, multiple overlapping task annotations on the same text would permit the system to learn from the interactions and delineations of different annotations. System generalization has been successfully shown in the BioNLP'11 Shared Task, but has resulted in a number of separate extraction systems. It could well be that the future of event extraction requires also the generalization of corpus annotations.

As future directions, we intend to further improve the scope and usability of our event extraction system. We will also continue our work on PubMed-scale event extraction, possibly applying some of the new extraction targets introduced by the BioNLP'11 Shared Task.

[4] http://www.uniprot.org/docs/bacsu

[5] http://subtiwiki.uni-goettingen.de/

Acknowledgments

We thank the Academy of Finland for funding, CSC — IT Center for Science Ltd for computational resources and Filip Ginter and Sofie Van Landeghem for help with the manuscript.

References

Jari Björne, Juho Heimonen, Filip Ginter, Antti Airola, Tapio Pahikkala, and Tapio Salakoski. 2009. Extracting complex biological events with rich graph-based feature sets. In *Proceedings of the BioNLP 2009 Workshop Companion Volume for Shared Task*, pages 10–18, Boulder, Colorado. Association for Computational Linguistics.

Jari Björne, Filip Ginter, Sampo Pyysalo, Jun'ichi Tsujii, and Tapio Salakoski. 2010. Scaling up biomedical event extraction to the entire PubMed. In *Proceedings of the 2010 Workshop on Biomedical Natural Language Processing*, pages 28–36, Uppsala, Sweden, July. Association for Computational Linguistics.

Jari Björne, Juho Heimonen, Filip Ginter, Antti Airola, Tapio Pahikkala, and Tapio Salakoski. 2011. Extracting contextualized complex biological events with rich graph-based feature sets. *Computational Intelligence, Special issue on Extracting Bio-molecular Events from Literature*. To appear, accepted in 2009.

Ekaterina Buyko, Erik Faessler, Joachim Wermter, and Udo Hahn. 2009. Event extraction from trimmed dependency graphs. In *Proceedings of the BioNLP 2009 Workshop Companion Volume for Shared Task*, pages 19–27. ACL.

Eugene Charniak and Mark Johnson. 2005. Coarse-to-fine n-best parsing and maxent discriminative reranking. In *Proceedings of the 43rd Annual Meeting of the Association for Computational Linguistics (ACL'05)*, pages 173–180. Association for Computational Linguistics.

Marie-Catherine de Marneffe, Bill MacCartney, and Christopher Manning. 2006. Generating typed dependency parses from phrase structure parses. In *Proceedings of LREC-06*, pages 449–454.

J. P. Euzéby. 1997. List of bacterial names with standing in nomenclature: a folder available on the internet. *Int J Syst Bacteriol*, 47(2):590–592.

Christiane Fellbaum, editor. 1998. *WordNet: an electronic lexical database*. MIT Press.

Juho Heimonen, Jari Björne, and Tapio Salakoski. 2010. Reconstruction of semantic relationships from their projections in biomolecular domain. In *Proceedings of the 2010 Workshop on Biomedical Natural Language Processing*, pages 108–116, Uppsala, Sweden, July. Association for Computational Linguistics.

Halil Kilicoglu and Sabine Bergler. 2009. Syntactic dependency based heuristics for biological event extraction. In *Proceedings of the BioNLP 2009 Workshop Companion Volume for Shared Task*, pages 119–127. ACL.

Jin-Dong Kim, Tomoko Ohta, Sampo Pyysalo, Yoshinobu Kano, and Jun'ichi Tsujii. 2009. Overview of BioNLP'09 Shared Task on Event Extraction. In *Proceedings of the BioNLP 2009 Workshop Companion Volume for Shared Task*, pages 1–9, Boulder, Colorado, June. Association for Computational Linguistics.

Jin-Dong Kim, Sampo Pyysalo, Tomoko Ohta, Robert Bossy, and Jun'ichi Tsujii. 2011. Overview of BioNLP Shared Task 2011. In *Proceedings of the BioNLP 2011 Workshop Companion Volume for Shared Task*, Portland, Oregon, June. Association for Computational Linguistics.

Marie-Catherine de Marneffe and Christopher Manning. 2008. The Stanford typed dependencies representation. In *COLING Workshop on Cross-framework and Cross-domain Parser Evaluation*.

David McClosky. 2010. *Any Domain Parsing: Automatic Domain Adaptation for Natural Language Parsing*. Ph.D. thesis, Department of Computer Science, Brown University.

Makoto Miwa, Sampo Pyysalo, Tadayoshi Hara, and Jun'ichi Tsujii. 2010a. A comparative study of syntactic parsers for event extraction. In *Proceedings of the 2010 Workshop on Biomedical Natural Language Processing*, BioNLP '10, pages 37–45, Stroudsburg, PA, USA. Association for Computational Linguistics.

Makoto Miwa, Rune Sætre, Jin-Dong Kim, and Jun'ichi Tsujii. 2010b. Event extraction with complex event classification using rich features. *J Bioinform Comput Biol*, 8:131–146.

Martin F. Porter. 1980. An algorithm for suffix stripping. *Program*, 14(3):130–137.

Sebastian Riedel, Hong-Woo Chun, Toshihisa Takagi, and Jun'ichi Tsujii. 2009. A markov logic approach to bio-molecular event extraction. In *Proceedings of the Workshop on Current Trends in Biomedical Natural Language Processing: Shared Task*, BioNLP '09, pages 41–49, Stroudsburg, PA, USA. Association for Computational Linguistics.

Ioannis Tsochantaridis, Thorsten Joachims, Thomas Hofmann, and Yasemin Altun. 2005. Large margin methods for structured and interdependent output variables. *Journal of Machine Learning Research (JMLR)*, 6(Sep):1453–1484.

Sofie Van Landeghem, Yvan Saeys, Bernard De Baets, and Yves Van de Peer. 2009. Analyzing text in search of bio-molecular events: a high-precision machine learning framework. In *Proceedings of the BioNLP 2009 Workshop Companion Volume for Shared Task*, pages 128–136. ACL.

Author Index